Human Genetics

Human Genetics

Chuck Armstrong

Callisto Reference

www.callistoreference.com

Callisto Reference,
118-35 Queens Blvd., Suite 400,
Forest Hills, NY 11375, USA

Visit us on the World Wide Web at:
www.callistoreference.com

ISBN: 978-1-64116-239-5 (Hardback)

Printed in China.

Cataloging-in-Publication Data

Human genetics / Chuck Armstrong.
p. cm.
Includes bibliographical references and index.
ISBN 978-1-64116-239-5
1. Human genetics. 2. Genetics. I. Armstrong, Chuck.
QH431 .H86 2019
599.935--dc23

Table of Contents

	Preface	**VII**
Chapter 1	**Human Genetics and its Variations**	**1**
	• Human Genetic Variation	15
	• Human Genome Structural Variation	22
	• Genetic Distance	33
	• Selective Sweep	39
	• Epigenetics	42
	• Variome	45
	• Heritability	50
Chapter 2	**Role of Genetics in Human Evolution**	**57**
	• Human Evolutionary Genetics	57
	• Ancient DNA	65
	• Archaeogenetics	69
	• Gene-centered View of Evolution	73
	• Adaptive Evolution in the Human Genome	83
	• Robustness	92
Chapter 3	**DNA: Sequencing, Damage and Repair**	**99**
	• Bases of DNA	133
	• DNA Sequencing	139
	• DNA Damage	142
	• DNA Repair	145
	• DNA Profiling	155
Chapter 4	**Human Chromosomes**	**157**
	• Centromere	170
	• Telomere	173
	• Autosome	175
	• Allosome	178
	• Nucleosome	198
	• Chromatin	219
Chapter 5	**Genetic Testing**	**224**
	• Carrier Testing	227
	• Preimplantation Genetic Diagnosis	229

- Predictive Testing 241
- DNA Paternity Testing 248
- Genealogical DNA Test 255

Permissions

Index

Preface

Human genetics is the study of inheritance in humans. It is approached in a multidisciplinary manner from the diverse fields of genetics such as classical, molecular and biochemical genetics, cytogenetics, developmental genetics, etc. The study of human genetics is vital to the understanding of human diseases and development of effective treatments. Besides these, it also presents valuable insights into human nature and personality. The concepts of autosomal dominant and recessive inheritance, X-linked and Y-linked inheritance are vital to this field. Pedigree chart analysis allows the determination of the parent who contributes to the transmission of a specific trait. Human genetics is an upcoming field of science that has undergone rapid development over the past few decades. Some of the diverse topics covered in this book address the varied branches that fall under this category. In this book, constant effort has been made to make the understanding of the difficult concepts of human genetics, as easy and informative as possible, for the readers.

A short introduction to every chapter is written below to provide an overview of the content of the book:

Chapter 1- The study of inheritance in humans is known as human genetics. Genetic variation refers to the change in biological systems in individuals and populations, generated due to mutation. This is an introductory chapter, which will discuss human genetics and genetic variation, human genome structural variation, selective sweep, epigenetics, heritability, etc.;

Chapter 2- Human evolution is a process that has led to the emergence of the modern human or Homo sapiens as a unique species within the hominid family. Genetics is at the core of this evolutionary process. This chapter has been carefully written to examine the role of genetics in human evolution, through the elucidation of topics such as human evolutionary genetics, ancient DNA, archaeogenetics and adaptive evolution in the human genome, among others;

Chapter 3- A molecule that is made of two strands of nucleotides forming a double helix carrying the genetic information vital for the functioning, growth and development of organisms is called the DNA. The aim of this chapter is to provide an insight into DNA sequencing, damage and repair;

Chapter 4- A chromosome is a DNA molecule, which contains the genetic information of an organism. Humans have 23 pairs of chromosomes. They are of two types, autosomes and allosomes. This chapter has been carefully written to provide a comprehensive study of human chromosomes through topics such as centromere, telomere, autosome, chromatin, etc.;

Chapter 5- The determination of human bloodlines and the diagnosis of the vulnerabilities to hereditary diseases is possible through genetic testing. This chapter discusses in extensive detail the varied forms of genetic testing such as carrier testing, predictive testing, DNA paternity testing, genealogical DNA test and preimplantation genetic diagnosis.

Finally, I would like to thank my fellow scholars who gave constructive feedback and my family members who supported me at every step.

Chuck Armstrong

Chapter 1

Human Genetics and its Variations

The study of inheritance in humans is known as human genetics. Genetic variation refers to the change in biological systems in individuals and populations, generated due to mutation. This is an introductory chapter, which will discuss human genetics and genetic variation, human genome structural variation, selective sweep, epigenetics, heritability, etc.

Human genetics is the study of the inheritance of characteristics by children from parents. Inheritance in humans does not differ in any fundamental way from that in other organisms.

The study of human heredity occupies a central position in genetics. Much of this interest stems from a basic desire to know who humans are and why they are as they are. At a more practical level, an understanding of human heredity is of critical importance in the prediction, diagnosis, and treatment of diseases that have a genetic component. The quest to determine the genetic basis of human health has given rise to the field of medical genetics. In general, medicine has given focus and purpose to human genetics, so the terms medical genetics and human genetics are often considered synonymous.

Through the study of human genetics, medical professionals and scientists develop new drugs, therapies and treatments for diseases that affect humanity. Several branches of study exist within human genetics, including clinical, biochemical and molecular. Scholars explore principles of disease, gene testing, gene therapy, genetic counseling, chromosome activity and medical genetics. Other professional options include positions as medical scientists, genetic engineers, genetic counselors, medical geneticists or biological scientists.

Human Genome

Human genome is all of the approximately three billion base pairs of deoxyribonucleic acid (DNA) that make up the entire set of chromosomes of the human organism. The human genome includes the coding regions of DNA, which encode all the genes (between 20,000 and 25,000) of the human organism, as well as the noncoding regions of DNA, which do not encode any genes. By 2003 the DNA sequence of the entire human genome was known.

The human genome, like the genomes of all other living animals, is a collection of long polymers of DNA. These polymers are maintained in duplicate copy in the form of chromosomes in every human cell and encode in their sequence of constituent bases (guanine [G], adenine [A], thymine[T], and cytosine [C]) the details of the molecular and physical characteristics that form the corresponding organism. The sequence of these polymers, their organization and structure, and the chemical modifications they contain not only provide the machinery needed to express the information held within the genome but also provide the genome with the capability to replicate, repair, package, and otherwise maintain itself. In addition, the genome is essential for the survival of the human organism; without it no cell or tissue could live beyond a short period of time. For example, red blood cells (erythrocytes), which live for only about 120 days, and skin cells, which on average live for only about 17 days, must be renewed to maintain the viability of the human body, and it is within the genome that the fundamental information for the renewal of these cells, and many other types of cells, is found.

The human genome is not uniform. Excepting identical (homozygous) twins, no two humans on Earth share exactly the same genomic sequence. Further, the human genome is not static. Subtle and sometimes not so subtle changes arise with startling frequency. Some of these changes are neutral or even advantageous; these are passed from parent to child and eventually become commonplace in the population. Other changes may be detrimental, resulting in reduced survival or decreased fertility of those individuals who harbor them; these changes tend to be rare in the population. The genome of modern humans, therefore, is a record of the trials and successes of the generations that have come before. Reflected in the variation of the modern genome is the range of diversity that underlies what are typical traits of the human species. There is also evidence in the human genome of the continuing burden of detrimental variations that sometimes lead to disease.

Knowledge of the human genome provides an understanding of the origin of the human species, the relationships between subpopulations of humans, and the health tendencies or disease risks of individual humans. Indeed, in the past 20 years knowledge of the sequence and structure of the human genome has revolutionized many fields of study, including medicine, anthropology, and forensics. With technological advances that enable inexpensive and expanded access to genomic information, the amount of and the potential applications for the information that is extracted from the human genome is extraordinary.

Completeness of the Human Genome Sequence

Although the human genome has been completely sequenced for all practical purposes, there are still hundreds of gaps in the sequence. A recent study noted more than 160 achromatic gaps of which 50 gaps were closed. However, there are still numerous gaps in the heterochromatic parts of the genome, which is much harder to sequence due to numerous repeats and other intractable sequence features.

Information Content

The human reference genome (GRC v38) has been successfully compressed to ~5.2-fold (marginal less than 550 MB) in 155 minutes using a desktop computer with 6.4 GB of RAM.

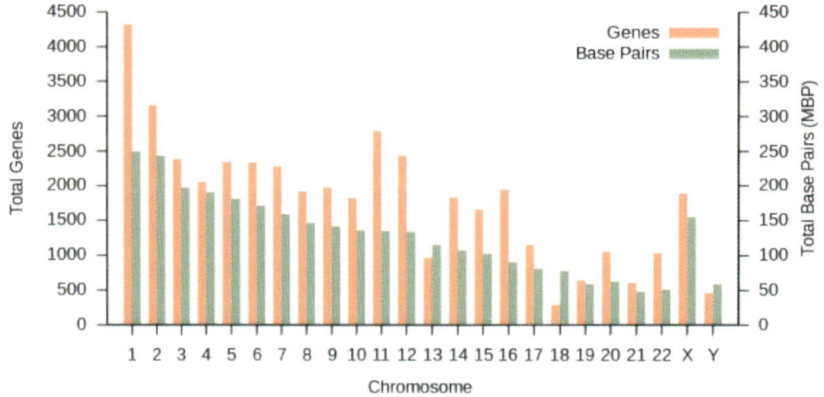

Diagram showing the number of base pairs on each chromosome in green.

The haploid human genome (23 chromosomes) is about 3.1 billion base pairs long and contains more than 20,000 distinct genes. Since 2 bits can code every base pair, this is about 800 megabytes of data. An individual somatic (diploid) cell contains twice this amount, that is, about 6.2 billion base pairs. Men have somewhat less than women because the Y chromosome is about 57 million base pairs whereas the X is about 156 million, but in terms of information men have more because the second X contains almost the same information as the first. Since individual genomes vary in sequence by less than 1% from each other, the variations of a given human's genome from a common reference can be losslessly compressed to roughly 4 megabytes.

The entropy rate of the genome differs significantly between coding and non-coding sequences. It is close to the maximum of 2 bits per base pair for the coding sequences (about 45 million base pairs), but less for the non-coding parts. It ranges between 1.5 and 1.9 bits per base pair for the individual chromosome, except for the Y-chromosome, which has an entropy rate below 0.9 bits per base pair.

Coding Versus Noncoding DNA

The content of the human genome is commonly divided into coding and noncoding DNA sequences. Coding DNA is defined as those sequences that can be transcribed into mRNA and translated into proteins during the human life cycle; these sequences occupy only a small fraction of the genome (<2%). Noncoding DNA is made up of all of those sequences (ca. 98% of the genome) that are not used to encode proteins.

Some noncoding DNA contains genes for RNA molecules with important biological functions (noncoding RNA, for example ribosomal RNA and transfer RNA). The exploration of the function and evolutionary origin of noncoding DNA is an important goal of contemporary genome research, including the ENCODE (Encyclopedia of DNA Elements) project, which aims to survey the entire human genome, using a variety of experimental tools whose results are indicative of molecular activity.

Because non-coding DNA greatly outnumbers coding DNA, the concept of the sequenced genome has become a more focused analytical concept than the classical concept of the DNA-coding gene.

Coding Sequences (Protein-coding Genes)

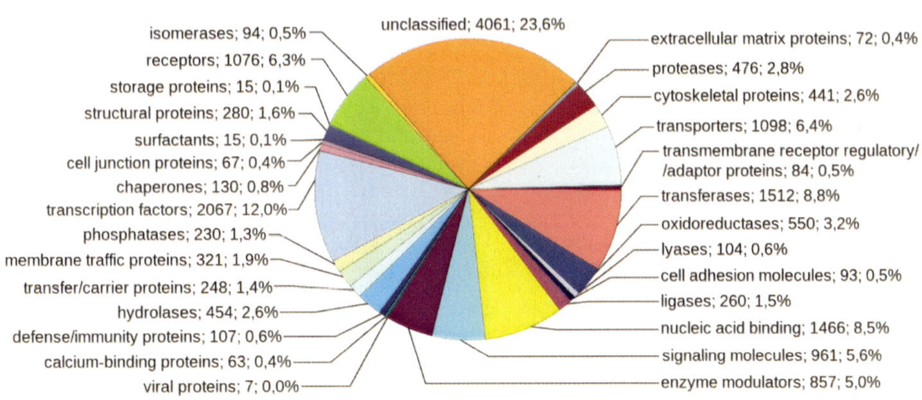

Figure: Human genes categorized by function of the transcribed proteins, given both as number of encoding genes and percentage of all genes.

Protein-coding sequences represent the most widely studied and best understood component of the human genome. These sequences ultimately lead to the production of all human proteins, although several biological processes (e.g. DNA rearrangements and alternative pre-mRNA splicing) can lead to the production of many more unique proteins than the number of protein-coding genes.

The complete modular protein-coding capacity of the genome is contained within the exome, and consists of DNA sequences encoded by exons that can be translated into proteins. Because of its biological importance, and the fact that it constitutes less than 2% of the genome, sequencing of the exome was the first major milepost of the Human Genome Project.

Number of protein-coding genes: About 20,000 human proteins have been annotated in databases such as Uniprot. Historically, estimates for the number of protein genes have varied widely, ranging up to 2,000,000 in the late 1960s, but several researchers pointed out in the early 1970s that the estimated mutational load from deleterious mutations placed an upper limit of approximately 40,000 for the total number of functional loci (this includes protein-coding and functional non-coding genes).

The number of human protein-coding genes is not significantly larger than that of many less complex organisms, such as the roundworm and the fruit fly. This difference may result from the extensive use of alternative pre-mRNA splicing in humans, which provides the ability to build a very large number of modular proteins through the selective incorporation of exons.

Protein-coding capacity per chromosome: Protein-coding genes are distributed unevenly across the chromosomes, ranging from a few dozen to more than 2000, with an especially high gene density within chromosomes 19, 11, and 1. Each chromosome contains various gene-rich and gene-poor regions, which may be correlated with chromosome bands and GC-content. The significance of these nonrandom patterns of gene density is not well understood.

Size of protein-coding genes: The size of protein-coding genes within the human genome shows enormous variability. The median size of a protein-coding gene is 26,288 bp (mean = 66,577 bp). For example, the gene for histone H1a (HIST1HIA) is relatively small and simple, lacking introns and encoding mRNA sequences of 781 nt and a 215 amino acid protein (648 nt open reading frame). Dystrophin (DMD) is the largest protein-coding gene in the human reference genome, spanning a total of 2.2 MB, while Titin (TTN) has the longest coding sequence (114,414 bp), the largest number of exons (363), and the longest single exon (17,106 bp). Over the whole genome, the median size of an exon is 122 bp (mean = 145 bp), the median number of exons is 7 (mean = 8.8), and the median coding sequence encodes 367 amino acids (mean = 447 amino acids).

Protein	Chrome	Gene	Length	Exons	Exon length	Intron length	Alt splicing
Breast cancer type 2 susceptibility protein	13	BRCA2	83,736	27	11,386	72,350	yes
Cystic fibrosis trans membrane conductance regulator	7	CFTR	202,881	27	4,440	198,441	yes
Cytochrome b	MT	MTCYB	1,140	1	1,140	0	no

Dystrophin	X	DMD	2,220,381	79	10,500	2,209,881	yes
Glyceraldehyde-3-phosphate dehydrogenase	12	GAPDH	4,444	9	1,425	3,019	yes
Hemoglobin beta subunit	11	HBB	1,605	3	626	979	no
Histone H1A	6	HIST1H1A	781	1	781	0	no
Titin	2	TTN	281,434	364	104,301	177,133	yes

Table: Examples of human protein-coding genes: Chrome, chromosome, Alt splicing, alternative pre-mRNA splicing.

Recently, a systematic meta-analysis of updated data of the human genome found that the largest protein-coding gene in the human reference genome is RBFOX1 (RNA binding protein, fox-1 homolog 1), spanning a total of 2.47 MB. Over the whole genome, considering a curated set of protein-coding genes, the median size of an exon is currently estimated to be 133 bp (mean = 309 bp), the median number of exons is currently estimated to be 8 (mean = 11), and the median coding sequence is currently estimated to encode 425 amino acids (mean = 553 amino acids).

Noncoding DNA (ncDNA)

Noncoding DNA is defined as all of the DNA sequences within a genome that are not found within protein-coding exons, and so are never represented within the amino acid sequence of expressed proteins. By this definition, more than 98% of the human genomes is composed of ncDNA.

Numerous classes of noncoding DNA have been identified, including genes for noncoding RNA (e.g. tRNA and rRNA), pseudo genes, introns, untranslated regions of mRNA, regulatory DNA sequences, repetitive DNA sequences, and sequences related to mobile genetic elements.

Numerous sequences that are included within genes are also defined as noncoding DNA. These include genes for noncoding RNA (e.g. tRNA, rRNA), and untranslated components of protein-coding genes (e.g. introns, and 5' and 3' untranslated regions of mRNA).

Protein-coding sequences (specifically, coding exons) constitute less than 1.5% of the human genome. In addition, about 26% of the human genome is introns. Aside from genes (exons and introns) and known regulatory sequences (8–20%), the human genome contains regions of noncoding DNA. The exact amount of noncoding DNA that plays a role in cell physiology has been hotly debated. Recent analysis by the ENCODE project indicates that 80% of the entire human genome is either transcribed, binds to regulatory proteins, or is associated with some other biochemical activity.

It "however" remains controversial whether all of this biochemical activity contributes to cell physiology, or whether a substantial portion of this is the result transcriptional and biochemical noise, which must be actively filtered out by the organism. Excluding protein-coding sequences, introns, and regulatory regions, much of the non-coding DNA is composed of: Many DNA sequences that do not play a role in gene expression have important biological functions. Comparative genomics studies indicate that about 5% of the genome contains sequences of noncoding DNA that are highly conserved, sometimes on time-scales representing hundreds of millions of years, implying that these noncoding regions are under strong evolutionary pressure and positive selection.

Many of these sequences regulate the structure of chromosomes by limiting the regions of heterochromatin formation and regulating structural features of the chromosomes, such as the telomeres and centromeres. Other noncoding regions serve as origins of DNA replication. Finally several regions are transcribed into functional noncoding RNA that regulate the expression of protein-coding genes, mRNA translation and stability, chromatin structure (including histone modifications), DNA methylation, DNA recombination, and cross-regulate other noncoding RNAs. It is also likely that many transcribed noncoding regions do not serve any role and that this transcription is the product of non-specific RNA Polymerase activity.

Pseudo Genes

Pseudo genes are inactive copies of protein-coding genes, often generated by gene duplication, that have become nonfunctional through the accumulation of inactivating mutations. The number of pseudo genes in the human genome is on the order of 13,000, and in some chromosomes is nearly the same as the number of functional protein-coding genes. Gene duplication is a major mechanism through which new genetic material is generated during molecular evolution.

For example, the olfactory receptor gene family is one of the best-documented examples of pseudo genes in the human genome. More than 60 percent of the genes in this family are non-functional pseudo genes in humans. By comparison, only 20 percent of genes in the mouse olfactory receptor gene family are pseudo genes. Research suggests that this is a species-specific characteristic, as the most closely related primates all have proportionally fewer pseudo genes. This genetic discovery helps to explain the less acute sense of smell in humans relative to other mammals.

Genes for Noncoding RNA (ncRNA)

Noncoding RNA molecules play many essential roles in cells, especially in the many reactions of protein synthesis and RNA processing. Noncoding RNA include tRNA, ribosomal RNA, microRNA, snRNA and other non-coding RNA genes including about 60,000 long non coding RNAs (lncRNAs). Although the number of reported lncRNA genes continues to rise and the exact number in the human genome is yet to be defined, many of them are argued to be non-functional.

Many ncRNAs are critical elements in gene regulation and expression. Noncoding RNA also contributes to epigenetics, transcription, RNA splicing, and the translational machinery. The role of RNA in genetic regulation and disease offers a new potential level of unexplored genomic complexity.

Introns and Untranslated Regions of mRNA

In addition to the ncRNA molecules that are encoded by discrete genes, the initial transcripts of protein coding genes usually contain extensive noncoding sequences, in the form of introns, 5'-untranslated regions (5'-UTR), and 3'-untranslated regions (3'-UTR). Within most protein-coding genes of the human genome, the length of intron sequences is 10- to 100-times the length of exon sequences.

Regulatory DNA Sequences

The human genome has many different regulatory sequences, which are crucial to controlling gene expression. Conservative estimates indicate that these sequences make up 8% of the genome, however extrapolations from the ENCODE project give that 20-40% of the genome is gene regulatory sequence. Some types of non-coding DNA are genetic "switches" that do not encode proteins, but do regulate when and where genes are expressed (called enhancers).

Regulatory sequences have been known since the late 1960s. The first identification of regulatory sequences in the human genome relied on recombinant DNA technology. Later with the advent of genomic sequencing, the identification of these sequences could be inferred by evolutionary conservation. The evolutionary branch between the primates and mouse, for example, occurred 70–90 million years ago. So computer comparisons of gene sequences that identify conserved non-coding sequences will be an indication of their importance in duties such as gene regulation.

Other genomes have been sequenced with the same intention of aiding conservation-guided methods, for exampled, the puffer fish genome. However, regulatory sequences disappear and re-evolve during evolution at a high rate.

As of 2012, the efforts have shifted toward finding interactions between DNA and regulatory proteins by the technique ChIP-Seq, or gaps where the DNA is not packaged by histones (DNase hypersensitive sites), both of which tell where there are active regulatory sequences in the investigated cell type.

Repetitive DNA Sequences

Repetitive DNA sequences comprise approximately 50% of the human genome.

About 8% of the human genome consists of tandem DNA arrays or tandem repeats, low complexity repeat sequences that have multiple adjacent copies (e.g. "CAGCAGCAG..."). The tandem sequences may be of variable lengths, from two nucleotides to tens of nucleotides. These sequences are highly variable, even among closely related individuals, and so are used for genealogical DNA testing and forensic DNA analysis.

Repeated sequences of fewer than ten nucleotides (e.g. the dinucleotide repeat (AC)n) are termed microsatellite sequences. Among the microsatellite sequences, trinucleotide repeats are of particular importance, as sometimes occur within coding regions of genes for proteins and may lead to genetic disorders. For example, Huntington's disease results from an expansion of the trinucleotide repeat (CAG) within the Huntingtin gene on human chromosome 4. Telomeres (the ends of linear chromosomes) end with a microsatellite hexanucleotide repeat of the sequence (TTAGGG)n.

Tandem repeats of longer sequences (arrays of repeated sequences 10–60 nucleotides long) are termed minisatellites.

Mobile Genetic Elements (Transposons) and their Relics

Transposable genetic elements, DNA sequences that can replicate and insert copies of themselves at other locations within a host genome, are an abundant component in the human genome. The most abundant transposon lineage, Alu, has about 50,000 active copies, and can be inserted into intragenic and intergenic regions. One other lineage, LINE-1, has about 100 active copies per genome (the number varies between people). Together with non-functional relics of old transposons, they account for over half of total human DNA. Sometimes called "jumping genes", transposons have played a major role in sculpting the human genome. Some of these sequences represent endogenous retroviruses, DNA copies of viral sequences that have become permanently integrated into the genome and are now passed on to succeeding generations.

Mobile elements within the human genome can be classified into LTR retrotransposons (8.3% of total genome), SINEs (13.1% of total genome) including Alu elements, LINEs (20.4% of total genome), SVAs and Class II DNA transposons (2.9% of total genome).

Genomic Variation in Humans

Human Reference Genome

With the exception of identical twins, all humans show significant variation in genomic DNA sequences. The human reference genome (HRG) is used as a standard sequence reference.

There are several important points concerning the human reference genome:

- The HRG is a haploid sequence. Each chromosome is represented once.
- The HRG is a composite sequence, and does not correspond to any actual human individual.
- The HRG is periodically updated to correct errors and ambiguities.
- The HRG in no way represents an "ideal" or "perfect" human individual. It is simply a standardized representation or model that is used for comparative purposes.

Measuring Human Genetic Variation

Most studies of human genetic variation have focused on single-nucleotide polymorphisms (SNPs), which are substitutions in individual bases along a chromosome. Most analyses estimate that SNPs occur 1 in 1000 base pairs, on average, in the achromatic human genome, although they do not occur at a uniform density. Thus follows the popular statement that "we are all, regardless of race, genetically 99.9% the same", although this would be somewhat qualified by most geneticists. For example, a much larger fraction of the genome is now thought to be involved in copy number variation. A large-scale collaborative effort to catalog SNP variations in the human genome is being undertaken by the International Hap Map Project.

The genomic loci and length of certain types of small repetitive sequences are highly variable from person to person, which is the basis of DNA fingerprinting and DNA paternity testing technologies. The heterochromatic portions of the human genome, which total several hundred million base pairs, are also thought to be quite variable within the human population (they are so repetitive and so long that they cannot be accurately sequenced with current technology). These regions contain few genes, and it is unclear whether any significant phenotypic effect results from typical variation in repeats or heterochromatin.

Most gross genomic mutations in gamete germ cells probably result in in viable embryos; however, a number of human diseases are related to large-scale genomic abnormalities. Down syndrome, Turner Syndrome, and a number of other diseases result from nondisjunction of entire chromosomes. Cancer cells frequently have aneuploidy of chromosomes and chromosome arms, although a cause and effect relationship between aneuploidy and cancer has not been established.

Mapping Human Genomic Variation

Whereas a genome sequence lists the order of every DNA base in a genome, a genome map identifies the landmarks. A genome map is less detailed than a genome sequence and aids in navigating around the genome.

An example of a variation map is the Hap Map being developed by the International Hap Map Project. The Hap Map is a haplotype map of the human genome, "which will describe the common patterns of human DNA sequence variation." It catalogs the patterns of small-scale variations in the genome that involve single DNA letters, or bases.

Researchers published the first sequence-based map of large-scale structural variation across the human genome in the journal. Large-scale structural variations are differences in the genome among people that range from a few thousand to a few million DNA bases; some are gains or losses of stretches of genome sequence and others appear as re-arrangements of stretches of sequence. These variations include differences in the number of copies individuals have of a particular gene, deletions, translocations and inversions.

SNP Frequency Across the Human Genome

Single-nucleotide polymorphisms (SNPs) do not occur homogeneously across the human genome. In fact, there is enormous diversity in SNP frequency between genes, reflecting different selective pressures on each gene as well as different mutation and recombination rates across the genome. However, studies on SNPs are biased towards coding regions, the data generated from them are unlikely to reflect the overall distribution of SNPs throughout the genome. Therefore, the SNP Consortium protocol was designed to identify SNPs with no bias towards coding regions and the Consortium's 100,000 SNPs generally reflect sequence diversity across the human chromosomes. The SNP Consortium aims to expand the number of SNPs identified across the genome to 300 000 by the end of the first quarter of 2001.

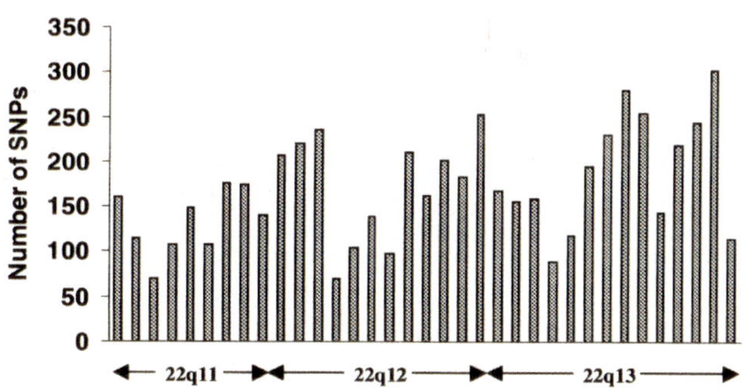

Figure: 'TSC' SNP distribution along the long arm of chromosome 22

Each column represents a 1 Mb interval; the approximate cytogenetic position is given on the x-axis. Clear peaks and troughs of SNP density can be seen, possibly reflecting different rates of mutation, recombination and selection.

Changes in non-coding sequence and synonymous changes in coding sequence are generally more common than non-synonymous changes, reflecting greater selective pressure reducing diversity at positions dictating amino acid identity. Transitional changes are more common than trans versions, with CpG dinucleotides showing the highest mutation rate, presumably due to deamination.

Personal Genomes

A personal genome sequence is a (nearly) complete sequence of the chemical base pairs that make up the DNA of a single person. Because medical treatments have different effects on different people due to genetic variations such as single-nucleotide polymorphisms (SNPs), the analysis of personal genomes may lead to personalized medical treatment based on individual genotypes.

The first personal genome sequence to be determined was that of Craig Venter in 2007. Personal genomes had not been sequenced in the public Human Genome Project to protect the identity of volunteers who provided DNA samples. That sequence was derived from the DNA of several volunteers from a diverse population. However, early in the Venter-led Celera Genomics genome sequencing effort the decision was made to switch from sequencing a composite sample to using DNA from a single individual later revealed to have been Venter himself. Thus the Celera human genome sequence released in 2000 was largely that of one man. Subsequent replacement of the early composite-derived data and determination of the diploid sequence, representing both sets of chromosomes, rather than a haploid sequence originally reported, allowed the release of the first personal genome. In April 2008, that of James Watson was also completed. Since then hundreds of personal genome sequences have been released, including those of Desmond Tutu, and of a Paleo-Eskimo. In November 2013, a Spanish family made their personal genomics data publicly available under a Creative Commons public domain license. The work was led by Manuel Corpas and the data obtained by direct-to-consumer genetic testing with 23 and Me and the Beijing Genomics Institute). This is believed to be the first such public genomics dataset for a whole family.

The sequencing of individual genomes further unveiled levels of genetic complexity that had not been appreciated before. Personal genomics helped reveal the significant level of diversity in the human genome attributed not only to SNPs but structural variations as well. However, the application of such knowledge to the treatment of disease and in the medical field is only in its very beginnings. Exome sequencing has become increasingly popular as a tool to aid in diagnosis of genetic disease because the exome contributes only 1% of the genomic sequence but accounts for roughly 85% of mutations that contribute significantly to disease.

Human Knockouts

In humans, gene knockouts naturally occur as heterozygous or homozygous loss-of-function gene knockouts. These knockouts are often difficult to distinguish, especially within heterogeneous genetic backgrounds. They are also difficult to find as they occur in low frequencies.

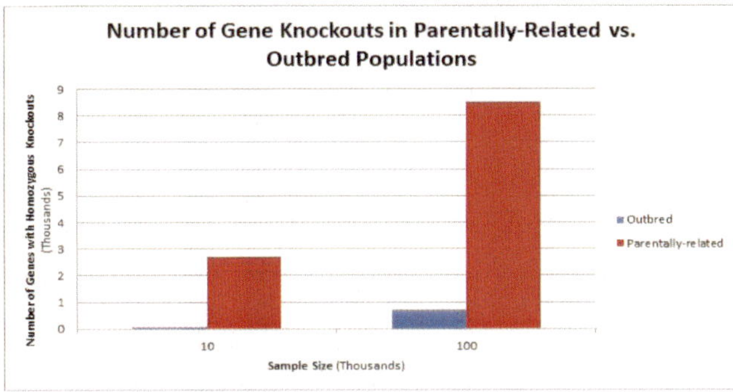

Populations with a high level of parental-relatedness result in a larger number of homozygous gene knockouts as compared to outbred populations.

Populations with high rates of consanguinity, such as countries with high rates of first-cousin marriages, display the highest frequencies of homozygous gene knockouts. Such populations include Pakistan, Iceland, and Amish populations. These populations with a high level of parental-relatedness have been subjects of human knock out research which has helped to determine the function of specific genes in humans. By distinguishing specific knockouts, researchers are able to use phenotypic analyses of these individuals to help characterize the gene that has been knocked out.

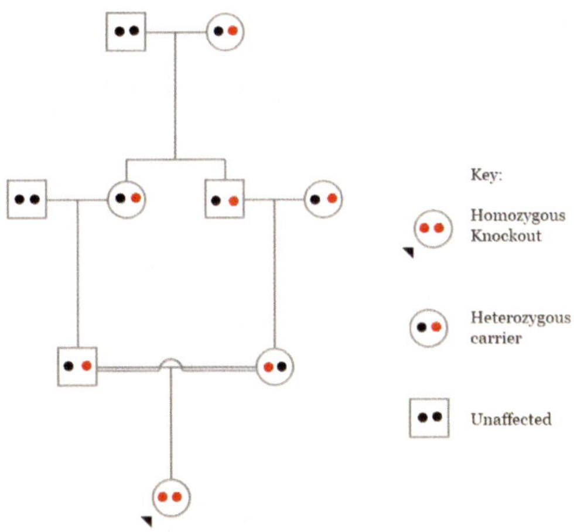

Figure: A pedigree displaying a first-cousin mating (carriers both carrying heterozygous knockouts mating as marked by double line)

Knockouts in specific genes can cause genetic diseases, potentially have beneficial effects, or even result in no phenotypic effect at all. However, determining a knockout's phenotypic effect and in humans can be challenging. Challenges to characterizing and clinically interpreting knockouts include difficulty calling of DNA variants, determining disruption of protein function (annotation), and considering the amount of influence mosaicism has on the phenotype.

One major study that investigated human knockouts is the Pakistan Risk of Myocardial Infarction study. It was found that individuals possessing a heterozygous loss-of-function gene knockout for the APOC3 gene had lower triglycerides in the blood after consuming a high fat meal as compared to individuals without the mutation. However, individuals possessing homozygous loss-of-function gene knockouts of the APOC3 gene displayed the lowest level of triglycerides in the blood after the fat load test, as they produce no functional APOC3 protein.

Human Genetic Disorders

Most aspects of human biology involve both genetic (inherited) and non-genetic (environmental) factors. Some inherited variation influences aspects of our biology that are not medical in nature (height, eye color, ability to taste or smell certain compounds, etc). Moreover, some genetic disorders only cause disease in combination with the appropriate environmental factors (such as diet). With these caveats, genetic disorders may be described as clinically defined diseases caused by genomic DNA sequence variation. In the most straightforward cases, the disorder can be associated with variation in a single gene. For example, cystic fibrosis is caused by mutations in the CFTR gene, and is the most common recessive disorder in Caucasian populations with over 1,300 different mutations known.

Disease-causing mutations in specific genes are usually severe in terms of gene function, and are fortunately rare, thus genetic disorders are similarly individually rare. However, since there are many genes that can vary to cause genetic disorders, in aggregate they constitute a significant component of known medical conditions, especially in pediatric medicine. Molecularly characterized genetic disorders are those for which the underlying causal gene has been identified, currently there are approximately 2,200 such disorders annotated in the OMIM database.

Studies of genetic disorders are often performed by means of family-based studies. In some instances population based approaches are employed, particularly in the case of so-called founder populations such as those in Finland, French-Canada, Utah, Sardinia, etc. Diagnosis and treatment of genetic disorders are usually performed by a geneticist-physician trained in clinical/medical genetics. The results of the Human Genome Project are likely to provide increased availability of genetic testing for gene-related disorders, and eventually improved treatment. Parents can be screened for hereditary conditions and counseled on the consequences, the probability it will be inherited, and how to avoid or ameliorate it in their offspring.

As noted above, there are many different kinds of DNA sequence variation, ranging from complete extra or missing chromosomes down to single nucleotide changes. It is generally presumed that much naturally occurring genetic variation in human populations is phenotypically neutral, i.e. has little or no detectable effect on the physiology of the individual (although there may be fractional differences in fitness defined over evolutionary time frames). Genetic disorders can be caused by any or all known types of sequence variation. To molecularly characterize a new genetic disorder, it is necessary to establish a causal link between a particular genomic sequence variant and the clinical disease under investigation. Such studies constitute the realm of human molecular genetics.

With the advent of the Human Genome and International Hap Map Project, it has become feasible to explore subtle genetic influences on many common disease conditions such as diabetes, asthma, migraine, schizophrenia, etc. Although some causal links have been made between genomic sequence variants in particular genes and some of these diseases, often with much publicity in the general media, these are usually not considered to be genetic disorders per se as their causes are complex, involving many different genetic and environmental factors. Thus there may be disagreement in particular cases whether a specific medical condition should be termed a genetic disorder. The categorized table below provides the prevalence as well as the genes or chromosomes associated with some human genetic disorders.

Disorder	Prevalence	Chromosome or gene involved
	Chromosomal conditions	
Down syndrome	1:600	Chromosome 21
Klinefelter syndrome	1:500–1000 males	Additional X chromosome
Turner syndrome	1:2000 females	Loss of X chromosome
Sickle cell anemia	1 in 50 births in parts of Africa; rarer elsewhere	β-globin (on chromosome 11)
	Cancers	
Breast/Ovarian cancer (susceptibility)	~5% of cases of these cancer types	BRCA1, BRCA2
FAP (hereditary nonpolyposis coli)	1:3500	APC
Lynch syndrome	5–10% of all cases of bowel cancer	MLH1, MSH2, MSH6, PMS2
	Neurological conditions	
Huntington disease	1:20000	Huntingtin
Alzheimer disease - early onset	1:2500	PS1, PS2, APP
	Other conditions	
Cystic fibrosis	1:2500	CFTR
Duchenne muscular dystrophy	1:3500 boys	Dystrophin

Evolution

Comparative genomics studies of mammalian genomes suggest that approximately 5% of the human genome has been conserved by evolution since the divergence of extant lineages approximately 200 million years ago, containing the vast majority of genes. The published chimpanzee genome differs from that of the human genome by 1.23% in direct sequence comparisons. Around 20% of this figure is accounted for by variation within each species, leaving only ~1.06% consistent sequence divergence between humans and chimps at shared genes. This nucleotide by nucleotide difference is dwarfed, however, by the portion of each genome that is not shared, including around 6% of functional genes that are unique to either humans or chimps.

In other words, the considerable observable differences between humans and chimps may be due as much or more to genome level variation in the number, function and expression of genes rather than DNA sequence changes in shared genes. Indeed, even within humans, there has been found to be a previously unappreciated amount of copy number variation (CNV) which can make up as much as 5 – 15% of the human genome. In other words, between humans, there could be +/- 500,000,000 base pairs of DNA, some being active genes, others inactivated, or active at different levels. The full significance of this finding remains to be seen. On average, a typical human protein-coding gene differs from its chimpanzee ortholog by only two amino acid substitutions; nearly one third of human genes have exactly the same protein translation as their chimpanzee orthologs. A major difference between the two genomes is human chromosome 2, which is equivalent to a fusion product of chimpanzee chromosomes 12 and 13. (later renamed to chromosomes 2A and 2B, respectively).

Humans have undergone an extraordinary loss of olfactory receptor genes during our recent evolution, which explains our relatively crude sense of smell compared to most other mammals. Evolutionary evidence suggests that the emergence of color vision in humans and several other primate species has diminished the need for the sense of smell.

In September 2016, scientists reported that, based on human DNA genetic studies, all non-Africans in the world today can be traced to a single population that exited Africa between 50,000 and 80,000 years ago.

Mitochondrial DNA

The human mitochondrial DNA is of tremendous interest to geneticists, since it undoubtedly plays a role in mitochondrial disease. It also sheds light on human evolution; for example, analysis of variation in the human mitochondrial genome has led to the postulation of a recent common ancestor for all humans on the maternal line of descent.

Due to the lack of a system for checking for copying errors, mitochondrial DNA (mtDNA) has a more rapid rate of variation than nuclear DNA. This 20-fold higher mutation rate allows mtDNA to be used for more accurate tracing of maternal ancestry. Studies of mtDNA in populations have allowed ancient migration paths to be traced, such as the migration of Native Americans from Siberia or Polynesians from southeastern Asia. It has also been used to show that there is no trace of Neanderthal DNA in the European gene mixture inherited through purely maternal lineage. Due to the restrictive all or none manner of mtDNA inheritance, this result (no trace of Neanderthal mtDNA) would be likely unless there were a large percentage of Neanderthal ancestry, or there was strong positive selection for that mtDNA (for example, going back 5 generations, only 1 of your 32 ancestors contributed to your mtDNA, so if one of these 32 was pure Neanderthal you would expect that ~3% of your autosomal DNA would be of Neanderthal origin, yet you would have a ~97% chance to have no trace of Neanderthal mtDNA).

Epigenome

Epigenetics describes a variety of features of the human genome that transcend its primary DNA sequence, such as chromatin packaging, histone modifications and DNA methylation, and which are important in regulating gene expression, genome replication and other cellular processes. Epigenetic markers strengthen and weaken transcription of certain genes but do not affect the actual sequence of DNA nucleotides. DNA methylation is a major form of epigenetic control over gene expression and one of the most highly studied topics in epigenetics. During development, the human DNA methylation profile experiences dramatic changes. In early germ line cells, the genome has very low methylation levels. These low levels generally describe active genes. As development progresses, parental imprinting tags lead to increased methylation activity.

Epigenetic patterns can be identified between tissues within an individual as well as between individuals themselves. Identical genes that have differences only in their epigenetic state are called epialleles. Epialleles can be placed into three categories: those directly determined by an individual's genotype, those influenced by genotype, and those entirely independent of genotype. The epigenome is also influenced significantly by environmental factors. Diet, toxins, and hormones impact the epigenetic state. Studies in dietary manipulation have demonstrated that methyl-deficient diets are associated with hypomethylation of the epigenome. Such studies establish epigenetics as an important interface between the environment and the genome.

Human Genetic Variation

Genetic variation is a term used to describe the variation in the DNA sequence in each of our genomes. Genetic variation is what makes us all unique, whether in terms of hair color, skin color or even the shape of our faces.

- Individuals of a species have similar characteristics but they are rarely identical, the difference between them is called variation.
- Genetic variation is a result of subtle differences in our DNA.
- Single nucleotide polymorphisms (SNPs, pronounced 'snips') are the most common type of genetic variation amongst people.
- Each single nucleotide polymorphism represents a difference in a single DNA base, A, C, G or T, in a person's DNA. On average they occur once in every 300 bases and are often found in the DNA between genes.
- Genetic variation results in different forms, or alleles, of genes. For example, if we look at eye color, people with blue eyes have one allele of the gene for eye color, whereas people with brown eyes will have a different allele of the gene.
- Eye color, skin tone and face shape are all determined by our genes so any variation that occurs will be due to the genes inherited from our parents.
- In contrast, although weight is partly influenced by our genetics, it is strongly influenced by our environment. For example, how much we eat and how often we exercise.
- Genetic variation can also explain some differences in disease susceptibility and how people react to drugs.
- Genetic variation is important in evolution. Evolution relies on genetic variation that is passed down from one generation to the next. Favorable characteristics are 'selected' for, survive and are passed on. This is known as natural selection.

Types of Variations

Variations are classified variously according to:

(i) Affected Trait:

Morphological, physiological, cytological and behavioristic.

(ii) Impact:

Useful, harmful and neutral or indifferent.

(iii) Parts:

Meristic (number of parts and their geometrical relations) and substantive (appearance),

(iv) Degree:

Continuous and discontinuous,

(v) Cells Affected:

Somatic and germinal,

(vi) Phenotypic (observable) and genotypic (constitutional).

I. Somatic or Somatogenic Variations

They are variations which affect the somatic or body cells of the organisms. They are also called modifications or acquired characters because they are got by an individual during its life time. Lamarck based his theory of evolution on the inheritance of acquired characters. However, as proved by Weismann, somatic variations generally die with the death of the individual and are hence non-inheritable. They are caused by three factors environment, use and disuse of organs and conscious efforts.

Environmental Factors

The environmental factors are medium, light, temperature, nutrition, wind, water supply, etc. The environmental factors bring about changes in phenotype of the individual. Different changes in the phenotype in response to different environmental factors, are called phenotype plasticity. A specific phenotype developed in response to a particular ecological condition is called Eco phenotype.

There are only slight modifications in animals but in plants the modifications are much more conspicuous. This is due to the environmental effect on the meristems of various parts. A slight change in the meristematic activity can have permanent effect on the plant. Environment can also change the amount of flowering and bring about non-inheritable changes in the floral parts. Some of the more important environmental factors are:

1. Medium

Some amphibious plants show heterophylly with dissected leaves inside water and entire leaves outside, e.g., Ranunculus aquatilis. Stockard placed eggs of fish Fundulus in sea water containing magnesium chloride. The young ones reared in such medium possessed one median eye instead of the two usual lateral eyes. Hydrangea bears blue flowers in acidic soil and pinkish flowers in alkaline soil.

2. Light

In the absence of light the plants remain etiolated. Shade produces elongated intemodes and thinner and broader leaves. It increases the softness of many vegetables. Strong light, on the contrary, helps in the production of more mechanical tissue and smaller and thicker leaves. Palisade parenchyma becomes multilayered under strong light but remains single layered under moderate intensities of light (e.g., Peach).

The effect of light has also been observed by Cunningham in flat fish Solea. The fish habitually rests on left side. It develops pigmentation and eyes on right side, the side exposed to light. If left side is exposed to light in the young fish, both eyes and pigmentation develop on that side.

3. Temperature

Temperature directly affects the metabolic activity of the organisms and rate of transpiration in plants. Plants growing in hot area show nanism of the aerial parts and greater growth of the root system. Strong sunlight and high temperature bring about sun-tanning of human skin by production of more melanin for protection against excessive insulation and ultra-violet radiations.

4. Nutrition

The individual provided with optimum nutrition grows best while the under-nourished one shows stunted growth. The abundance or deficiency of a mineral salt produces various types of deformities in plants. A larva of honey bee fed on royal jelly grows into queen while the one fed on the bee bread develops into worker.

5. Water

Plants growing in soils deficient in water or in areas with little rainfall show modifications in order to reduce transpiration and retain water, e.g., succulence, spines, reduced leaves, thick coating, sunken stomata, etc. Those growing in humid and moist area produce luxuriant growth.

Use and Disuse of Organs

This is mostly observed in higher animals. The organ which is put to continuous use develops more while the organ less used develops little. A wrestler or a player who performs daily exercise develops a stronger and more muscular body than another man who does not do any exercise. A lion, tiger or bear kept in a zoo is weaker than the one living in jungle.

Conscious Efforts

Modifications due to conscious efforts are observed only in those animals which have intelligence. Receiving education, training of some pets, slim bodies, boring of pinna, long neck, small feet, mutilations in pets, bonsai, etc. are some of the examples of conscious efforts.

II. Germinal or Blastogenic Variations

They are produced in the germ cells of an organism and are inheritable. They may be already present in ancestors or may be formed suddenly. Accordingly, the germinal variations are of two types, continuous and discontinuous.

1. Continuous Variations:

They are also called fluctuating variations because they fluctuate on either side (both plus and minus) of a mean or average for the species. Continuous variations are typical of quantitative characteristics. They show differences from the average which are connected with it through small intermediate forms.

If plotted as a graph, the mean or normal characteristic will be found to be possessed by maximum number of individuals. The number of individuals will decrease with the increase in degree of fluctuation. The graph wills appear to be bell shaped. Continuous variations are already present in different organisms or races of a species.

They are produced by:

i. Chance separation or segregation of chromosomes at the time of gamete or spore formation.

ii. Crossing over or exchange of segments between homologous chromosomes during meiosis.

iii. Chance combination of chromosomes during fertilization. Therefore, these variations are also known by the name of recombination's.

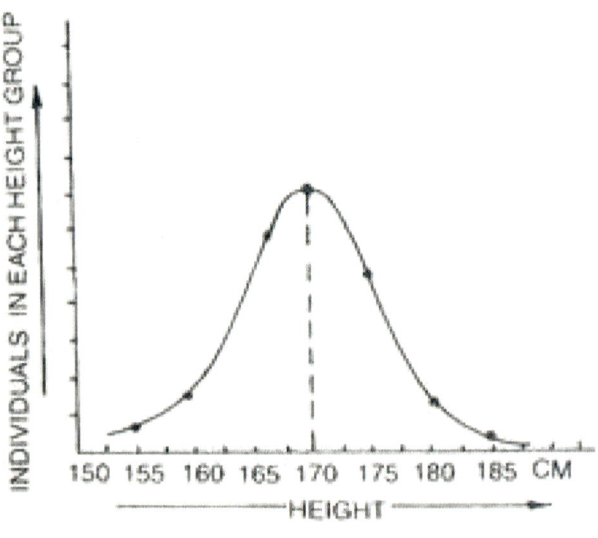

Figure: Continuous varaiations or fluctuations in the height of adult human beings

They make an organism better fitted to struggle for existence in a particular environment. They also enable human beings to improve the races of important plants and animals. However, they are unable to form a new species though Darwin based his evolution theory of natural selection on continuous variations. Continuous variations are of two types:

(a) Substantive

They influence appearance including shape, size, weight and color of a part or whole of the organism, e.g., height, shape of nose, skin color, color of eyes, hair, length of fingers or toes, yield of milk, eggs, etc.

(b) Meristic

They influence the number of parts, e.g., multiple alleles in blood groups, number of grains in an ear of wheat, number of epicalyx segments in Althaea, tentacles in Hydra or segments in earthworm, etc.

2. Discontinuous Variations:

They are also called sports, saltations or mutations. They are sudden unpredictable inheritable departures from the normal without any intermediate stage. The organism in which a mutation occurs is called a mutant. Discontinuous variations form the basis of mutation theory of evolution proposed by de Vries.

Discontinuous variations or mutations are caused by:

a) Chromosomal aberrations like deletion, duplication, inversion and translocation,

b) Change in chromosome number through aneuploidy and polyploidy,

c) Change in gene structure and expression due to addition, deletion or change in nucleotides.

The discontinuous variations are of two types:

a) Substantive they affect the shape, size and color, e.g., short legged Ancon sheep, hornless or polled cattle, hairless cats, piebald patching in man, brachydactyly, syndactyly, etc.

b) Meristic. They affect the number of parts, e.g., polydactyly (six or more digits) in humans.

Measures of Variation

Genetic variation among humans occurs on many scales, from gross alterations in the human karyotype to single nucleotide changes. Chromosome abnormalities are detected in 1 of 160 live human births. Apart from sex chromosome disorders, most cases of aneuploidy result in death of the developing fetus (miscarriage); the most common extra autosomal chromosomes among live births are 21, 18 and 13.

Nucleotide diversity is the average proportion of nucleotides that differ between two individuals. As of 2004, the human nucleotide diversity was estimated to be 0.1% to 0.4% of base pairs. In 2015, the 1000 Genomes Project, which sequenced one thousand individuals from 26 human populations, found that "a typical [individual] genome differs from the reference human genome at 4.1 million to 5.0 million sites affecting 20 million bases of sequence." Nearly all (>99.9%) of these sites are small differences, either single nucleotide polymorphisms or brief insertion-deletions in the genetic sequence, but structural variations account for a greater number of base-pairs than the SNPs and indels.

As of 2017, the Single Nucleotide Polymorphism Database (dbSNP), which lists SNP and other variants, listed 324 million variants found in sequenced human genomes.

Single Nucleotide Polymorphisms

A single nucleotide polymorphism (SNP) is a difference in a single nucleotide between members of one species that occurs in at least 1% of the population. The 2,504 individuals characterized by the 1000 Genomes Project had 84.7 million SNPs among them. SNPs are the most common type of sequence variation, estimated in 1998 to account for 90% of all sequence variants. Other sequence variations are single base exchanges, deletions and insertions. SNPs occur on average about every 100 to 300 bases and so are the major source of heterogeneity.

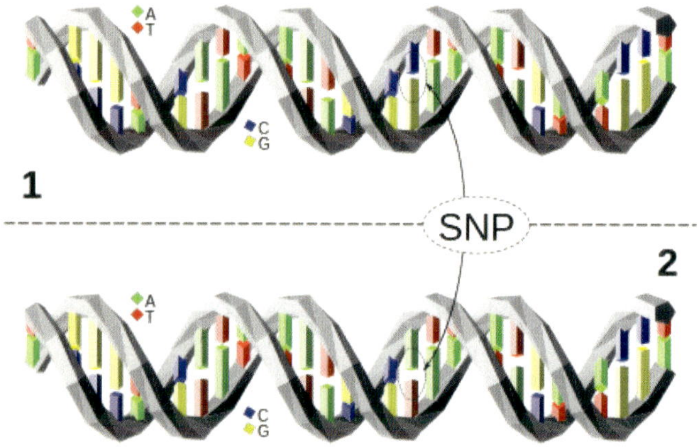

Figure: DNA molecule 1 differs from DNA molecule 2 at a single base-pair location (a C/T polymorphism).

A functional, or non-synonymous, SNP is one that affects some factor such as gene splicing or messenger RNA, and so causes a phenotypic difference between members of the species. About 3% to 5% of human SNPs are functional. Neutral, or synonymous SNPs are still useful as genetic markers in genome-wide association studies, because of their sheer number and the stable inheritance over generations.

A coding SNP is one that occurs inside a gene. There are 105 Human Reference SNPs that result in premature stop codons in 103 genes. This corresponds to 0.5% of coding SNPs. They occur due to segmental duplication in the genome. These SNPs result in loss of protein, yet all these SNP alleles are common and are not purified in negative selection.

Structural Variation

Structural variation is the variation in structure of an organism's chromosome. Structural variations, such as copy-number variation and deletions, inversions, insertions and duplications, account for much more human genetic variation than single nucleotide diversity. This was concluded in 2007 from analysis of the diploid full sequences of the genomes of two humans: Craig Venter and James D. Watson. This added to the two haploid sequences which were amalgamations of sequences from many individuals, published by the Human Genome Project and Celera Genomics respectively.

According to the 1000 Genomes Project, a typical human has 2,100 to 2,500 structural variations, which include approximately 1,000 large deletions,160 copy-number variants, 915 Alu insertions, 128 L1 insertions, 51 SVA insertions, 4 NUMTs, and 10 inversions.

Copy Number Variation

A copy-number variation (CNV) is a difference in the genome due to deleting or duplicating large regions of DNA on some chromosome. It is estimated that 0.4% of the genomes of unrelated humans differ with respect to copy number. When copy number variation is included, human-to-human genetic variation is estimated to be at least 0.5% (99.5% similarity). Copy number variations are inherited but can also arise during development.

Pratas et al. have built a visual map with the regions with high genomic variation of the modern-human reference assembly relatively to a Neanderthal of 50k.

Epigenetics

Epigenetic variation is variation in the chemical tags that attach to DNA and affect how genes get read. The tags, "epigenetic markings, act as switches that control how genes can be read." At some alleles, the epigenetic state of the DNA, and associated phenotype, can be inherited across generations of individuals.

Genetic Variability

Genetic variability is a measure of the tendency of individual genotypes in a population to vary (become different) from one another. Variability is different from genetic diversity, which is the amount of variation seen in a particular population. The variability of a trait is how much that trait tends to vary in response to environmental and genetic influences.

Clines

In biology, a cline is a continuum of species, populations, races, varieties, or forms of organisms that exhibit gradual phenotypic and/or genetic differences over a geographical area, typically as a result of environmental heterogeneity. In the scientific study of human genetic variation, a gene cline can be rigorously defined and subjected to quantitative metrics.

Haplogroups

In the study of molecular evolution, a haplogroup is a group of similar haplotypes that share a common ancestor with a single nucleotide polymorphism (SNP) mutation. Haplogroups pertain to deep ancestral origins dating back thousands of years.

The most commonly studied human haplogroups are: Y-chromosome (Y-DNA) haplogroups and mitochondrial DNA (mtDNA) haplogroups, both of which can be used to define genetic populations. Y-DNA is passed solely along the patrilineal line, from father to son, while mtDNA is passed down the matrilineal line, from mother to both daughter and son. The Y-DNA and mtDNA may change by chance mutation at each generation.

Variable Number Tandem Repeats

A variable number tandem repeat (VNTR) is the variation of length of a tandem repeat. A tandem repeat is the adjacent repetition of a short nucleotide sequence. Tandem repeats exist on many chromosomes, and their length varies between individuals. Each variant acts as an inherited allele, so they are used for personal or parental identification. Their analysis is useful in genetics and biology research, forensics, and DNA fingerprinting.

Short tandem repeats (about 5 base pairs) are called microsatellites, while longer ones are called minisatellites.

Human Genome Structural Variation

Genome structural variations (SVs) in the human genome are defined as DNA sequence polymorphisms of at least a few dozen or few hundred bases in length and include deletions, duplications, inversions, translocation, retro element insertions, and more complex rearrangements that could be thought of as consisting of multiple fragments from the just listed categories. More bases in a personal genome are affected by SVs than by single nucleotide polymorphisms (SNPs), suggesting that SV have a larger or comparable effect on personal phenotype than SNPs. SVs frequently occur in tumor genomes, with several tumor types (e.g., ovarian) having SVs as the dominant type of genomic alteration. Numerous de novo SVs have been linked to various diseases.

Because of their size and enrichment in repeat regions, these are the most challenging variants to discover and analyze. Even more challenging is the precise identification of SV breakpoints at a single base pair resolution. But reward is huge, as precise breakpoints hold invaluable information about the origin of each SV; i.e., about the mutational process that created it. The main mechanisms of SV mutagenesis are largely known or hypothesized based on existing evidence: Non-Allele Homologous Recombination (NAHR), Non-Homologous End Joining (NHEJ), Micro homology-Mediated End Joining (MMEJ), errors during replication (replicative mechanisms), and retro element insertions.

Characterization of Structural Variation

Germline and Somatic Structural Variation

Structural variants are important contributors to genome variation and consideration of these variants is necessary for disease association and cancer genetics studies. In this topic, we briefly review current knowledge about structural variation in human and cancer genomes.

Germline Structural Variation

Characterizing the DNA sequence differences that distinguish individuals is a major challenge in human genetics. Until a few years ago, the primary focus was to identify single nucleotide polymorphisms (SNPs), and projects such as Hap Map provide catalogs of common SNPs in several human populations. Recent whole-genome sequencing and microarray measurements have shown that structural variation, including duplications, deletions, and inversions of large blocks of DNA sequence, is common in the human genome. SVs include both copy number variants – duplications and deletions – that change the number of copies of a segment of the genome, and balanced rearrangements – such as inversions and translocations – that do not alter the copy number of the genome. The Database of Genomic Variants currently lists approximately 66 thousand copy number variants and approximately 900 inversion variants in the human genome, and this number continues to increase. Some of these entries are multiple reports of the same variant due to problems in merging SV predictions across different platforms/technologies. Nevertheless, SVs are extensive in human populations.

Germline SVs account for a greater share of the total nucleotide differences between two individual human genomes than SNPs. Copy number variants alone account for approximately 18% of genetic variation in gene expression, having little overlap with variation associated to SNPs, and can affect the expression of genes up to 300 kb away from the variant. Both common and rare SVs have recently been linked to several human diseases including autism and schizophrenia. In addition to SVs that cause disease, SVs segregating in a population perturb patterns of linkage disequilibrium and haplotype structure. Thus, it is essential to catalog SVs in order to understand their consequences for human population genetics. Incorrect identification of SVs in samples can lead to spurious genetic associations resulting from the undetected SVs, erroneous merging of distinct variants in different samples, and failure to recognize heterozygosity at a locus.

Finally, structural variants are also present in model organisms such as mouse and fruit fly. Identifying these variants is important for animal models of human diseases.

Somatic Structural Variation and Cancer

Cancer is a disease driven by somatic mutations that accumulate during the lifetime of an individual. The inheritance of mutations by daughter cells during mitosis and selection for advantageous mutations make cancer a "micro evolutionary process" within a population of cells. Decades of cytogenetic studies have shown that somatic structural variants are a feature of many cancer genomes. These early studies, particularly in leukemia's and lymphoma, identified a number of recurrent chromosomal rearrangements that are present in many patients with the same type of cancer. For example, a significant fraction of patients with chronic myelogenous leukemia (CML) exhibit a translocation between chromosomes 9 and 22. The breakpoints of this translocation lie in two genes, BCR and ABL, and the translocation results in the BCR-ABL fusion gene that is directly implicated in the development of this cancer. In addition to fusion genes, somatic SVs can also lead to altered expression of oncogenes and tumor suppressor genes due to both genetic and epigenetic mechanisms. For example, in Burkitt's lymphoma, a translocation activates the MYC oncogene by fusing it with a strong promoter.

In solid tumors, the situation is more complicated. Many solid tumors have genomes that are extensively rearranged compared to the normal healthy genome from which they were derived. These highly rearranged genomes are thought to be a product of genome instability resulting from mutations in the DNA repair machinery. This complex organization of cancer genomes obscures functional driver SVs in a background of passenger mutations. However, with the availability of higher-resolution genomics technologies, recurrent fusion genes are also being found in solid tumors, such as prostate and lung cancers. These results suggest that additional events remain to be discovered. Next-generation DNA sequencing technologies provide the opportunity to reconstruct the organization of cancer genomes at single nucleotide resolution. Projects including The Cancer Genome Atlas (TCGA) and International Cancer Genome Consortium (ICGC) are using these technologies to measure somatic mutations in thousands of cancer genomes.

Mechanisms of Structural Variation

As additional genetic and somatic structural variants are characterized, there is increasing opportunity to characterize the mechanisms that produce these variants. A distinguishing feature of the different mechanisms is the amount of sequence similarity, or homology, at the breakpoints of the structural variant. One extreme is little or no sequence similarity. These variants are thought to result from random (or near random) double-stranded breaks in DNA. These breaks might occur due to exposure to external DNA damaging agents. For example, ultraviolet radiation or various chemotherapy drugs produce double-stranded breaks. Aberrant repair of these breaks result in structural variants. This mechanism is termed non-homologous end-joining (NHEJ).

The opposite extreme is high sequence similarity at the breakpoints. This mechanism is termed non-allelic homologous recombination (NAHR). This mechanism is similar to the normal biological process of homologous recombination that occurs during meiosis and exchanges DNA between two homologous chromosomes. But as the name states, NAHR is a rearrangement that occurs between homologous sequences that are not the same allele on homologous chromosomes. Rather NAHR occurs between repetitive sequences on the genome. The human genome contains numerous repetitive sequences ranging from Alu elements of 300 bp to segmental duplications, also called low copy repeats, of tens to hundreds of kbp. Thus, there are numerous substrates for NAHR in the human genome, and not surprisingly numerous reported structural variants that result from NAHR. For example, the 1000 Genomes Project, a large NIH project to survey all classes of variation – SNPs through SV – in 1000 human genomes recently reported that approximately 23% of deletions were a result of NAHR. Importantly, due to technical limitations in discovering NAHR-mediated SVs, this percentage may be an underestimate.

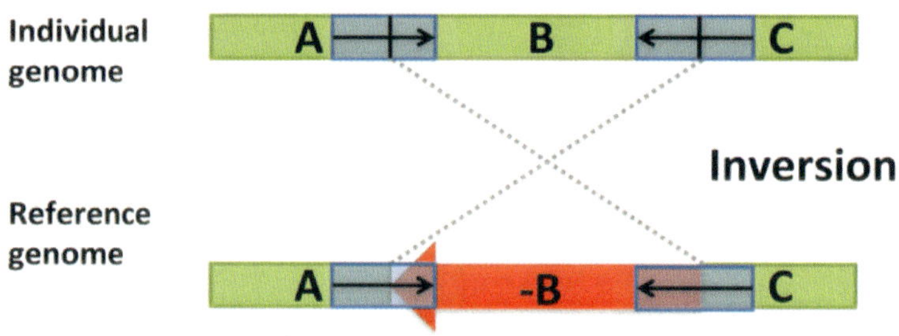

Figure: An inversion resulting from non-allelic homologous recombination (NAHR) between two nearly identical segmental duplications (blue boxes) with opposite orientations (arrows).

The inversion flips the orientation of the subsequence, or block, *B* in one genome relative to the other genome.

There are other mechanisms for the formation of SVs. The division between homology mediated and non-homologous mechanisms may not be so strict. NHEJ events sometimes have some degree of micro homology (e.g. 2–25 bp of similarity) at their breakpoints. Other mechanisms such as fork stalling and template switching (FoSTeS) have also been proposed. Some of these are reviewed in. Finally, the relative contribution of each of these mechanisms in generating germline SVs versus somatic SVs remains an active area of investigation, with conflicting reports about the importance of repetitive sequences in somatic structural variants found in cancer genomes.

Technologies for Measurement of Structural Variation

Structural variants vary widely in size and complexity, ranging from insertions/deletions of hundreds of nucleotides to large scale chromosomal rearrangements. Large structural variants can be visualized directly on chromosomes, through cytogenetic techniques such as chromosome painting, spectral karyotyping (SKY), or fluorescent in situ hybridization (FISH). In fact, Sturtevant and Dobzhansky studied inversion polymorphisms in Drosophila in the 1920's – well before the modern genomics era. However, SVs that are too small to be directly observed on chromosomes are generally more difficult to detect and to characterize than single nucleotide polymorphisms (SNPs). Much of the recent excitement surrounding structural variation stems from improvements in genomics technologies that allow more complete measurements of SVs of all types. These include microarrays and more recently next-generation DNA sequencing technologies.

Microarrays

The first genome-wide surveys of SVs in the human genome in 2004 utilized microarray-based techniques such as array comparative genomic hybridization (aCGH). In aCGH, differentially fluorescently labeled DNA from an individual, or test, genome and a reference genome are hybridized to an array of genomic probes derived from the reference genome. Measurements of test reference fluorescence ratio, called the copy number ratio, at each probe identifies locations of the test genome that are present in higher or lower copy in the reference genome. Microarrays containing hundreds of thousands of probes are available, and thus one obtains copy number ratios at hundreds of thousands of locations. Since individual copy number ratios are subject to various types of experimental error, computational techniques are needed to analyze aCGH data.

aCGH is equally applicable for measurement of germline SVs in normal genomes and somatic SVs in cancer genomes. In fact, aCGH was originally developed for cancer genomics applications. aCGH is now very affordable making it possible to detect copy number variants in large numbers of genomes at reasonable cost. However, aCGH has two important limitations. First, because aCGH measures only differences in the number of copies of a genomic region between a test and reference genome, aCGH detects only copy number variants. Thus, aCGH is blind to copy-neutral, or balanced, variants such as inversions, or reciprocal translocations. Moreover, aCGH requires that the genomic probes from the reference genome lie in non-repetitive regions, making it difficult to detect SVs with breakpoints in repetitive regions, such as NAHR events or the insertion/ deletion of repetitive sequences.

Next-generation DNA Sequencing Technologies

DNA sequencing technology has advanced dramatically in recent years, and several "next-generation" DNA sequencing technologies from companies such as Illumina, ABI, and 454 have significantly lowered the cost of sequencing DNA. However, these technologies, and the Sanger sequencing technique they are replacing, are severely limited in the length of a DNA molecule that can be sequenced. Present sequencing technologies produce short sequences of DNA, called reads, that range from 25–1000 nucleotides, or base pairs (bp), with the upper end of this range requiring technologies (e.g. Sanger and 454) that are considerably more expensive. Much of the recent excitement in DNA sequencing has been in short read DNA sequencers (e.g.llumina Genome Analyzer, Life Technologies Solid and Ion Torrent) that yield reads of only 25–150 nucleotides. These reads are much shorter than the one to two hundred million bp of a typical human chromosome. However, the large number of reads that are produced (hundreds of millions), results in a cost per nucleotide that is several orders of magnitude lower than Sanger sequencing.

Many DNA sequencing technologies employ a paired end, or mate pair, sequencing protocol to increase the effective read length. In this protocol two reads are generated from opposite ends of a longer DNA fragment, or insert. With earlier Sanger sequencing protocols, the sizes of these DNA fragments were dictated by the cloning vector that was used. Fragment, or insert, sizes of 2 kb–150 kb could be obtained by cloning into bacterial plasmids or bacterial artificial chromosomes (BACs). With next-generation technologies, a variety of techniques have been employed to generate paired reads. At present, the most efficient and effective techniques produce paired reads from fragments of only a few hundred bp, although fragments of 2–3 kb are available. Thus, next-generation sequencing technologies have both limited read lengths and limited insert sizes compared to Sanger sequencing.

There are two approaches to detecting SVs from next-generation DNA sequencing data. The first is de novo assembly. In this approach, sophisticated algorithms are used to reconstruct the genome sequence from overlaps between reads. The assembled genome sequence is then compared to the reference genome, or the assembled genomes of other individuals, to identify all types of variants. If the genome sequence is successfully assembled, this approach is the best for characterization of SVs. Unfortunately, assembling a human genome de novo – i.e. with no prior information – of sufficient quality for structural variation studies remains difficult with limited read lengths. Currently, human genome assemblies are highly fragmented, consisting of tens-hundreds of thousands of contigs, intermediate sized sequences of thousands to tens of thousands of nucleotides. Moreover, the associations between some structural variants and repetitive sequences implies that assemblies of finished (not draft quality) are necessary for comprehensive coverage of structural variation. Improving de novo assembly is a very active research area, but human genome assemblies of high enough quality for SV studies remain out of reach for inexpensive short-read technologies.

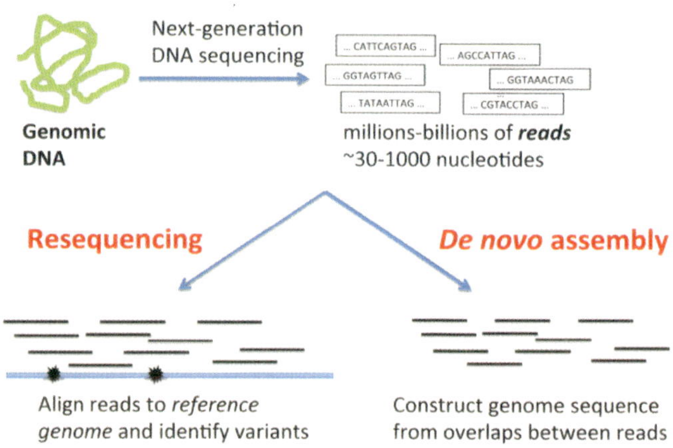

Figure: Major approaches to detect structural variants in an individual genome from next-generation sequencing data are de novo assembly and resequencing.

In de novo assembly, the individual genome sequence is constructed by examining overlaps between reads. In resequencing approaches, reads from the individual genome are aligned to a closely related reference genome. Examination of the resulting alignments reveals differences between the individual genome and the reference genome

The second approach to detect SVs in next-generation DNA sequencing data is a "resequencing" approach that leverages the extensive finishing efforts undertaken in the Human Genome Project. In a resequencing approach, one finds differences between an individual genome and a closely related reference genome whose sequence is known by aligning reads from the individual genome to the reference genome. Differences (variants) between the genomes correspond to differences between the aligned reads and the reference sequence. In the next section, we describe how to predict SVs using a resequencing approach.

New DNA Sequencing Technologies

Many of the challenges in reliable measurement of SVs described above are related to limitations in sequencing technologies. In particular, SVs with breakpoints in highly-repetitive sequences are beyond the abilities of current technologies. New "third-generation" and single-molecule technologies promise additional advantages for structural variation discovery. These advantages include longer read lengths, easier sample preparation, lower input DNA requirements, and higher throughput. For example, Pacific Biosciences recently released their Single-Molecule Real Time (SMRT) sequencing, a technology that measures in real time the incorporation of nucleotides by a single DNA polymerase molecule immobilized in a nanopore.

One application of this technology is strobe sequencing. A strobe read, or strobe, consists of multiple sub reads from a single contiguous molecule of DNA. These sub reads are separated by a number of "dark" nucleotides (called advances), whose identity is unknown. Thus far, Pacific Biosciences has demonstrated strobes of lengths up to 20 kb with 2–4 sub reads each of 50–400 bp. Additional improvements are expected as technology matures. Strobes generalize the concept of paired reads by including more than two reads from a single DNA fragment. Strobes provide long-range sequence information with low input DNA requirements, a feature missing from current sequencing technologies. This additional information is useful for detection and de novo assembly of complex SV that lie in highly repetitive regions, or contain multiple breakpoints in a small region. However, the advantages of strobes are reduced by higher single-nucleotide error rates. Thus, realizing the advantages of strobes requires new algorithms that exploit information from multiple, spaced subreads to overcome high single-nucleotide error rates.

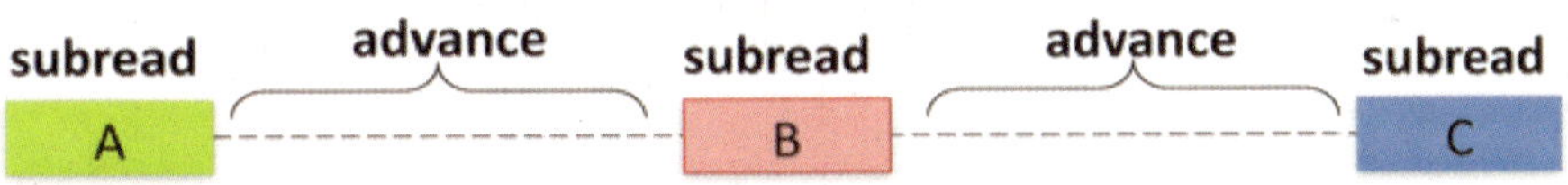

Figure: A strobe with 3 sub reads.

Sequencing technologies continues its rapid development. Improvements in the chemistry, imaging, and manufacture of existing technologies are increasing their read lengths, insert lengths, and throughput. Additional sequencing technologies are under active development. Nanopore-based technologies that directly read the nucleotides of long molecules of DNA hold promise for a dramatic shift in DNA sequencing where extremely long reads (tens of kb) are generated, making both de novo assembly and variant detection by resequencing straightforward problems.

Resequencing Strategies for Structural Variation

A resequencing strategy predicts SVs by alignments of sequence reads to the reference genome. There are two main steps in any resequencing strategy:

1. Alignments of reads;
2. Prediction of SVs from alignments.

Resequencing approaches are straightforward in principle, but in practice sensitive and specific detection of structural variation in human genomes is notoriously difficult. While some types of SVs are easy to detect with next-generation sequencing technologies, other complex SVs are refractory to detection. This is due to both technological limitations and biological features of SVs. DNA sequencing technologies produce reads with sequencing errors, have limited read lengths and insert sizes, and have other sampling biases (e.g. in GC-rich regions). Biologically, human SVs are: (i) enriched for repetitive sequences near their breakpoints; (ii) may overlap, have multiple states or complex architectures; and (iii) recurrent (but not identical) variants may exist at the same locus. These properties mean that the alignment of reads to the reference genome and the prediction of SVs from these alignments is not always an easy task. Algorithms are required to make highly sensitive and specific predictions of SVs.

We review the main issues in predicting SVs using a resequencing approach. We begin with read alignment. Then we describe the three major approaches that are used to identify structural variants from aligned reads: (i) split reads; (ii) depth of coverage analysis; and (iii) paired-end mapping.

Read Alignment

Alignment of reads to a reference genome is a special case of sequence alignment, one of the most researched problems in bioinformatics. However, the specialized task of aligning millions-billions of individual short reads led to the development of new software programs tailored to this task, such as Maq, BWA, Bowtie/Bowtie2, BFAST, mrsFAST, etc. A key decision in read alignment for SV detection is whether to consider only reads with a single, best alignment to the reference genome, or to also include reads with multiple high-quality alignments. Some read alignment programs will output only a single alignment for each read, in some cases choosing an alignment randomly if there are multiple alignments of equal score. If one uses only reads with a unique alignment, then there is limited power to detect SVs whose breakpoints lie in repetitive regions, such as SVs resulting from NAHR. On the other hand, if one allows reads whose alignment is ambiguous, then the problem of SV prediction requires an algorithm to distinguish among the multiple possible alignments for each read. Many SV prediction algorithms analyze only unique alignments, although several recent algorithms use ambiguous alignments. A few of these are noted below.

Split Reads

A direct approach to detect structural variants from aligned reads is to identify reads whose alignments to the reference genome are in two parts. These so called split reads contain the breakpoint of the structural variant. To reduce false positive predictions of structural variants, one requires the presence of multiple split reads sharing the same breakpoint. Because the two parts of a split read align independently to the reference genome, these alignments must be long enough to be aligned uniquely (or with little ambiguity) to the reference. Thus, split read analysis is a feasible strategy only when the reads are sufficiently long. For example, if one has a 36 bp read containing the breakpoint of an SV at its midpoint, one must align the two 18 bp halves of the read to the reference genome. Finding unique alignments of an 18 bp sequence is often not possible. There are no reports of successful prediction of structural variants from split reads alone using next generation DNA sequencing reads less that 50 bp in length. Instead, split read methods have been proposed that use paired reads, and require that one read in the pair has a full length alignment to the reference. This alignment of the read from one end of the fragment is used to anchor the search for alignments of the other split read of the fragment.

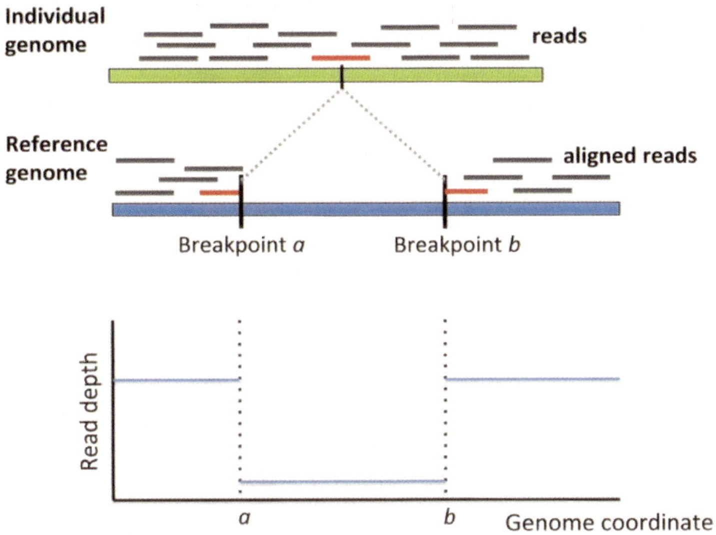

Figure: Identification of a deletion in an individual genome by split read analysis (middle), and by depth of coverage analysis (bottom).

Depth of Coverage

Depth of coverage (also called read depth) analysis detects differences in the number of reads that align to intervals in the reference genome. Assuming that reads are sampled uniformly from the genome sequence, the number of reads that contain a given nucleotide of the reference is, on average, $c = \frac{NL}{G}$, where N is the number of reads, L is the length of each read, and G is the length of the genome. This is the Lander-Waterman model, and the parameter c, called the coverage, is a key parameter in a sequencing experience. For example, recent cancer sequencing projects with Illumina technology have used "30X coverage" which means that the number of reads and length of reads are chosen such that $c = 30$.

Now, if the individual genome contained a deletion of a segment of the human reference genome, the coverage of this segment would be reduced by half – if the deletion was heterozygous or reduced to zero – if the deletion was homozygous. Similarly, if an interval of the reference genome was duplicated, or amplified, in the individual genome, the coverage of this interval would increase in proportion to the number of copies. Thus, the observed coverage of an interval of the reference genome, the depth of coverage, gives an indication of the number of copies of this interval in the individual genome. Of course, there are numerous additional factors to consider beyond this simple analysis. For example, since reads are sampled at random from the genome, coverage is not constant, but rather follows a distribution with mean c. A Poisson distribution is typically used as an approximation to this distribution, although other distributions sometimes provide a better fit to the data. In addition, repetitive sequences in the reference genome and biases in sequencing (e.g. different coverage of GC-rich regions) also affect depth of coverage calculations. Nevertheless, there are several computational methods for depth of coverage analysis. Many of these are largely similar to those used to analyze microarray copy number data.

Paired-end Sequencing and Mapping

The most common approach for resequencing SVs is paired-end mapping (PEM). Paired-end mapping was used to identify somatic SVs in cancer genomes and the same idea has been applied to identify germline structural variants. While the early paired-end mapping studies used older clone-based sequencing, paired-end mapping is now possible using various next-generation sequencing technologies.

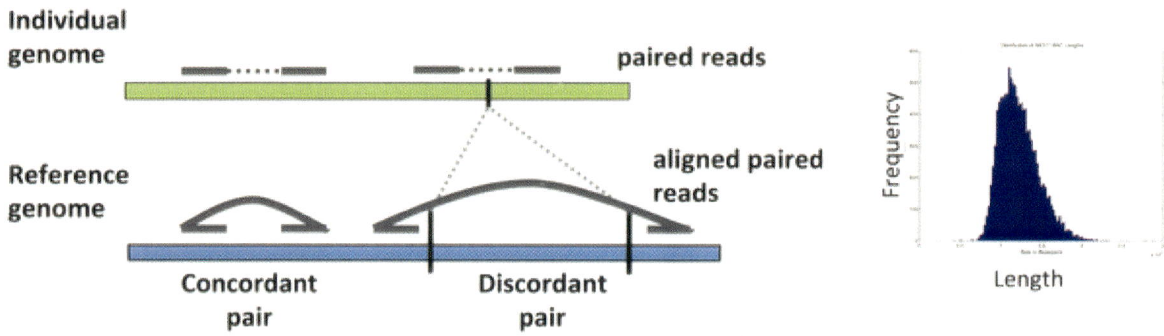

Figure: Paired end mapping (PEM).

Fragments from an individual genome are sequenced from both ends and the resulting paired reads are aligned to a reference genome. Most paired reads correspond to concordant pairs, where the distance between the alignment of each read agrees with the distribution of fragment lengths (right). The remaining discordant pairs suggest structural variants (here a deletion) that distinguish the individual and reference genome

In PEM, a paired-end sequencing protocol is used to obtain paired reads from opposite ends of a larger DNA fragment, or clone, from a individual genome. These paired reads are then aligned to a reference genome. Most paired reads result in concordant pairs where the distance between aligned reads is equal to the fragment length. In contrast, discordant pairs have alignments with abnormal distance or that lie on different chromosomes. These suggest the presence of an SV or a sequencing error. For example, a discordant pair whose distance between alignments is too long suggests a deletion in the individual genome, while a discordant pair whose alignments are on different chromosomes suggests a translocation. Other types of discordant pairs identify inversions, transpositions, or duplications that distinguish the individual genome from the reference genome. Note that in general the length of any particular sequenced fragment is not known. Rather, during the preparation of genomic DNA for sequencing, the DNA is fragmented and fragments are size-selected to an appropriate target length. It is desirable for this size selection to be as strict as possible, so that only fragments near the target length are sequenced. However, in practice the size selection procedure produces fragments whose lengths vary around the target length. Typically, the distribution of fragment lengths is obtained empirically by examining the distances between all aligned paired reads, as most fragments will correspond to a concordant pair.

To distinguish real SVs from sequencing errors, one looks for clusters of discordant pairs that indicate the same SV. Numerous algorithms have been developed to predict SVs by finding clusters of discordant pairs. Early algorithms used only those paired reads whose alignments to the reference genome were non-ambiguous; i.e. there was only a single "best alignment". More sophisticated algorithms use paired reads with multiple ambiguous alignments to the reference genome and use a variety of combinatorial and statistical techniques to select among these alignments. Finally, some approaches model the fact that the human genome is diploid to avoid making inconsistent structural variant predictions.

All of the approaches above rely on predicting structural variants that are supported by multiple paired reads. Some, but not all, of them are careful when determining whether a group of paired reads genuinely support the same variant. A key feature of GASV is that it records both the information that the paired reads reveal about the boundaries (breakpoints) of the structural variant and the uncertainty associated with this measurement. Most types of SV, including deletions, inversions, and translocations have two breakpoints a and b where the reference genome is cut. The segments adjacent to these coordinates are then pasted together in a way that is particular to the type of SV. For example, a deletion is defined by coordinates $a$and b in the reference genome such that the nucleotide at position a is joined to the nucleotide at position b in the individual genome. Note that this is a simplification of the underlying biology, as there are sometimes small insertions or deletions at breakpoints, but these small changes have limited effect on the analysis of larger structural variants.

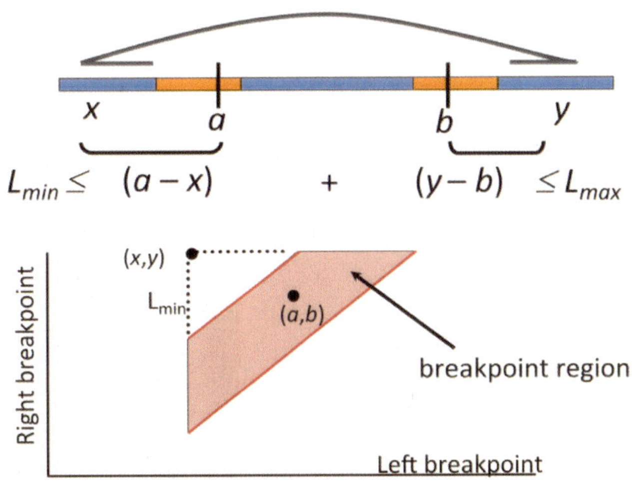

(Top) A discordant pair (arc) indicates a deletion with unknown breakpoints a and b located in orange blocks. Positions x,y and the minimum L_{min} and maximum L_{max} length of end-sequenced fragments constrain breakpoints (a,b) to lie within the indicated orange blocks, and are governed by the indicated linear inequalities. (Bottom) A polygon in 2D genome space expresses the linear dependency between breakpoints a and b and records the uncertainty in the location of the breakpoints.

Now the discordant pairs that indicate an SV have the property that the locations of the read alignments are near the breakpoints a and b. However, a paired read does not give independent information about the breakpoint a and the breakpoint b. Rather, the breakpoints a and b are related by a linear inequality that defines a polygon in 2D genome space called the breakpoint region. For example, suppose that the pair of reads from a single fragment align to the same chromosome of the reference genome such that the read with lower coordinate starts at position x in the reference and the read with higher coordinate ends at position y in the reference. (For simplicity, we ignore the fact that the sequence of a read can align to either strand (forward or reverse) of the reference genome. The strand of an alignment gives additional information about the location of the breakpoint.) If the sequenced fragment has length L then the breakpoints a and b satisfy the equation $(a-x)+(y-b)=L$. As described above, the size of any particular fragment is typically unknown.

Rather, one defines a minimum size L_{min} and maximum size L_{max} of a sequenced fragment, perhaps according to the empirical fragment length distribution. Thus, we have the inequality

$$L_{min} \leq (a-x)+(y-b) \leq L_{max}.$$

This equation defines the unknown breakpoints a and b in terms of the known coordinates x and y of the aligned reads and the length of sequenced fragments. The pairs of breakpoints (a,b) that satisfy this equation form a polygon (specifically a trapezoid) in two-dimensional genome space. We define the breakpoint region B of discordant pair (x,y) to be the breakpoints $(a\ b)$ satisfying the above equation.

This geometric representation provides a principled way to combine information across multiple paired-reads: multiple paired-reads indicate the same variant if their corresponding breakpoint regions intersect. The geometric representation also provides precise breakpoint localization by multiple paired reads; separates multiple measurements of the same variant from measurements of nearby or overlapping variants; and facilitates robust comparisons across multiple samples and measurement technologies. Finally, the approach is computationally efficient as it relies on computational geometry algorithms for polygon intersection. These scale to millions of discordant pairs that result from next-generation sequencing platforms.

While the algorithms above consider many of the issues in prediction of structural variants, there remains room for improvement. Most notably, many algorithms still use only one of the possible signals of structural variants: read depth, split reads, or paired reads. Improvements in specificity are likely possible by integrating these multiple signals into a single prediction algorithm.

Representation of Structural Variants

Next generation DNA sequencing technologies are dramatically reducing the cost of sequence-based surveys of structural variants, while oligonucleotide aCGH techniques are now used in studies profiling tens of thousands of genomes. Large projects like the 1000 Genomes Project and The Cancer Genome Atlas (TCGA) are performing paired-end sequencing and aCGH of many human genomes, and matched tumor and normal genomes, respectively. At the same time, smaller or single investigator projects are using a variety of paired-end sequencing approaches and/or microarray-based techniques with different trade-offs in cost-per-sample vs. measurement resolution. Thus, in the near future there will be an enormous number of measurements of SVs, but using a wide range of technologies of varying resolution, sensitivity, and specificity. This diversity of approaches will likely continue for some time as investigators explore tradeoffs between the cost of measuring variants in one sample with high confidence versus surveying variants in many samples with lower confidence per sample. For example, in cancer genome studies the goal of finding recurrent mutations demands the survey of many genomes and thus large sample sizes might be preferred over high coverage sequencing of one sample.

The problem of comparing variants across samples and/or measurement platforms is less studied than the problem of detecting variants in a single sample. Standard practice remains to use heuristics that merge predicted structural variants into the same variant in they overlap by a significant fraction (e.g. 50–70%) on the reference genome. For example, the Database of

Genomic Variations (DGV), arguably the most comprehensive repository of measured human structural variants, merges structural variant predictions whose coordinates overlap by $\geq 70\,\%$ on the reference genome. Such heuristics are typically the only approach available to databases of human structural variants because many early studies did not report information on the uncertainty (i.e. "error bars") in the boundaries (breakpoints) of the variant. This situation makes it difficult to explicitly separate multiple measurements of the same variant from measurements of nearby variants or overlapping variants. This situation is now improving, and more recent software records both the information that the measurement reveals about the breakpoints of the structural variant and the uncertainty associated with this measurement. Software that uses this uncertainty to classify and compare SVs across samples and measurement platforms is also now available. Such precision provides increased confidence in associations between a structural variant and a disease, helps separate germline from somatic structural variants in cancer genome sequencing projects, and aids in the study of rare recurrent variants that might occur on a variety of genetic backgrounds.

Genetic Distance

Genetic distance is the degree of genetic difference (genomic difference) between species or populations that is measured by some numerical method. Thus, the average number of codon or nucleotide differences per gene is a measure of genetic distance. There are various molecular data that can be used for measuring genetic distance.

In the comparison of closely related species or populations, however, the effect of polymorphism cannot be neglected, and one has to examine many proteins or genes. For this reason, it is customary to measure the genetic distance between populations in terms of a function of allele frequencies for many genetic loci. All proposed distances are bounded between 0 and 1 and have the property that the distance between an individual and itself is 0.

Genetic distances based on a model of DNA substitution can only be estimated for aligned sequences. Distance estimations are therefore unreliable when the sequences to be compared are too heterogeneous to align. There are, however, a number of methods for deriving distances from non-aligned sequence data; they are particularly useful for whole-genome data. A common approach is to assess the frequencies with which all possible strings of a fixed length (e.g., all possible "words" that can be formed with six nucleotides) occur in each sequence to be compared; these frequency vectors can then be used to define a distance function between the sequences. It has been shown, by using both simulated and empirical datasets, that the analysis of alignment-free distances results in phylogenetic trees with higher overall reconstruction accuracies if variation of substitution rates within the sequences is high.

The Nci's standard genetic distance is defined as follows. Consider two populations, X and Y. Let x_{ik} and y_{ik} be the frequencies of the k-th alleles ($i = 1, \cdots, N$, $k \in \{1, 2\}$ in a binary coded GA) in X and Y, respectively. The probability of identity of two randomly chosen genes is $j_{xi} = x_{i1}^2 + x_{i2}^2$ in the population X, while it is $j_{yi} = y_{i1}^2 + y_{i2}^2$ in the population Y. The probability of identity of a gene from X and a gene from Y is $j_{xyi} = x_{i1} y_{i1} + x_{i2} y_{i2}$.

The normalized identity of genes between X and Y with respect to a locus is defined as,

$$I_i = \frac{j_{xyi}}{\sqrt{j_{xi}}\sqrt{j_{yi}}},$$

Where, Ii = 1.0 if the two populations have the same alleles in identical frequencies, and Ii = 0.0 if they have no common alleles. The normalized identity of genes between X and Y with respect to the average in all loci is defined as,

$$I = \frac{J_{XY}}{\sqrt{J_X}\sqrt{J_Y}},$$

Where $J_X = \sum_{i=1}^{N} j_{xi} \,/ N$, $J_Y = \sum_{i=1}^{N} j_{yi} / N$ and $J_{XY} = \sum_{i=1}^{N} j_{xyi} \,/ N$. The genetic distance between X and Y is defined as,

$$D = -\log_e I,$$

under the assumption that the mutation rate per locus is sufficiently small. However, the above definition cannot be applied to the standard GA directly, because it is assumed that a new allele always appears on a locus when a mutation occurs, while "back mutations " frequently occur in the standard GA, due to the binary coding scheme. Therefore, the genetic distance between the population at the initial generation and the one at the last generation is calculated as:

$$D_{final} = \sum_{1}^{T-1} D_{t,t+1}$$

Where T is the number of the last generation and $D_{t,t+1}$ is the genetic distance between the population in the t-th and the (t + 1)-th generation. The rate of gene substitution is defined as the genetic distance per generation.

Cavalli-Sforza Chord Distance

In 1967, Luigi Luca Cavalli-Sforza and A. W. F. Edwards published this measure. It assumes that genetic differences arise due to genetic drift only. One major advantage of this measure is that the populations are represented in a hyper sphere, the scale of which is one unit per gene substitution. The chord distance in the hyper dimensional sphere is given by;

$$D_{CH} = \frac{2}{\pi}\sqrt{2(1 - \sum_l \sum_u \sqrt{X_u Y_u})}$$

Some authors drop the factor $\frac{2}{\pi}$ to simplify the formula at the cost of losing the property that the scale is one unit per gene substitution.

Reynolds, Weir and Cockerham's Genetic Distance

In 1983, this measure was published by John Reynolds, B.S. Weir and C. Clark Cockerham. This measure assumes that genetic differentiation occurs only by genetic drift without mutations. It estimates the coancestry coefficient Θ which provides a measure of the genetic divergence by:

$$\Theta_w = \sqrt{\frac{\sum_l \sum_u (X_u - Y_u)^2}{2\sum_l (1 - \sum_u X_u Y_u)}}$$

Many other measures of genetic distance have been proposed with varying success.

Nei's DA Distance

This distance assumes that genetic differences arise due to mutation and genetic drift, but this distance measure is known to give more reliable population trees than other distances particularly for microsatellite DNA data.

$$D_A = 1 - \sum_l \sum_u \sqrt{X_u Y_u} / L$$

Euclidean Distance

$$D_{EU} = \sqrt{\sum_u (X_u - Y_u)^2}$$

Goldstein Distance

It was specifically developed for microsatellite markers and is based on the stepwise-mutation model (SMM). μ_X and μ_Y are the means of the allele sizes in population X and Y.

$$(\delta\mu)^2 = \sum_l (\mu_X - \mu_Y)^2 / L$$

Nei's Minimum Genetic Distance

This measure assumes that genetic differences arise due to mutation and genetic drift.

$$D_m = \frac{(J_X + J_Y)}{2} - J_{XY}$$

Roger's Distance

$$D_R = \frac{1}{L}\sqrt{\frac{\sum_u (X_u - Y_u)^2}{2}}$$

Fixation Index

A commonly used measure of genetic distance is the fixation index, which varies between 0 and 1. A value of 0 indicates that two populations are genetically identical (minimal or no genetic diversity between the two populations) whereas a value of 1 indicates that two populations are genetically different (maximum genetic diversity between the two populations). No mutation is assumed. Large populations between which there is much migration, for example, tend to be little differentiated whereas small populations between which there is little migration tend to be greatly differentiated. Fst is a convenient measure of this differentiation, and as a result Fst and related statistics are among the most widely used descriptive statistics in population and evolutionary genetics. But Fst is more than a descriptive statistic and measure of genetic differentiation. Fst is directly related to the Variance in allele frequency among populations and conversely to the degree of resemblance among individuals within populations. If Fst is small, it means that allele frequencies within each population are very similar; if it is large, it means that allele frequencies are very different.

Comparison of Genetic and Physical Distances

Genetic distances between loci are determined by the frequency with which recombination events occur between the genes. Genetic distances are expressed in centimorgans (cM) where 1 cM corresponds to a recombination frequency of 1%. It is generally assumed that homologous recombination events can occur at any point in the DNA sequence and therefore genetic distances reflect physical distances. The ability to compare the relationship between genetic and physical distances more directly is now possible because of the technological advances in gene mapping allowing determination of the physical distance between two loci.

In humans, a genetic distance of 1 cM is calculated to be approximately 1200 kb (the average size of the human genome is about 2700 cM and contains approximately 3.3 × 106 kb of DNA). Within the HLA complex, there are regions where genetic and physical distances approximate calculated values and there are other regions where there are marked discrepancies. The physical distance (~1000 kb) between HLA-DR and HLA-B corresponds to the genetic distance (1 cM). By contrast, a recombination frequency of 3% (3 cM) has been observed between the HLA-DB and DQ sub regions; these loci are physically separated by only ~500 kb. Therefore, there is a significantly increased frequency of recombination between HLA-DP and HLA-DQ compared to the physical distance that separates them. These data imply that the properties of any given DNA segment may vary from those of another.

Display Techniques

Genetic Maps

Several techniques reduce distance matrices into a more manageable two dimensional graphical representation. The first of these, principal coordinates analysis, is the method most commonly used to produce the familiar two-dimensional genetic "maps." This technique has been reviewed elsewhere. With the exception of new interpolation techniques and color graphic. Little new methodological work appears to have been done since these reviews were published. The principal coordinates approach is mathematically sound, and the frequency of its use in the published literature indicates a high degree of consumer satisfaction.

Multidimensional scaling also generates genetic maps and can be performed on either metric or nonmetric data. It has the advantage of statistical robustness. Lalouel reviews the technique extensively and discusses its application to both symmetric and asymmetric matrices.

Evolutionary Trees

In marked contrast to the two-dimensional maps, a great deal of work has been done on estimating the structure of evolutionary trees during the past several years. This reflects the fact that trees attempt to convey more information about relationships between populations than do maps of a substantial body of effort, it is still debatable whether tree methodologies really achieve this goal.

Many tree methods are based on the criterion of maximum parsimony: the tree which requires the fewest number of gene substitutions between branching points is found. This is equivalent to the "minimum evolution" method first used by Edwards & Cavalli-Sforza. The approach works best with populations that have diverged relatively recently, since back mutations and convergent evolution violate the assumptions of the model.

Other tree methodologies are based on the compatibility principle. Here an attempt is made to find the tree that is compatible with the largest number of variables (e.g. gene frequencies). The compatibility and parsimony approaches have been reviewed thoroughly and critically by Felsenstein.

A substantial amount of work has been done on maximum like hood techniques for phylogeny estimation. The tree is chosen in which the statistical likelihood of a given data set is maximized, given the particular phylogenetic model under consideration. Fleckenstein is the leading exponent of this approach, and he has developed a "restricted" maximum likelihood algorithm which avoids some of the pitfalls of earlier endeavors. In particular, the problem of singularities in the likelihood surface is overcome. This approach has the advantages of a solid statistical basis and the ability to estimate standard errors of branch lengths. Also, the likelihoods of various topologies are calculated, allowing a well-defined statistical evaluation of which topology is "best." Disadvantages include relatively slow computation and dependence on the assumptions of this parametric method. Also, it is possible to obtain only a local maximum when using this technique; thus it does not guarantee that the "best" tree will be found.

Templeton has proposed a highly original approach to tree estimation. His technique is intended for use with restriction site and/or DNA sequence data (which will be discussed below). A maximum parsimony tree is generated for each restriction enzyme used (these correspond roughly to loci or small sets of loci). Then the compatibility criterion is applied in order to select a single tree that corresponds most closely with the largest number of individual parsimony trees. A nonparametric sign test can be used to determine whether two trees differ significantly. This approach has the advantages of distribution free robustness and relative ease of computation. If convergent evolution due to parallel mutations is probable (i.e. when λ_t.05, where λ is the gene substitution rate and t is the time since divergence between subpopulations), the method does not work well. In this case, Templeton advises resorting to the maximum likelihood approach cited above. An application of these methods to mitochondrial DNA data from humans and great apes has yielded a provocative phylogeny. The human line splits before the chimp-gorilla divergence, which leads Templeton to conclude that bipedalism (or at least an incipient version of it) may have been the primitive state in the human-chimp gorilla ancestor, while knuckle-walking is a derived state which evolved after humans split from the chimp-gorilla line.

Nei et al have compared several tree-making algorithms, including Farris's maximum parsimony approach and the unweighted paired-group method (UPGMA), a simple agglomerative clustering technique. The techniques were applied to simulated data where the true phylogeny was known. Interestingly, the UPGMA technique, which consumes very\' little computer time, gave the most accurate results. Recently, a method has been derived to estimate the standard errors of branching points in UPGMA trees, allowing statistical inference. It would be interesting to conduct a similar comparison between UPGMA and the maximum likelihood approach.

One of the most persistent problems in applying tree methods to human data is the omnipresence of gene flow between populations after their divergence. This vitiates any estimate of divergence times. A partial solution to this problem has been proposed by Lathrop. who presents a method that simultaneously calculates maximum likelihood estimates of phylogenies and admixture rates between populations. A critical assumption is that the duration of admixture is short relative to the period of isolation between two subpopulations. Often, not enough is known about the migration history of human populations to ensure that this assumption will be fulfilled. Still, the method should be useful in some cases.

As in any genetic distance methodology, tree estimation is more reliable when a large number of loci are used. Estimated trees are nearly always erroneous (i.e. the topological arrangement will be wrong) if the number of loci is less than 30. if populations are closely related (as in many anthropological studies), even trees based on 100 loci will usually be incorrect. This is because the error variance of genetic distances increases when the distances arc small. In addition, adding more populations to the analysis increases the probability that the tree will be incorrect. Thus, it is best to use as few populations as possible. One good rule of thumb is to use at least three to four times as many independent characters as populations.

Maps or Trees

In deciding whether to estimate genetic maps or trees, the researcher must weigh the advantages and disadvantages of each. Maps offer computational case and good statistical definition. Unlike comparisons at the species level, comparisons of human populations sometimes do not involve any type of hierarchical structure. Since trees imply hierarchy, they tend to be inappropriate in such situations. The ubiquity and complexity of human migration patterns guarantee that branching points in trees will nearly always be suspect, in spite of promising new advances in incorporating admixture into tree estimation. On the other hand, many tree methods now allow the estimation of confidence limits for branching points and branch lengths, while statistical error estimation is absent from map methods.

Statistical confidence has remained sadly neglected in nearly ail distance studies. Whether by using tree methods that allow statistical inference, or by examining the standard errors of genetic distances themselves and using maps, more attention needs to be paid to statistical significance. It could well be that many controversies in human microevolution involve topologies (either maps or trees) that are not statistically different from one another.

Autocorrelation

A novel development in the display of genetic relationships is the application of spatial autocorrelation methods to gene frequency data. An autocorrelation is the correlation of a series of frequencies, in this case arranged along a spatial coordinate system, with itself. At a displacement distance of zero, the autocorrelation is of course 1.0. But it is expected to decrease as the displacement distance increases. For example, the correlation between the series and itself, displaced by 50 kilometers, might be 0.5. Autocorrelations at regular geographic distances are plotted against distance (displacement) in a correlogram. In a pure drift system, there should be no spatial pattern of gene frequencies, so all autocorrelations at non-zero displacement values should be zero. Autocorrelation patterns corresponding to environmental gradients could indicate natural selection, while others (especially if they are similar for all loci) could reflect patterns of gene flow. This method has been applied to data from Bougainville and indicates, in agreement with earlier studies, that gene flow is the primary' determinant of genetic variation here. It has also been applied to pan-European gene frequencies to test hypotheses regarding the spread of farming during the Neolithic period.

Selective Sweep

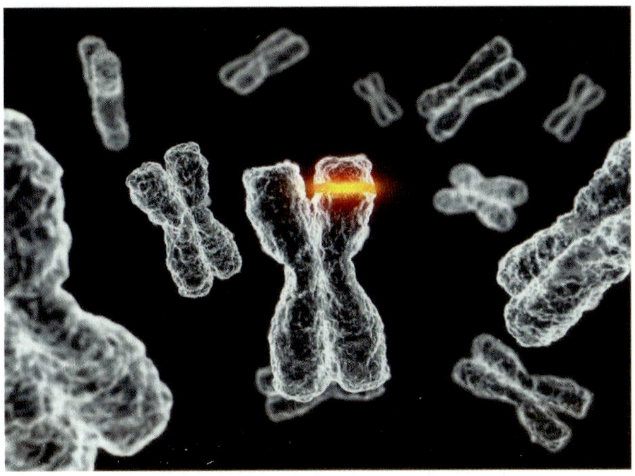

Selective sweep, or genetic hitchhiking, is a genetics and evolution term that explains how alleles for favorable adaptations, and their associated alleles near them on chromosomes, become more frequently seen in a population due to natural selection.

Strong Alleles

Natural selection works to choose the most favorable alleles for an environment in order to keep a species passing down those traits generation after generation. The more favorable the allele for the environment, the more likely the individuals that possess that allele will be to live long enough to reproduce and pass that desirable trait down to their offspring. Eventually, undesirable traits will be bred out of the population and only the strong alleles will be left to continue on.

Proccess of a Selective Sweep

The selection of these preferred traits can be very strong. After a particularly strong selection for a trait that is the most desirable, a selective sweep will happen. Not only will the genes that code for the favorable adaptation increase in frequency and be seen more often in the population, other traits that are controlled by alleles that are close in proximity to those favorable alleles will also be selected for, whether they are good or bad adaptations.

Also called "genetic hitchhiking", these extra alleles come along for the selection ride. This phenomenon may be the reason why some seemingly undesirable traits get passed down, even if it does not make the population the "fittest". One major misconception of how natural selection works is the idea that if only the desirable traits are selected for, then all other negatives, such as genetic diseases, should be bred out of the population. Yet, these not so favorable characteristics seem to persist. Some of this could be explained by the idea of selective sweep and genetic hitchhiking.

Examples of Selective Sweep in Humans

Do you know someone who is lactose intolerant? People who suffer from lactose intolerance are unable to fully digest milk or milk products like cheese and ice cream. Lactose is a type of sugar that is found in milk that requires the enzyme lactase in order to be broken down and digested. Human infants are born with lactase and can digest the lactose. However, by the time they reach adulthood, a large percentage of the human population loses the ability to produce lactase and therefore can no longer handle drinking or eating milk products.

Looking Back at our Ancestors

About 10,000 years ago, our human ancestors learned the art of agriculture and subsequently started to domesticate animals. The domestication of cows in Europe allowed these people to use cow's milk for nutrition. Over time, those individuals who had the allele to make lactase possessed the favorable trait over those who could not digest the cow's milk.

A selective sweep occurred for the Europeans and the ability to get nutrition from milk and milk products was highly positively selected. Therefore, the majority of Europeans possessed the ability to make lactase. Other genes hitchhiked along with this selection. In fact, researchers estimate that about a million base pairs of DNA hitchhiked along with the sequence that coded for the lactase enzyme.

Another Example is Skin Color

Another example of a selective sweep in humans is skin color. As human ancestors moved from Africa where dark skin is a necessary protection against the direct ultraviolet rays of the sun, less direct sunlight meant that the dark pigments were no longer necessary for survival. Groups of these early humans moved north to Europe and Asia and gradually lost the dark pigmentation in favor of a lighter coloring for the skin.

Not only was this lack of dark pigmentation favored and selected, nearby alleles that controlled the rate of metabolism hitchhiked along. Metabolic rates have been studied for different cultures all over the world and have been found to correlate very closely to the type of climate where the individual lives, much like the skin coloring genes. It is proposed that the skin pigmentation gene and the metabolic rate gene were involved in the same selective sweep in the early human ancestors.

Detection

Whether or not a selective sweep has occurred can be investigated in various ways. One method is to measure linkage disequilibrium, i.e., whether a given haplotype is overrepresented in the population. Under neutral evolution, genetic recombination will result in the reshuffling of the different alleles within a haplotype, and no single haplotype will dominate the population. However, during a selective sweep, selection for a positively selected gene variant will also result in selection of neighboring alleles and less opportunity for recombination. Therefore, the presence of strong linkage disequilibrium might indicate that there has been a recent selective sweep, and can be used to identify sites recently under selection.

There have been many scans for selective sweeps in humans and other species, using a variety of statistical approaches and assumptions.

In maize, a recent comparison of yellow and white corn genotypes surrounding Y1—the phytoene synthetase gene responsible for the yellow endosperm color, shows strong evidence for a selective sweep in yellow germplasm reducing diversity at this locus and linkage disequilibrium in surrounding regions. White maize lines had increased diversity and no evidence of linkage disequilibrium associated with a selective sweep.

Relevance to Disease

Because selective sweeps allow for rapid adaptation, they have been cited as a key factor in the ability of pathogenic bacteria and viruses to attack their hosts and survive the medicines we use to treat them. In such systems, the competition between host and parasite is often characterized as an evolutionary "arms race", so the more rapidly one organism can change its method of attack or defense, the better. This has elsewhere been described by the Red Queen hypothesis. Needless to say, a more effective pathogen or a more resistant host will have an adaptive advantage over its conspecifics, providing the fuel for a selective sweep.

One example comes from the human influenza virus, which has been involved in an adaptive contest with humans for hundreds of years. While antigenic drift (the gradual change of surface antigens) is considered the traditional model for changes in the viral genotype, recent evidence suggests that selective sweeps play an important role as well. In several flu populations, the time to the most recent common ancestor (TMRCA) of "sister" strains, an indication of relatedness, suggested that they had all evolved from a common progenitor within just a few years. Periods of low genetic diversity, presumably resultant from genetic sweeps, gave way to increasing diversity as different strains adapted to their own locales.

A similar case can be found in Toxoplasma gondii, a remarkably potent protozoan parasite capable of infecting warm-blooded animals. T. gondii was recently discovered to exist in only three clonal lineages in all of Europe and North America. In other words, there are only three genetically distinct strains of this parasite in all of the Old World and much of the New World. These three strains are characterized by a single monomorphic version of the gene Chr1a, which emerged at approximately the same time as the three modern clones. It appears then, that a novel genotype emerged containing this form of Chr1a and swept the entire European and North American population of Toxoplasma gondii, bringing with it the rest of its genome via genetic hitchhiking. The

South American strains of T. gondii, of which there are far more than exist elsewhere, also carry this allele of Chr1a.

Epigenetics

Epigenetics is the study of heritable changes in gene expression (active versus inactive genes) that do not involve changes to the underlying DNA sequence — a change in phenotype without a change in genotype — which in turn affects how cells read the genes. Epigenetic change is a regular and natural occurrence but can also be influenced by several factors including age, the environment/lifestyle, and disease state. Epigenetic modifications can manifest as commonly as the manner in which cells terminally differentiate to end up as skin cells, liver cells, brain cells, etc. Or, epigenetic change can have more damaging effects that can result in diseases like cancer. At least three systems including DNA methylation, histone modification and non-coding RNA (ncRNA)-associated gene silencing are currently considered to initiate and sustain epigenetic change. New and ongoing research is continuously uncovering the role of epigenetics in a variety of human disorders and fatal diseases.

Epigenetics and the Environment: How Lifestyle can Influence Epigenetic Change from One Generation to the Next

The field of epigenetics is quickly growing and with it the understanding that both the environment and individual lifestyle can also directly interact with the genome to influence epigenetic change. These changes may be reflected at various stages throughout a person's life and even in later generations. For example, human epidemiological studies have provided evidence that prenatal and early postnatal environmental factors influence the adult risk of developing various chronic diseases and behavioral disorders. Studies have shown that children born during the period of the Dutch famine from 1944-1945 have increased rates of coronary heart disease and obesity after maternal exposure to famine during early pregnancy compared to those not exposed to famine. Less DNA methylation of the insulin-like growth factor II (IGF2) gene, a well-characterized epigenetic locus, was found to be associated with this exposure. Likewise, adults that were prenatally exposed to famine conditions have also been reported to have significantly higher incidence of schizophrenia.

Lifestyle Affect on Individual Epigenetics and Health

Although our epigenetic marks are more stable during adulthood, they are still thought to be dynamic and modifiable by lifestyle choices and environmental influence. It is becoming more apparent that epigenetic effects occur not just in the womb, but over the full course of a human life span, and that epigenetic changes could be reversed. There are numerous examples of epigenetics that show how different lifestyle choices and environmental exposures can alter marks on top of DNA and play a role in determining health outcomes.

The environment is being investigated as a powerful influence on epigenetic tags and disease susceptibility. Pollution has become a significant focus in this research area as scientists are finding that air pollution could alter methyl tags on DNA and increase one's risk for neurodegenerative disease. Interestingly, B vitamins may protect against harmful epigenetic effects of pollution and may be able to combat the harmful effects that particular matter has on the body.

Researchers have found that a ketogenic diet – consuming high amounts of fat, adequate protein, and low carbohydrates – increases an epigenetic agent naturally produced by the body. Diet has also been shown to modify epigenetic tags in significant ways. The field of nutriepigenomics explores how food and epigenetics work together to influence health and wellbeing. For example, a study found that a high fat, low carb diet could open up chromatin and improve mental ability via HDAC inhibitors. Other studies have found that certain compounds within the foods we consume could protect again cancer by adjusting methyl marks on oncogenes or tumor suppressor genes. Ultimately, an epigenetic diet may guide people toward the optimal food regimen as scientific studies reveal the underlying mechanisms and impact that different foods have on the epigenome and health.

Links to Disease

Among all the epigenetics research conducted so far, the most extensively studied disease is cancer, and the evidence linking epigenetic processes with cancer is becoming "extremely compelling," says Peter Jones, director of the University of Southern California's Norris Comprehensive Cancer Center. Halfway around the world, Toshikazu Ushijima is of the same mind. The chief of the Carcinogenesis Division of Japan's National Cancer Center Research Institute says epigenetic mechanisms are one of the five most important considerations in the cancer field, and they account for one-third to one-half of known genetic alterations.

Many other health issues have drawn attention. Epigenetic immune system effects occur, and can be reversed. The team says it's the first to establish a specific link between aberrant histone modification and mechanisms underlying lupus-like symptoms in mice, and they confirmed that a drug in the research stage, trichostatin A, could reverse the modifications. The drug appears to reset the aberrant histone modification by correcting hypo acetylation at two histone sites.

Lupus has also been a focus of Bruce Richardson, chief of the Rheumatology Section at the Ann Arbor Veterans Affairs Medical Center and a professor at the University of Michigan Medical School. In studies published in the May–August 2004 issue of International Reviews of Immunology and the October 2003 issue of Clinical Immunology, he noted that pharmaceuticals such as the heart drug pro-cainamide and the antihypertensive agent hydralazine cause lupus in some people, and demonstrated that lupus-like disease in mice exposed to these drugs is linked with DNA methylation alterations and interruption of signaling pathways similar to those in people.

Substantial Changes

Most epigenetic modification, by whatever mechanism, is believed to be erased with each new generation, during gameto-genesis and after fertilization. However, one of the more startling reports published in 2005 challenges this belief and suggests that epigenetic changes may endure in at least four subsequent generations of organisms.

Michael Skinner, a professor of molecular biosciences and director of the Center for Reproductive Biology at Washington State University, and his team described how they briefly exposed pregnant rats to individual relatively high levels of the insecticide methoxychlor and the fungicide vinclozolin, and documented effects such as decreased sperm production and increased male infertility in the male pups. Digging for more information, they found altered DNA methylation of two genes. As they continued the experiment, they discovered the adverse effects lasted in about 90% of the males in all four subsequent generations they followed, with no additional pesticide exposures.

The findings are not known to have been reproduced. If they are reproducible, however, it could "provide a new paradigm for disease etiology and basic mechanisms in toxicology and evolution not previously appreciated," says Skinner. He and his colleagues are conducting follow-up studies, assessing many other genes and looking at other effects such as breast and skin tumors, kidney degeneration, and blood defects.

Other studies have found that epigenetic effects occur not just in the womb, but over the full course of a human life span. Manel Esteller, director of the Cancer Epigenetics Laboratory at the Spanish National Cancer Center in Madrid, and his colleagues evaluated 40 pairs of identical twins, ranging in age from 3 to 74, and found a striking trend. Younger twin pairs and those who shared similar lifestyles and spent more years together had very similar DNA methylation and histone acetylation patterns. But older twins, especially those who had different lifestyles and had spent fewer years of their lives together, had much different patterns in many different tissues, such as lymphocytes, epithelial mouth cells, intra-abdominal fat, and selected muscles.

As one example, the researchers found four times as many differentially expressed genes between a pair of 50-year-old twins compared to 3-year-old twins, and the 50-year-old twin with more DNA hypomethylation and histone hyperacetylation (the epigenetic changes usually associated with transcriptional activity) had the higher number of overexpressed genes. The degree of epigenetic change therefore was directly linked with the degree of change in genetic function.

Sometimes the effects of epigenetic mechanisms show up in living color. Changes in the pigmentation of mouse pup fur, ranging from yellow to brown, were directly tied to supplementation of the pregnant mother's diet with vitamin B12, folic acid, choline, and betaine. The color changes were directly linked to alterations in DNA methylation. In a study forthcoming, Jirtle and his colleagues also induced these alterations through maternal ingestion of genistein, the major phytoestrogen in soy, at doses comparable to those a human might receive from a high-soy diet. The methylation changes furthermore appeared to protect the mouse offspring against obesity in adulthood, although there are hints that genistein may also cause health problems, via additive or synergistic effects on DNA methylation, when it interacts with other substances such as folic acid.

Variome

The term "variome" refers to the sum of all the genetic variations found in different populations of the same species. The variome of the various populations of Homo sapiens is remarkably similar from one continent to another, suggesting that our species has evolved over the past ten thousand years from a small original gene pool.

Human Variome Project

The Human Variome Project (HVP) acts as an umbrella organization, actively engaging with partners and stakeholders in each country to ensure that genetic variation information, generated during routine diagnostic and predictive testing, is collected and shared. The HVP is also instrumental in establishing and maintaining the standards, systems, and infrastructure that will embed the sharing of this knowledge in routine clinical practice.

The United Nations Educational, Scientific and Cultural Organization (UNESCO) serves as an important channel for the involvement of developing countries in the HVP, as it did during the Human Genome Project. One of the main goals of UNESCO is the development of international science that meets social needs in health, food, education, and other standards of living.

This goal has become increasingly relevant in the Post-2015 Development Agenda, which aims to address these global challenges, including the burden of diseases on the performance and growth of many nations, particularly in developing countries where issues of public health are of major concern. Formed at the end of World War II, UNESCO was one of numerous initiatives for international scientific cooperation undertaken by the nascent United Nations. These scientific cooperation initiatives were seen as diplomatic opportunities to promote collaborations among nations in hopes of fostering peace and development. The same sentiment is true today with the HVP, one of the latest efforts by UNESCO to promote collaboration.

The looming need to collect and share human genetic data and the problems surrounding this was seen as far back as 1994 by a group of geneticists, bioinformaticians, and scientists who met and established the Mutation Database Initiative under the auspices of the Human Genome Organization (HUGO-MDI). This marked the beginning of the evolutionary pathway that eventually became the HVP. The HUGO-MDI was founded to encourage the creation of new locus-specific databases; it was successful enough that a small, but highly active, community of database curators was brought together. This community organized themselves into the Human Genome Variation Society (HGVS) in 2001 and soon began publishing recommendations on how to improve the quality of efforts to create databases of genetic variations.

One landmark achievement was the formalization on the now globally accepted HGVS nomenclature for the naming and describing of sequence variations. Richard Cotton, a world-renowned specialist in genetic variation from the University of Melbourne, was fundamental in the establishment and development of both the HUGO-MDI and HGVS, and he led the organizations for many years.

The early successes of the HUGO-MDI and HGVS highlighted the enormous challenges that need to be overcome to achieve the complete collection of information on all genetic variations from all countries. Motivated by the knowledge that collaboration across disciplines and cultures would be the only way to ensure enough data could be pooled to produce better and cheaper results for patients with genetic diseases, Cotton convened the HVP in 2006. The meeting was held in Melbourne, Australia, with financial support from Tony Abbott (the current Australian prime minister was the health minister at the time) and the Victorian government. It brought together leading geneticists, diagnosticians, researchers, and bioinformaticians from thirty countries, as well as representatives of UNESCO, the World Health Organization, the Organization for Economic Co-operation and Development, and the European Commission.

The project sparked the immediate interest of those present, but the participants realized that to expand its global reach the project would need a formal structure and commitment at the governmental level. This led to the founding in 2010 of the Human Variome Project International Limited, a nonprofit Australian public company. The nonprofit would centralize coordination of the project. This structure allowed UNESCO to establish official relations with the organization in 2011 and to approach governments to ascertain their interest in the project—similar to the role UNESCO had fulfilled for the Human Genome Project in the late 1980s. Many governments, recognizing the benefits of global collaboration, reacted with enthusiasm to the potential of this new project in improving the diagnoses of diseases and patient care.

Unlike the Human Genetic Diversity Project, which is trying to work out the genetic differences among different ethnic populations, the HVP provides a central repository hub for genetic information with direct application to improving health. An editorial in Nature Genetics describes the HVP as the successor to the Human Genome Project and pointed out the importance of this function: "much of the necessary work is currently happening across the globe—but is just insufficiently coordinated."

The project aims to provide opportunities for training, education, and capacity building, especially in developing countries. The organization updates consortium members on its plans and progress through biennial meetings. At its fourth meeting in June 2012, hosted by UNESCO at its Paris headquarters in the framework of its International Basic Sciences Programme, the HVP's Project Road map to 2016 was presented. Among its goals are the completion of high-quality gene- and disease-specific databases for at least three thousand genes by 2016 and a further five thousand genes, thus making it a total of eight thousand genes by 2022. The Project Roadmap also sets a target of forty countries—double the current total—sharing information with these international databases by 2016.

Sharing of genetic and genomic data, particularly when they are linked to patient clinical data, is almost always subject to local laws, regulations, and professional codes of practice. This makes developing a single standard approach to data collection, storage, access, and transfer almost impossible. Since public health issues transcend both domestic and international policies to encourage greater adoption of data-sharing practices while retaining local control over data and their use, the HVP works with stakeholders within individual countries, including national health systems, ministries of health, and national societies of human genetics, to establish what it calls HVP Country Nodes. An HVP Country Node acts as a national focal point for genomic data-sharing activities and has a specific role in connecting all the laboratories in a country that provide genetic testing services. Each node is managed and financed locally by a committee or organization that represents a sufficient number of national stakeholder groups, and the node enjoys the backing or support of the country's human genetics society or similar professional body.

So far, nodes have been established in twenty countries: Australia, Austria, Belgium, China, Cyprus, the Czech Republic, Egypt, Italy, Kuwait, Malaysia, Mexico, Nepal, the Netherlands, Nigeria, the Republic of Korea, Spain, the United Kingdom, the United States, Venezuela, and Vietnam. The HVP Country Nodes do not operate in isolation. As part of an international consortium, they are active in HVP activities, participating in the development of HVP Standards and Guidelines and sharing their knowledge and experience with other HVP Country Nodes. Continuing membership of the HVP Consortium and recognition as an HVP Country Node is at all times subject to the HVP Code of Conduct.

An HVP Country Node consists of three components. One is a repository or linked network of databases where information on a genetic variation within a country is collected and stored. This repository enables the sharing of the information both nationally and internationally. The second is a governance structure that ensures that the work of the node is both sustainable in the long term and consistent with all relevant national and international ethical, legal, and social requirements and considerations. The third is a set of policies and procedures that ensures that the repository is operated and maintained in a responsible and accountable manner that is consistent with both national standards and the HVP's Standards and Guidelines.

Action in the third component is driven within the consortium by interest groups that are formed around broad topical areas, such as ethics, phenotype, and pathogenicity. Consortium members are divided into working groups around very narrow topics to produce standards and specify the systems and infrastructure required to address particular issues. This process is overseen by the HVP's International Scientific Advisory Committee, which leads the HVP in matters of strategic scientific direction for current and future activities. The committee is also responsible for managing the development and publication of all HVP Standards and Guidelines, as well as the arbitration of the dispute resolution process. Voting members of the committee are elected by the two advisory councils; one is the representative body for gene- and disease-specific databases and the other for HVP Country Nodes. Nodes in certain regions can assume a supporting, coordinating, and developing role for neighboring countries to become Regional Nodes.

The overall activities and the international management of the HVP are facilitated by the small staff of the International Coordinating Office. It is organized in a manner that ensures that the core scientific focus of the project is maintained, while retaining the necessary commercial and organizational skills to manage the project.

New Partners, New Frontiers

UNESCO is the only agency within the UN system that deals with fundamental science. Its unique combination of cultural and scientific interests makes it an ideal forum for interdisciplinary discussion and the promotion of understanding. By acting as a bridge among governments, governmental organizations, and nongovernmental agencies such as the HVP, UNESCO is able to facilitate effective international cooperation.

Through UNESCO, nations not currently involved in the HVP may gain access to it and ultimately make some contribution of their own. In its dealings with the HVP specifically, UNESCO provides a focal point for the exchange of data, technology, and samples relevant to genomic research and also for debate among scientists from different disciplines or from widely separated countries.

UNESCO's creation of the International Bioethics Committee, a body committed to ensuring respect for human dignity and freedom in the field of life sciences and its applications, with special attention to patient confidentiality, confers to the organization the legitimacy to work in such a sensitive field as the collection of genetic data. In working under UNESCO's umbrella, the HVP consequently is in compliance with the Universal Declaration on the Human Genome and Human Rights, the International Declaration on Human Genetic Data, and the Universal Declaration on Bioethics and Human Rights.

A tremendous step forward for the project took place in 2013 with the establishment of the HVP South East Asian Node. It represented significant recognition by Malaysian authorities that genetic and genomic healthcare is an important part of a well-developed health system, and it flagged their serious engagement with initiatives to provide these services to their citizens, as well as working closely within the region to address common challenges.

The launch of the HVP South East Asian Node was officiated by Tan Sri Muhyiddin Bin Yassin, Malaysia's deputy prime minister and minister for education, and Omar Osman, vice chancellor of University Saints Malaysia. The node will assist Brunei, Malaysia, Singapore, Thailand, and Vietnam in their national efforts to share information on genetic variations in Southeast Asian populations among associated states and the rest of the world.

Nodes also are being developed in Portugal (by the Pediatric Hospital of Coimbra) and Brazil (by the University of São Paulo). UNESCO is working with both institutions to develop the HVP through the Community of Portuguese Language Countries formed by Angola, Brazil, Cape Verde, East Timor, Equatorial Guinea, Guinea-Bissau, Mozambique, Portugal, and São Tomé and Príncipe (known by its Portuguese acronym, CPLP, for Comunidade dos Países de Língua Portuguesa). The basic idea is to use the soft power of a common language as a diplomatic instrument to exchange science and technology across continents, fostering North-South and South-South cooperation. Recently, the Eduardo Mondlane University in Maputo, Mozambique, started discussions with UNESCO and the HVP to establish a country node.

Current methods of data sharing from research are suboptimal because they are greatly distributed across different databases.

The fifth biennial meeting of the HVP Consortium highlighted the diverse range of human genomics research projects being undertaken by African institutions and researchers involved in the Human Heredity and Health in Africa (H3Africa) initiative.

The projects presented ran the gamut of human genomics research from identifying the genetic variants involved in the development of diseases primarily affecting individuals of African descent to investigating the underlying genomic components of susceptibility to infectious diseases such as trypanosomiasis, or sleeping sickness. Researchers also reported on the development of a continent-spanning bioinformatics capability and a network of state-of-the-art bio repositories.

The Nigerian Country Node focal point, from the National Biotechnology Development Agency in Abuja, asserted that Africa now has the capacity to support the research that needs to be done and African scientists should be able to collaborate globally from Africa rather than subjecting the continent to further brain drain. He emphasized that the continent is now a fertile environment for science and technology development.

Challenges persist in the effort to expand the HVP in Africa, the Middle East, Asia, and Latin America, regions that need to be better represented in the HVP Consortium. Furthermore, HVP Standards and Guidelines for mutation detection methods, data collection, data curation, variation nomenclature, and genetic counseling must be applied in all the countries that join the consortium. That will mean ensuring that the infrastructure and working methods of the institutions in each of the countries that sets up a country (or regional) node would be of a level that complies with what is required in order to apply the standards and guidelines of the HVP.

There also is a need to dissipate any hesitation some scientists in the developed world may have regarding the social and political stability of countries in the developing world by familiarizing them with local conditions there.

To achieve both aims, exploratory missions have been held on-site by delegations consisting of UNESCO staff and members of the HVP's International Coordinating Office. Their goal is to establish preliminary contacts with local institutions and to discuss and negotiate the kind of interventions required in loco.

An accurate analysis of existing infrastructure and facilities and an open dialogue among scientists, both local and international, and local authorities has been key toward building confidence among all the stakeholders. Creating effective person-to-person communication centered on the exchange of information, ideas, and perspectives has made it possible to promote training and empowerment in developing countries and to establish a much more realistic picture of the local situation. In the process, some biased ideas, often based on misleading knowledge, have been eradicated.

Separately, national and regional issues needed to be addressed. National differences among the internal organizations of the scientific community have made a big difference in identifying the leading research group. Paradoxically, it is proving easier to proceed in countries that lack a national association of human genetics. In those countries, direct contact between permanent delegations and national commissions to UNESCO and central authorities, such as ministries of science and technology or national research institutions, typically leads to intervention at a national level that swiftly identifies an HVP focal point.

By contrast, in countries that have a national human genetics society, existing rivalries between different institutions have hindered the identification of a single national focal point. In those instances, after consultation with the central organization of the national society, missions to participate at the annual congress of the local association have been organized. UNESCO representatives presented the project to the governing body and led a process to reach consensus through participative meetings that included all the major national stakeholders in the field of human genetics. Conducting these collegial consultations has proved to be a successful strategy for UNESCO.

Once a country node is established at a national level, the challenge is to create regional awareness on the project in order to use the sharing of genetic information as a diplomatic instrument. Identifying a country that is able to play a leading role ended up being crucial in the leverage of interest at a regional level. The idea was to go through a detailed country analysis and geopolitical study of the region. The process needs to take into consideration several factors: the economic and financial condition, the political and societal landscape, the education level, and the quality of existing infrastructures, as well as some important cultural and historical elements and regional tensions. As an example, the use of the common language in the Community of Portuguese Language Countries was a successful idea to connect the scientific communities from countries across three continents and open the dialogue for a fruitful cooperation.

Ultimately, for the HVP to truly achieve its scientific aim of sharing data to promote better treatment and healthcare around the world, the nature of the work both requires and fosters diplomacy. To create more channels of communication among communities promoting free access to scientific information, to strengthen science education, and to shape international dialogue, the HVP relies on a diverse set of national and international partners. The close collaboration between the HVP and UNESCO ensures that the scientific efforts to understand human genetics fosters cooperation at the regional level, builds capacity at the national level, and addresses common challenges at the global level.

Heritability

Heritability is the amount of phenotypic (observable) variation in a population that is attributable to individual genetic differences. Heritability, in a general sense, is the ratio of variation due to differences between genotypes to the total phenotypic variation for a character or trait in a population. The concept typically is applied in behaviour genetics and quantitative genetics, where heritability estimates are calculated by using either correlation and regression methods or analysis of variance (ANOVA) methods.

Types of Heritability

Broad-Sense Heritability

Our starting definition of heritability as "the proportion of variation in a trait explained by inherited genetic variants" refers to this most general version of heritability. Mathematically, we'd define the broad-sense heritability as:

$$H^2 = \frac{\sigma_G^2}{\sigma_P^2}$$

where σ_G^2 is the variance in the trait explained by genetics (G), and σ_P^2 is the total variance of the trait in the population.

We make three important observations about this definition. First, it's entirely flexible about how specific genetic effects contribute to σ_G^2. The broad-sense H^2 doesn't care whether σ_G^2 comes from a single Mendelian variant in just one gene, or the small additive effects from variants in 100 different genes, or complex interactions between every variant in the whole genome. We'll see below that this is an important distinction between broad-sense H^2 and some of the other types of heritability.

Second, broad-sense H^2 is entirely flexible about how σ_G^2 relates to σ_P^2. We could choose to assume that the effects of genes and environment are independent and thus write:

$$H^2 = \frac{\sigma_G^2}{\sigma_P^2 + \sigma_E^2}$$

but that assumption isn't required. By simply writing the denominator as σ_P^2 we allow for the possibility that genetic and environmental factors are correlated or interact in some way. This is important since it highlights that the effect of environment on the trait isn't simply the "remainder" after accounting for all the genetic effects, instead they can overlap and interact in complex ways.

Narrow-sense Heritability

In practice, the flexibility of broad-sense H^2 makes it very hard to estimate without making strong assumptions. Allowing for effects of all possible interactions of all possible genetic variants means having a functionally infinite space of possible effects. One useful way to simplify this is to think of the total variance explained by genetics as a combination of additive effects, dominant/recessive effects, and interaction effects between different variants.

$$\sigma_G^2 = \sigma_A^2 + \sigma_D^2 + \sigma_I^2$$

For a number of reasons, we might expect the variance explained by additive genetic effects σ_A^2 to be the largest and most immediately useful portion of the total σ_G^2. Focusing on just this additive genetic component leads us to the definition of the narrow-sense heritability h^2:

$$h^2 = \frac{\sigma_A^2}{\sigma_P^2}$$

If there are no dominant/recessive or interaction effects (i.e. $\sigma_D^2 = \sigma_I^2 = 0$) then the narrow-sense and broad-sense heritability are the same ($h^2 = H^2$). Otherwise the narrow-sense heritability will be smaller ($h^2 < H^2$) since it excludes these other types of genetic effects.

Historically, most scientific discussion of the heritability of different traits has focused on h^2. One of the nice features of h^2 is that it implies a simple relationship between how genetically related two people are and how similar the trait will be for those two people. We can use this relationship to estimate h^2 in twin and family studies.

In the simplest case, we can compare monozygotic twins (often called "identical" or MZ twins) to dizygotic ("fraternal" or DZ) twins. MZ twins shared all of their DNA, while DZ twins share half of their DNA on average. Twins also largely share the same environment regardless of whether they are MZ or DZ. So to estimate h^2 we can observe how correlated a trait is between pairs of MZ twins and how correlated the trait is between DZ twins and see if those correlations are different. If the MZ twins pairs, with their higher genetic similarity, are more strongly correlated than the DZ twin pairs, that suggests that genetics explains some of the variance in the trait.

SNP-heritability

The above flavors of heritability have referred to "genetic effects" conceptually without requiring any consideration of specific genetic variants and their association with the trait. Now that advances in genetics have made it possible to actually collect data on these specific variants, there's the opportunity to evaluate how much each of these observed variants contribute to heritability.

In particular, we can consider one type of genetic variant called a single nucleotide polymorphism (SNP), which is a change of a single base pair of DNA at a specific location in the genome. For example, some people may have an A at that location, while other people have a G. There are millions of these locations in the genome that commonly vary between different people, and much of the current research in human genetics is focused on understanding the effects of these variants.

So given a set "S" of SNPs that we've observed, how much of the variance in the trait can they explain? That leads us to define the SNP-heritability h_g^2, the proportion of variance explained by additive effects of the observed SNPs, which we could write as:

$$h_g^2 = \frac{\sigma^2_{SNP_{s \in S}}}{\sigma_P^2}$$

If we compare this to the above definitions, it's evident that $h_g^2 \leq h^2 \leq H^2$ since h_g^2 is limited to additive effects from only a subset of genetic variants.

This definition of h_g^2 still hides the effects of individual SNPs though, so it's useful to introduce an alternate version. If we call our trait y, and say each SNP x_j has an additive effect β_j, then we can write:

$$y = \sum_{SNP_{s \in S}} x_j \beta_j + \in$$

Where $\in$ is a residual term for effects not explained by the sum of the SNP effects. We can then define h_g^2 based on the variance of this sum of SNP effects compared to the total variance of the trait:

$$h_g^2 = \frac{var\left(\sum_{SNPs \in S} x_j \beta_j\right)}{var(y)}$$

It's worth highlighting two key features of h_g^2. First, you might notice that we've defined h_g^2 based on some set of SNPs "S". In practice, this set of SNPs is going to depend on (a) the SNP data that has been observed and (b) the method used for estimating h_g^2. This makes it tricky to compare values of h_g^2 between different methods and different studies, though in most cases it's safe to at least assume it refers to commonly-occurring SNPs. Second, the variance explained by SNPs may or may not reflect the effects of those particular SNPs as opposed to the effects of other genetic variants the SNPs are correlated with. This is just an extension of our previous discussion above about the meaning of variance "explained", but worth reiterating since it would be easy to misinterpret SNP-heritability as fully excluding the causal effects of other types of genetic variation.

There are a couple of different methods that have been developed for estimating h_g^2 from observed SNPs. In practice, we don't know the true βj so we have to use other tricks. The first approach, known as GREML (Genomic relatedness matrix Restricted Maximum Likelihood; commonly implemented in GCTA), uses SNPs to estimate the genetic similarity between random individuals and compare that to their trait similarity. This is conceptually similar to the twin-based estimation described above, but uses the observed low-level genetic similarity in SNP data from individuals who aren't directly related. You can read about the statistical details here with a more recent review here.

A second approach is called linkage disequilibrium (LD) score regression, implemented in ldsc. This is the method we are applying to the UK Bio bank data set. LD score regression depends on the key observation that some SNPs are correlated with (i.e. in LD with) other genetic variants, so observing that SNP in turn "tags" information about the effects of other variants. The basic idea then is that if there are lots and lots of small genetic effects spread across the genome (i.e. the trait is "polygenic"), then the strength of the relationship between each individual SNP and the trait should be (on average) proportional to how much total genetic variation that SNP tags.

Making Sense of Heritability

Heritability estimates have a value between 0 and 1. These values are sometimes represented as percentages, for instance "depression is 70% heritable" would correspond to an h^2 of 0.7.

However, this does not mean that 70% of an individual's depression is genetic, with the environment making up the other 30%. It also does not mean that 70% of depressed individuals are so because of their genes.

To make sense of this, imagine that we found that height was 80% heritable. It seems obvious that this could not mean that only 80% of people have their height genetically influenced. It also strange to think that my particular height of 165cm can be broken down in to 132cm of genetically caused growth and 33cm of environmentally caused growth.

The interplay of genes and environment for individual traits is recognized by geneticists, and cannot be broken down in to percentage values.

For my Labrador Bob to have brown fur, he needs to have particular genes to code for the brown pigment expressed in his coat. He also needs an environment in which to develop, in order to be able to grow fur at all.

Figure: The color of Labrador fur

In an individual case, it does not make sense to say how much is genetically or environmentally caused. It does not make sense to talk about "how genetic" blond hair, short stature, or Bobs brown fur is.

This is why heritability estimates can only be applied to populations.

Causes of Trait Variation

So what do these numbers refer to?

Heritability concerns how much variation in traits is caused by variation in genes.

If we looked a population of people and measured their height we are likely find variation between them – some short, some tall, and some in-between. Heritability tells us if this variation occurs because people have different genes or because they live in different environments.

In the Labrador case – it seems that dogs with particular genes are golden, while others with different genes are black or brown. Thus, there is variation in genes between the three colored groups.

If we looked at the environments in which they were raised, you would find that no matter what environment these dogs are raised in, their coat color is not affected.

So, Labrador coat color variation is caused by variation in the genes, and is highly heritable. As changes in the environment have no effect at all, the heritability would be 100%, or $h^2 = 1$.

Human skin color however, has a lower heritability estimate.

Although we know that some variation in color can be explained by differences in genes, we also know that variation in the environment – such as sun exposure, can affect the color of peoples' skin.

So, variation in skin color is caused partially by variation in genes and partially by variation in environment.

Skin color is influenced by genes as well as the environment, like being out in the sun.

Some Strange Consequences of Heritability

As heritability is a measure of the causes of variation in traits, things which we ordinarily think of as having a genetic basis can turn out to have low heritability.

For instance, "walking on two legs" is a human trait which does not vary much. When it does vary, this is usually due to environmental variations, such as accidents where people lose the function of one or both legs.

As a consequence, "walking on two legs" has an h^2 close to 0. This does not mean that genes are not necessary for humans to walk on two legs. What it means is that variation in this trait is caused by primarily non-genetic factors.

Another strange consequence of heritability is that the estimate depends upon which population you examine.

For example, the heritability of hair color in a Chinese population would be quite low, yet in Australia would be quite high. This is because in China there is little "natural" variation in hair color – variation that is genetically caused. As such, any large variations are usually due to environmental factors, such as artificial dyes.

So while heritability does measure the causal impact of genes, it does so in a very specific and limited way.

References

- Pierce BA (2012). Genetics : a conceptual approach (4th ed.). New York: W.H. Freeman. pp. 538–540. ISBN 978-1-4292-3250-0.
- Human-genetics, science: britannica.com, Retrieved 11 May 2018
- Bennett EA, Keller H, Mills RE, Schmidt S, Moran JV, Weichenrieder O, Devine SE (Dec 2008). "Active Alu retrotransposons in the human genome". Genome Research. 18 (12): 1875–83. doi:10.1101/gr.081737.108. PMC 2593586. PMID 18836035
- What-is-genetic-variation: yourgenome.org, Retrieved 29 March 2018

- NCBI_user_services (29 March 2004). "Mapping Factsheet". Ncbi.nlm.nih.gov. Archived from the original on 19 July 2010. Retrieved 31 May 2009
- Genetic-variation-types-and-importance-of-genetic-variations-12073: yourarticlelibrary.com, Retrieved 26 June 2018
- Driscoll DA, Gross S (June 2009). "Clinical practice. Prenatal screening for aneuploidy". The New England Journal of Medicine. 360 (24): 2556–62. doi:10.1056/NEJMcp0900134. PMID 19516035
- What-is-selective-sweep-1224718: thoughtco.com, Retrieved 30 June 2018
- Barton NH, Briggs DE, Eisen JA, Goldstein DB, Patel NH (2007). Evolution. Cold Spring Harbor, NY: Cold Spring Harbor Laboratory Press. ISBN 978-0-87969-684-9
- Human-variome-project: sciencediplomacy.org, Retrieved 17 April 2018
- Liang KH, Yeh CT (2013). "A gene expression restriction network mediated by sense and antisense Alu sequences located on protein-coding messenger RNAs". BMC Genomics. 14: 325. doi:10.1186/1471-2164-14-325. PMC 3655826. PMID 23663499
- Explainer-what-is-heritability-21334: theconversation.com, Retrieved 27 April 2018
- "2008 Release: Researchers Produce First Sequence Map of Large-Scale Structural Variation in the Human Genome". genome.gov. Retrieved 2009-05-31

Chapter 2

Role of Genetics in Human Evolution

Human evolution is a process that has led to the emergence of the modern human or Homo sapiens as a unique species within the hominid family. Genetics is at the core of this evolutionary process. This chapter has been carefully written to examine the role of genetics in human evolution, through the elucidation of topics such as human evolutionary genetics, ancient DNA, archaeogenetics and adaptive evolution in the human genome, among others.

Human Evolutionary Genetics

Genetics has played an increasingly important role in studies of the last two million years of human evolution. Evolutionary genetics is concerned with the mechanisms that explain the existence and maintenance of genetic variation in traits. All else equal, one would expect selection to deplete genetic variation in heritable traits related to fitness eventually. However, such genetic variation is ubiquitous and underlies stable individual differences that play prominent roles in psychological theories, be it as traits under intersexual (e.g. attractiveness, agreeableness, intelligence) and intrasexual selection (masculinity, aggressiveness), life history traits, formidability in recalibration theory, sociometer sensitivity, perceived vulnerability to infection in the behavioral immune system, attachment security, or the tendency to show strong reciprocity in cooperation. Though these theories ascribe adaptive roles to individual differences, more or less explicitly linking them to fitness, their genetic variation is often taken for granted.

Evolutionary genetics can help evolutionary psychologists unearth clues to the ultimate reasons behind e.g. humans' cognitive faculties that go beyond what can gleaned through paleontology and archaeology. This information can have very practical implications, e.g. helping to understand how natural and sexual selection, when altered through changing mores or policy, will affect certain traits.

Genetic Architecture

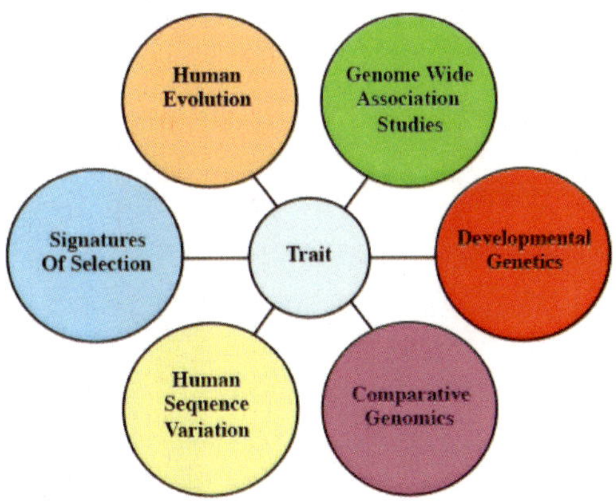

Some research in molecular genetics has been carried out with the aim of characterizing the genetic architecture of traits, sometimes also called the genotype-phenotype map. The genetic architecture of a trait can provide important clues to the evolutionary history and the mechanisms that govern the maintenance of genetic variation in the trait. Characterizing the genetic architecture of a quantitative trait would ideally involve its robustness to mutations (canalization) as well as its evolvability. It would also imply gauging its degree of pleiotropy (whether the genes involved also have simultaneous other effects) and the importance of non-additive genetic variation (i.e. epistasis and dominance, variation that does not breed true to the next generation). Unfortunately, many examinations of the genetic architecture are limited to estimates of the number and effect size of involved genetic variants.

Origin of Apes

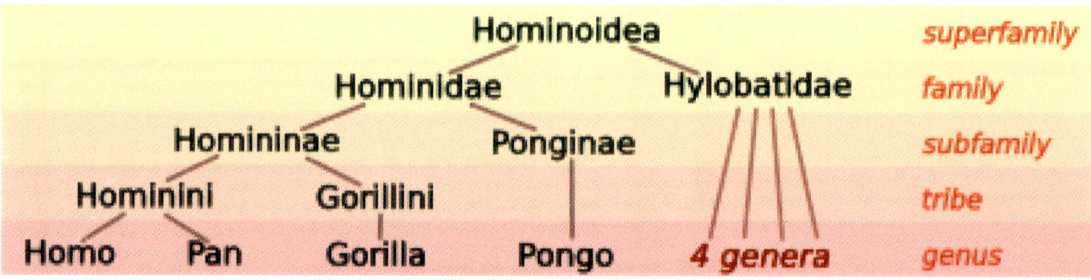

Taxonomic relationships of hominoids.

Biologists classify humans, along with only a few other species, as great apes (species in the family Hominidae). The living Hominidae include two distinct species of chimpanzee (the bonobo, *Pan paniscus*, and the common chimpanzee, *Pan troglodytes*), two species of gorilla (the western gorilla, *Gorilla gorilla*, and the eastern gorilla, *Gorilla graueri*), and two species of orangutan (the Bornean orangutan, *Pongo pygmaeus*, and the Sumatran orangutan, *Pongo abelii*). The great apes with the family Hylobatidae of gibbons form the superfamily Hominoidea of apes.

Apes, in turn, belong to the primate order (>400 species), along with the Old World monkeys, the New World monkeys, and others. Data from both mitochondrial DNA (mtDNA) and nuclear DNA (nDNA) indicate that primates belong to the group of Euarchontoglires, together with Rodentia, Lagomorpha, Dermoptera, and Scandentia. This is further supported by Alu-like short interspersed nuclear elements (SINEs), which have been found only in members of the Euarchontoglires.

Cladistics

A phylogenetic tree is usually derived from DNA or protein sequences from populations. Often, mitochondrial DNA or Y chromosome sequences are used to study ancient human demographics. These single-locus sources of DNA do not recombine and are almost always inherited from a single parent, with only one known exception in mtDNA. Individuals from closer geographic regions generally tend to be more similar than individuals from regions farther away. Distance on a phylogenetic tree can be used approximately to indicate:

1. Genetic distance: The genetic difference between humans and chimps is less than 2%, or twenty times larger than the variation among modern humans.

2. Temporal remoteness of the most recent common ancestor: The mitochondrial most recent common ancestor of modern humans lived roughly 200,000 years ago, latest common ancestors of humans and chimps between four and seven million years ago.

Speciation of Humans and the African Apes

The separation of humans from their closest relatives, the non-human apes (chimpanzees and gorillas), has been studied extensively for more than a century. Five major questions have been addressed:

- Which apes are our closest ancestors?
- When did the separations occur?
- What was the effective population size of the common ancestor before the split?
- Are there traces of population structure (subpopulations) preceding the speciation or partial admixture succeeding it?
- What were the specific events (including fusion of chromosomes 2a and 2b) prior to and subsequent to the separation?

General Observations

Different parts of the genome show different sequence divergence between different hominoids. It has also been shown that the sequence divergence between DNA from humans and chimpanzees varies greatly. For example, the sequence divergence varies between 0% to 2.66% between non-coding, non-repetitive genomic regions of humans and chimpanzees. Additionally gene trees, generated by comparative analysis of DNA segments, do not always fit the species tree. Summing up:

- The sequence divergence varies significantly between humans, chimpanzees and gorillas.
- For most DNA sequences, humans and chimpanzees appear to be most closely related, but some point to a human-gorilla or chimpanzee-gorilla clade.
- The human genome has been sequenced, as well as the chimpanzee genome. Humans have 23 pairs of chromosomes, while chimpanzees, gorillas, and orangutans have 24. Human chromosome 2 is a fusion of two chromosomes 2a and 2b that remained separate in the other primates.

Divergence Times

The divergence time of humans from other apes is of great interest. One of the first molecular studies, published in 1967 measured immunological distances (IDs) between different primates. Basically, the study measured the strength of immunological response that an antigen from one species (human albumin) induces in the immune system of another species (human, chimpanzee, gorilla and Old World monkeys). Closely related species should have similar antigens and therefore weaker immunological response to each other's antigens. The immunological response of a species to its own antigens (e.g. human to human) was set to be 1.

The ID between humans and gorillas was determined to be 1.09, that between humans and chimpanzees was determined as 1.14. However, the distance to six different Old World monkeys was on average 2.46, indicating that the African apes are more closely related to humans than to monkeys. The authors consider the divergence time between Old World monkeys and hominoids to be 30 million years ago (MYA), based on fossil data, and the immunological distance was considered to grow at a constant rate. They concluded that divergence time of humans and the African apes to be roughly ~5 MYA. That was a surprising result. Most scientists at that time thought that humans and great apes diverged much earlier (>15 MYA).

The gorilla was, in ID terms, closer to human than to chimpanzees; however, the difference was so slight that the trichotomy could not be resolved with certainty. Later studies based on molecular genetics were able to resolve the trichotomy: chimpanzees are phylogenetically closer to humans than to gorillas. However, some divergence times estimated later (using much more sophisticated methods in molecular genetics) do not substantially differ from the very first estimate in 1967, but a recent paper puts it at 11–14 MYA.

Divergence Times and Ancestral Effective Population Size

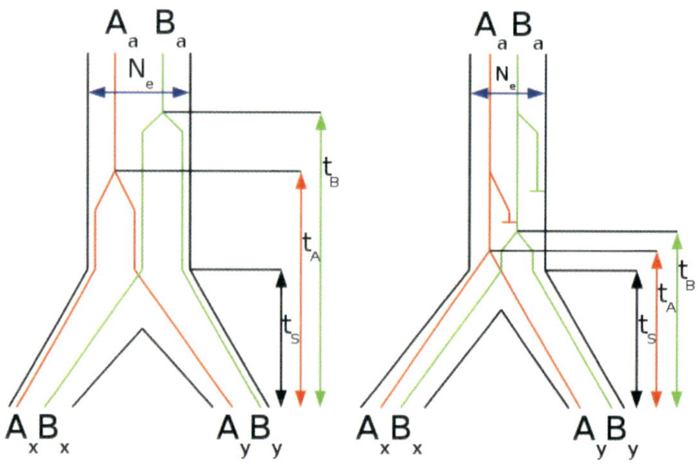

Figure: The sequences of the DNA segments diverge earlier than the species

A large effective population size in the ancestral population (left) preserves different variants of the DNA segments (=alleles) for a longer period of time. Therefore, on average, the gene divergence times (tA for DNA segment A; tB for DNA segment B) will deviate more from the time the species diverge (tS) compared to a small ancestral effective population size (right).

Current methods to determine divergence times use DNA sequence alignments and molecular clocks. Usually the molecular clock is calibrated assuming that the orangutan split from the African apes (including humans) 12-16 MYA. Some studies also include some Old World monkeys and set the divergence time of them from hominoids to 25-30 MYA. Both calibration points are based on very little fossil data and have been criticized.

If these dates are revised, the divergence times estimated from molecular data will change as well. However, the relative divergence times are unlikely to change. Even if we can't tell absolute divergence times exactly, we can be pretty sure that the divergence time between chimpanzees and humans is about six fold shorter than between chimpanzees (or humans) and monkeys.

One study used 15 DNA sequences from different regions of the genome from human and chimpanzee and 7 DNA sequences from human, chimpanzee and gorilla. They determined that chimpanzees are more closely related to humans than gorillas. Using various statistical methods, they estimated the divergence time human-chimp to be 4.7 MYA and the divergence time between gorillas and humans (and chimps) to be 7.2 MYA.

Additionally, they estimated the effective population size of the common ancestor of humans and chimpanzees to be ~100,000. This was somewhat surprising since the present day effective population size of humans is estimated to be only ~10,000. If true that means that the human lineage would have experienced an immense decrease of its effective population size (and thus genetic diversity) in its evolution.

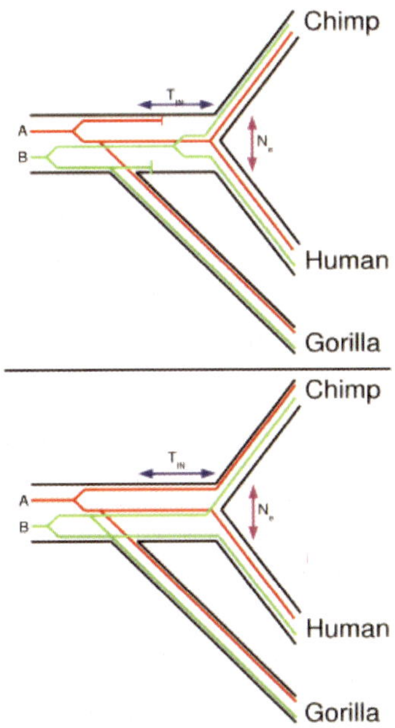

Figure: Analysis of DNA sequence of Human, Gorilla and Chimp.

A and B are two different loci. In the upper figure they fit to the species tree. The DNA that is present in today's gorillas diverged earlier from the DNA that is present in today's humans and chimps. Thus, both loci should be more similar between human and chimp than between gorilla and chimp or gorilla and human. In the lower graph, locus A has a more recent common ancestor in human and gorilla compared to the chimp sequence. Whereas chimp and gorilla have a more recent common ancestor for locus B. Here the gene trees are incongruent to the species tree.

Another study sequenced 53 non-repetitive, intergenic DNA segments from a human, a chimpanzee, a gorilla, and orangutan. When the DNA sequences were concatenated to a single long sequence, the generated neighbor-joining tree supported the *Homo-Pan* clade with 100% bootstrap (that is that humans and chimpanzees are the closest related species of the four). When three species are fairly closely related to each other (like human, chimpanzee and gorilla), the trees obtained from DNA sequence data may not be congruent with the tree that represents the speciation (species tree).

The shorter internodal time span (T_{IN}) the more common are incongruent gene trees. The effective population size (N_e) of the internodal population determines how long genetic lineages are preserved in the population. A higher effective population size causes more incongruent gene trees. Therefore, if the internodal time span is known, the ancestral effective population size of the common ancestor of humans and chimpanzees can be calculated.

When each segment was analyzed individually, 31 supported the *Homo-Pan* clade, 10 supported the *Homo-Gorilla* clade, and 12 supported the *Pan-Gorilla* clade. Using the molecular clock, the authors estimated that gorillas split up first 6.2-8.4 MYA and chimpanzees and humans split up 1.6-2.2 million years later (internodal time span) 4.6-6.2 MYA. The internodal time span is useful to estimate the ancestral effective population size of the common ancestor of humans and chimpanzees.

A parsimonious analysis revealed that 24 loci supported the *Homo-Pan* clade, 7 supported the *Homo-Gorilla* clade, 2 supported the *Pan-Gorilla* clade and 20 gave no resolution. Additionally, they took 35 protein coding loci from databases. Of these 12 supported the *Homo-Pan* clade, 3 the *Homo-Gorilla* clade, 4 the *Pan-Gorilla* clade and 16 gave no resolution. Therefore, only ~70% of the 52 loci that gave a resolution (33 intergenic, 19 protein coding) support the 'correct' species tree. From the fraction of loci which did not support the species tree and the internodal time span they estimated previously, the effective population of the common ancestor of humans and chimpanzees was estimated to be ~52 000 to 96 000. This value is not as high as that from the first study (Takahata), but still much higher than present day effective population size of humans.

A third study used the same dataset that Chen and Li used but estimated the ancestral effective population of 'only' ~12,000 to 21,000, using a different statistical method.

Genetic Differences between Humans and other Great Apes

The alignable sequences within genomes of humans and chimpanzees differ by about 35 million single-nucleotide substitutions. Additionally about 3% of the complete genomes differ by deletions, insertions and duplications.

Since mutation rate is relatively constant, roughly one half of these changes occurred in the human lineage. Only a very tiny fraction of those fixed differences gave rise to the different phenotypes of humans and chimpanzees and finding those is a great challenge. The vast majority of the differences are neutral and do not affect the phenotype.

Molecular evolution may act in different ways, through protein evolution, gene loss, differential gene regulation and RNA evolution. All are thought to have played some part in human evolution.

Gene Loss

Many different mutations can inactivate a gene, but few will change its function in a specific way. Inactivation mutations will therefore be readily available for selection to act on. Gene loss could thus be a common mechanism of evolutionary adaptation (the "less-is-more" hypothesis).

80 genes were lost in the human lineage after separation from the last common ancestor with the chimpanzee. 36 of those were for olfactory receptors. Genes involved in chemoreception and immune response are overrepresented. Another study estimated that 86 genes had been lost.

Hair Keratin Gene: KRTHAP1

A gene for type I hair keratin was lost in the human lineage. Keratins are a major component of hairs. Humans still have nine functional type I hair keratin genes, but the loss of that particular gene may have caused the thinning of human body hair. The gene loss occurred relatively recently in human evolution—less than 240,000 years ago.

Myosin Gene: MYH16

Stedman stated that the loss of the sarcomeric myosin gene MYH16 in the human lineage led to smaller masticatory muscles. They estimated that the mutation that led to the inactivation (a two base pair deletion) occurred 2.4 million years ago, predating the appearance of *Homo ergaster/ erectus* in Africa. The period that followed was marked by a strong increase in cranial capacity, promoting speculation that the loss of the gene may have removed an evolutionary constraint on brain size in the genus *Homo*.

Another estimate for the loss of the MYH16 gene is 5.3 million years ago, long before *Homo* appeared.

CASPASE12, a cysteinyl aspartate proteinase. The loss of this gene is speculated to have reduced the lethality of bacterial infection in humans.

Gene Addition

Segmental duplications (SDs or LCRs) have roles in creating new primate genes and shaping human genetic variation.

Human-specific DNA Insertions

When the human genome was compared to the genomes of five comparison primate species, including the chimpanzee, gorilla, orangutan, gibbon, and macaque, it was found that there are approximately 20,000 human-specific insertions believed to be regulatory. While most insertions appear to be fitness neutral, a small amount have been identified in positively selected genes showing associations to neural phenotypes and some relating to dental and sensory perception-related phenotypes. These findings hint at the seemingly important role of human-specific insertions in the recent evolution of humans.

Selection Pressures

Human accelerated regions are areas of the genome that differ between humans and chimpanzees to a greater extent than can be explained by genetic drift over the time since the two species shared a common ancestor. These regions show signs of being subject to natural selection, leading to the evolution of distinctly human traits. Two examples are HAR1F, which is believed to be related to brain development and HAR2 (HACNS1) that may have played a role in the development of the opposable thumb.

It has also been hypothesized that much of the difference between humans and chimpanzees is attributable to the regulation of gene expression rather than differences in the genes themselves.

Analyses of conserved non-coding sequences, which often contain functional and thus positively selected regulatory regions, address this possibility.

Sequence Divergence between Humans and Apes

When the draft sequence of the common chimpanzee (*Pan troglodytes*) genome was published, 2400 million bases (of ~3160 million bases) were sequenced and assembled well enough to be compared to the human genome. 1.23% of this sequenced differed by single-base substitutions. Of this, 1.06% or less was thought to represent fixed differences between the species, with the rest being variant sites in humans or chimpanzees. Another type of difference, called indels (insertions/deletions) accounted for many fewer differences (15% as many), but contributed ~1.5% of unique sequence to each genome, since each insertion or deletion can involve anywhere from one base to millions of bases.

A companion paper examined segmental duplications in the two genomes, whose insertion and deletion into the genome account for much of the indel sequence. They found that a total of 2.7% of achromatic sequence had been differentially duplicated in one or the other lineage.

Percentage sequence divergence between humans and other hominids			
Locus	Human-Chimp	Human-Gorilla	Human-Orangutan
Alu elements	2	-	-
Non-coding (Chr. Y)	1.68 ± 0.19	2.33 ± 0.2	5.63 ± 0.35
Pseudogenes (autosomal)	1.64 ± 0.10	1.87 ± 0.11	-
Pseudogenes (Chr. X)	1.47 ± 0.17	-	-
Noncoding (autosomal)	1.24 ± 0.07	1.62 ± 0.08	3.08 ± 0.11
Genes (Ks)	1.11	1.48	2.98
Introns	0.93 ± 0.08	1.23 ± 0.09	-
Xq13.3	0.92 ± 0.10	1.42 ± 0.12	3.00 ± 0.18
Subtotal for X chromosome	1.16 ± 0.07	1.47 ± 0.08	-
Genes (Ka)	0.8	0.93	1.96

The sequence divergence has generally the following pattern: Human-Chimp < Human-Gorilla << Human-Orangutan, highlighting the close kinship between humans and the African apes. Alu elements diverge quickly due to their high frequency of CpG dinucleotides which mutate roughly 10 times more often than the average nucleotide in the genome. The mutation rate is higher in the male germ line, therefore the divergence in the Y chromosome—which is inherited solely from the father—is higher than in autosomes. The X chromosome is inherited twice as often through the female germ line as through the male germ line and therefore shows slightly lower sequence divergence. The sequence divergence of the Xq13.3 region is surprisingly low between humans and chimpanzees.

Mutations altering the amino acid sequence of proteins (K_a) are the least common. In fact ~29% of all orthologous proteins are identical between human and chimpanzee. The typical protein differs by only two amino acids. The measures of sequence divergence shown in the table only take the substitutional differences, for example from an A (adenine) to a G (guanine), into account. DNA sequences may however also differ by insertions and deletions (indels) of bases. These are usually stripped from the alignments before the calculation of sequence divergence is performed.

Genetic Differences between Modern Humans and Neanderthals

An international group of scientists completed a draft sequence of the Neanderthal genome in May 2010. The results indicate some breeding between modern humans (*Homo sapiens*) and Neanderthals (*Homo neanderthalensis*), as the genomes of non-African humans have 1-4% more in common with Neanderthals than do the genomes of sub-Saharan Africans. Neanderthals and most modern humans share a lactose-intolerant variant of the lactase gene that encodes an enzyme that is unable to break down lactose in milk after weaning. Modern humans and Neanderthals also share the FOXP2 gene variant associated with brain development and with speech in modern humans, indicating that Neanderthals may have been able to speak. Chimps have two amino acid differences in FOXP2 compared with human and Neanderthal FOXP2.

Genetic Differences among Modern Humans

Homo sapiens is thought to have emerged about 300,000 years ago. It dispersed throughout Africa, and after 70,000 years ago throughout Eurasia and Oceania. A 2009 study identified 14 "ancestral population clusters", the most remote being the San people of Southern Africa.

With their rapid expansion throughout different climate zones, and especially with the availability of new food sources with the domestication of cattle and the development of agriculture, human populations have been exposed to significant selective pressures since their dispersal. For example, East Asians have been found to be separated from Europids by a number of concentrated alleles suggestive of selection pressures, including variants of the EDAR, ADH1B, ABCC1, and ALDH2genes. The East Asian types of ADH1B in particular are associated with rice domestication and would thus have arisen after the development of rice cultivation roughly 10,000 years ago. Several phenotypical traits of characteristic of East Asians are due to a single mutation of the EDAR gene, dated to c. 35,000 years ago.

As of 2017, the Single Nucleotide Polymorphism Database (dbSNP), which lists SNP and other variants, listed a total of 324 million variants found in sequenced human genomes. Nucleotide diversity, the average proportion of nucleotides that differ between two individuals, is estimated at between 0.1% and 0.4% for contemporary humans (compared to 2% between humans and chimpanzees). This corresponds to genome differences at a few million sites; the 1000 Genomes Project similarly found that "a typical [individual] genome differs from the reference human genome at 4.1 million to 5.0 million sites ... affecting 20 million bases of sequence."

Ancient DNA

Ancient DNA (aDNA) refers to the study of DNA extracted from specimens that died decades, hundreds or sometimes thousands of years ago. Examples include the analysis of DNA recovered from archaeological finds, museum specimens, fossil remains and other unusual specimens. In general these specimens were not preserved for the purpose of genetic and genomic studies. The techniques used in extracting aDNA are applicable to any situation where DNA has degraded to the extent that conventional fresh DNA extraction techniques cannot be used. Practically speaking, the term aDNA relates to the condition of the DNA, not necessarily the age.

Techniques

Different techniques are required to extract ancient DNA, and the extraction therefore needs to be handled at a specialist aDNA laboratory. A fresh DNA sample can be on the order of micrograms. If the lab is exposed to low levels of alien DNA on the order of nanograms or picograms, the contamination will not show up in the results. In contrast, an aDNA sample is typically on the order of nanograms or even picograms, so that extra nanograms or picograms of contamination could be fatal to the analysis.

The issue with aDNA extraction is simply that DNA is a very complex structure that degrades as soon as the organism dies due to bacteria that cause the corpse to decompose. This is accelerated if the DNA is exposed to "the elements", and by any chemicals that might be present (such as embalming fluid). The oldest specimens that have yielded aDNA tend to be found in cool dry climates at high altitudes that helped retard the bacterial action and kept the DNA away from heat and moisture.

The Y-chromosome is almost 60 million base pairs long and there is only one per cell. DNA analysis of the Y depends on extracting enough DNA from certain regions within those 60 million base pairs for analysis. For highly degraded remains, it's highly unlikely that enough of the right Y survives for analysis.

There is a much better chance of recovering enough mitochondrial DNA (mtDNA) for an identification. This makes it easier for the laboratory to extract usable DNA, but a lot harder on the genealogists looking for the family reference, as you have to follow the female line. There are up to 1,000 mitochondria per cell, each with five to ten copies of its own 16,569 base-pairs genome. Therefore, there can be as many as 10,000 copies per cell of the mtDNA genome. This results in a much higher probability of recovering mtDNA from severely degraded remains.

Embalming creates further problems. The formaldehyde found in embalming fluid not only denatures DNA, but also causes DNA strands to cross link to themselves and other strands of DNA, much like a wadded up ball of duct tape. The damage is permanent. The formaldehyde oxidizes to paraformaldehyde, which can inhibit the Proteinase K used during the extraction. So for embalmed remains, the extraction of aDNA must overcome the issues of degradation by bacterial action involved in decomposition and degradation due to exposure to the elements, in addition to the inhibition of the extraction process by the presence of oxidized formaldehyde. There have been protocols developed to try to break the cross-links formed by the formaldehyde. These involve microwaving and temperature cycling bone powder. Unfortunately, for very fragile specimens, this protocol can destroy the DNA as well. What has been more successful is to soak the bone powder in a PBS solution that allows the paraformaldehyde to float to the top, with the bone powder sinking to the bottom. Once the paraformaldehyde is removed, the remaining bone powder is dissolved with a demineralization process, releasing DNA that is hiding deep in the bone matrix that has not been affected by the embalming process. This can double the yield of aDNA.

The best place to look for aDNA is in the petrous bone. Teeth are also favoured for ancient DNA. Enamel is the hardest substance in the body and although it does not contain DNA, it provides physical protection to the dentine within it and helps to protect the DNA in the dentine. Any other dense compact bone, such as a femur or another long bone, can also be used.

History of Ancient DNA Studies

1980s

The first study of what would come to be called aDNA was conducted in 1984, when Russ Higuchi and colleagues at the University of California, Berkeley reported that traces of DNA from a museum specimen of the Quagga not only remained in the specimen over 150 years after the death of the individual, but could be extracted and sequenced. Over the next two years, through investigations into natural and artificially mummified specimens, Svante Pääbo confirmed that this phenomenon was not limited to relatively recent museum specimens but could apparently be replicated in a range of mummified human samples that dated as far back as several thousand years.

The laborious processes that were required at that time to sequence such DNA (through bacterial cloning) were an effective brake on the development of the field of ancient DNA (aDNA). However, with the development of the Polymerase Chain Reaction (PCR) in the late 1980s, the field began to progress rapidly. Double primer PCR amplification of aDNA (jumping-PCR) can produce highly skewed and non-authentic sequence artifacts. Multiple primer, nested PCR strategy was used to overcome those shortcomings.

1990s

The post-PCR era heralded a wave of publications as numerous research groups tried their hands at aDNA. A series of incredible findings had been published, claiming authentic DNA could be extracted from specimens that were millions of years old, into the realms of what Lindahl has labeled Antediluvian DNA. The majority of such claims were based on the retrieval of DNA from organisms preserved in amber. Insects such as stingless bees, termites, and wood gnats, as well as plant and bacterial sequences were extracted from Dominican amber dating to the Oligocene epoch. Still older sources of Lebanese amber-encased weevils, dating to within the Cretaceous epoch, reportedly also yielded authentic DNA. DNA retrieval was not limited to amber.

Several sediment-preserved plant remains dating to the Miocene were successfully investigated. Then, in 1994 and to international acclaim, Woodward *et al.* reported the most exciting results to date — mitochondrial cytochrome b sequences that had apparently been extracted from dinosaur bones dating to more than 80 million years ago. When in 1995 two further studies reported dinosaur DNA sequences extracted from a Cretaceous egg, it seemed that the field would revolutionize knowledge of the Earth's evolutionary past. Even these extraordinary ages were topped by the claimed retrieval of 250-million-year-old halobacterial sequences from halite.

Whole genome sequencing started to yield results in 1995.

2000s

Single primer extension (SPEX) amplification was introduced in 2007 to address postmortem DNA modification damage.

Ancient DNA Revolution

The field of aDNA-studies has been revolutionized, with the introduction of much cheaper research-techniques, leading to new insights in human migrations.

Problems and Errors

Degradation Processes

Due to degradation processes (including cross-linking, deamination and fragmentation) ancient DNA is of lower quality in comparison with modern genetic material. The damage characteristics and ability of aDNA to survive through time restricts possible analyses and places an upper limit on the age of successful samples. There is a theoretical relationship between time and DNA degradation, although differences in environmental conditions complicate things. Samples subjected to different conditions are unlikely to predictably align to a uniform age-degradation relationship. The environmental effects may even matter after excavation, as DNA decay rates may increase, particularly under fluctuating storage conditions. Even under the best preservation conditions, there is an upper boundary of 0.4-1.5 million years for a sample at around to contain sufficient DNA for contemporary sequencing technologies.

Research into the decay of mitochondrial and nuclear DNA in Moa bones has modeled mitochondrial DNA degradation to an average length of 1 base pair after 6,830,000 years at -5° C. The decay kinetics have been measured by accelerated aging experiments further displaying the strong influence of storage temperature and humidity on DNA decay. Nuclear DNA degrades at least twice as fast as mtDNA. As such, early studies that reported recovery of much older DNA, for example, from Cretaceous dinosaur remains, may have stemmed from contamination of the sample.

Age Limit

A critical review of ancient DNA literature through the development of the field highlights that few studies after about 2002 have succeeded in amplifying DNA from remains older than several hundred thousand years. A greater appreciation for the risks of environmental contamination and studies on the chemical stability of DNA have resulted in concerns being raised over previously reported results. The dinosaur DNA was later revealed to be human Y-chromosome, while the DNA reported from encapsulated halobacteria has been criticized based on its similarity to modern bacteria, which hints at contamination. A 2007 study also suggests that these bacterial DNA samples may not have survived from ancient times, but may instead be the product of long-term, low-level metabolic activity.

aDNA may contain a large number of postmortem mutations, increasing with time. Some regions of polynucleotide are more susceptible to this degradation, so sequence data can bypass statistical filters used to check the validity of data. Due to sequencing errors, great caution should be applied to interpretation of population size. Substitutions resulting from deamination cytosine residues are vastly over-represented in the ancient DNA sequences. Miscoding of C to T and G to A accounts for the majority of errors.

Contamination

Another problem with ancient DNA samples is contamination by modern human DNA and by microbial DNA (most of which is also ancient). New methods have emerged in recent years to prevent possible contamination of aDNA samples, including conducting extractions under extreme sterile conditions, using special adapters to identify endogenous molecules of the sample (over ones that may have been introduced during analysis), and applying bioinformatics to resulting sequences.

Human aDNA

Due to the considerable anthropological, archaeological, and public interest directed toward human remains, they have received considerable attention from the DNA community.

Due to the morphological preservation in mummies, many studies from the 1990s and 2000s used mummified tissue as a source of ancient human DNA. Examples include both naturally preserved specimens, for example, those preserved in ice, such as the Ötzi the Iceman, or through rapid desiccation, such as high-altitude mummies from the Andes, as well as various sources of artificially preserved tissue (such as the chemically treated mummies of ancient Egypt). However, mummified remains are a limited resource. The majority of human aDNA studies have focused on extracting DNA from two sources that are much more common in the archaeological record – bone and teeth. Several other sources have also yielded DNA, including paleofaeces and hair. Contamination remains a major problem when working on ancient human material.

Ancient pathogen DNA has been successfully retrieved from samples dating to more than 5,000 years old in humans and as long as 17,000 years ago in other species. In addition to the usual sources of mummified tissue, bones and teeth, such studies have also examined a range of other tissue samples, including calcified pleura, tissue embedded in paraffin, and formalin-fixed tissue. Efficient computational tools have been developed for pathogen and microorganism aDNA analyses in a small (QIIME) and large scale (FALCON).

Results

Taking preventative measures in their procedure against such contamination though, a 2012 study analyzed bone samples of a Neanderthal group in the El Sidrón cave, finding new insights on potential kinship and genetic diversity from the aDNA. In November 2015, scientists reported finding a 110,000-year-old tooth containing DNA from the Denisovan hominin, an extinct species of human in the genus Homo.

The research has added new complexity to the peopling of Eurasia. It has also revealed new in-formation about links between the ancestors of Central Asians and the indigenous peoples of the Americas. In Africa, older DNA degrades quickly due to the warmer tropical climate, although, in September 2017, ancient DNA samples, as old as 8,100 years old, have been reported.

Archaeogenetics

Archaeogenetics is the study of ancient DNA using various molecular genetic methods and DNA resources. The Greek word “arkhaios” meaning ancient; genetics meaning hereditary.

It is the reconstruction of ancient demography from patterns of gene differences in contemporary populations. Population size, population movements and subdivision into partially- isolated subpopulations, leave characteristic signatures in the DNA of the contemporary populations.

The field of genetic research has seen drastic changes since James Watson and Francis Crick first discovered the double helix structure of DNA in 1953. Thirty years after their pioneering breakthrough, the field of ancient DNA was born in the mid 1980's, with the extraction and sequencing of DNA from the quagga, an extinct South African equid, along with the extraction of DNA from an Egyptian mummy sample. Earlier attempts at DNA extraction were unfortunately often foiled by the lack of appropriate technology to allow scientists to distinguish between endogenous ancient DNA (known as aDNA) and outside sources of DNA contamination.

One of the key developments in genetic studies, the development of PCR (polymerase chain reaction) technology, has allowed geneticists to amplify genetic material for analysis. While revolutionary for the field of genetics, PCR did have serious problems for the study of ancient DNA, replicating not only the surviving ancient DNA, but also any contaminating DNA from other exogenous sources present in the sample. For this reason, many earlier reports of DNA extraction from ancient specimens, and all reports from specimens over a million years old – including all reports of dinosaur DNA – have been widely dismissed as being the amplified DNA of modern contaminants. The issue of contamination has remained a major issue, with Cooper and Poinar, openly criticising the lack of contamination control from many practitioners in the field. Simultaneously, they suggested a list of standards to be followed, such as having an isolated work area and an outside lab replicate results, which have laid much of the groundwork for modern aDNA standards.

Population Genetics

Both modern and ancient DNA can be used to analyse past population descent and migration. By observing the aforementioned changes in the genetic code, that are inherited, researchers have been able to analyse large scale population migrations, and develop theories as to how migrants interbred with local populations. Most notably, this was used to investigate the "out of Africa" theory of how anatomically modern humans originated and migrated from Africa. This model was later expanded upon, with additional analyses suggesting a level of genetic breeding between non-African humans and other species of hominin such as Neanderthals and Denisovans.

Much research in the past decades has focused on the origins and migration of prehistoric populations, and a particular "hot topic" of examination has been whether the spread of agricultural and other technologies associated with the Neolithic – the so called "Neolithic package" – better fits into the aforementioned cultural or demic model of diffusion. Related work has focused on both analysing the genetic spread of ancient human populations and the genetic signature of the domesticated animals that accompanied them, in order to test and develop new migration theories. Indeed, as genetic sequencing costs decrease, large scale population studies will become an increasingly lucrative area of research and it will be exciting to see what future results will uncover.

Fossil DNA Preservation

Fossil retrieval starts with selecting an excavation site. Potential excavation sites are usually identified with the mineralogy of the location and visual detection of bones in the area. However, there are more ways to discover excavation zones using technology such as field portable x-ray fluorescence and Dense Stereo Reconstruction. Tools used include knives, brushes, and pointed trowels which assist in the removal of fossils from the earth.

To avoid contaminating the ancient DNA, specimens are handled with gloves and stored in -20 °C immediately after being unearthed. Ensuring that the fossil sample is analyzed in a lab that has not been used for other DNA analysis could prevent contamination as well. Bones are milled to a powder and treated with a solution before the polymerase chain reaction (PCR) process. Samples for DNA amplification may not necessarily be fossil bones. Preserved skin, salt- preserved or air-dried, can also be used in certain situations.

DNA preservation is difficult because the bone fossilization degrades and DNA is chemically modified, usually by bacteria and fungi in the soil. The best time to extract DNA from a fossil is when it is freshly out of the ground as it contains six times the DNA when compared to stored bones. The temperature of extraction site also affects the amount of obtainable DNA, evident by a decrease in success rate for DNA amplification if the fossil is found in warmer regions. A drastic change of a fossil's environment also affects DNA preservation. Since excavation causes an abrupt change in the fossil's environment, it may lead to physiochemical change in the DNA molecule. Moreover, DNA preservation is also affected by other factors such as the treatment of the unearthed fossil like (e.g. washing, brushing and sun dring), pH, irradiation, the chemical composition of bone and soil, and hydrology. There are three perseveration digenetic phases. The first phase is bacterial putrefaction, which is estimated to cause a 15-fold degradation of DNA. Phase 2 is when bone chemically degrades, mostly by depurination. The third digenetic phase occurs after the fossil is excavated and stored, in which bone DNA degradation occurs most rapidly.

Methods of DNA Extraction

Once a specimen is collected from an archaeological site, DNA can be extracted through a series of processes. One of the more common methods utilizes silica and takes advantage of polymerase chain reactions in order to collect ancient DNA from bone samples.

There are several challenges that add to the difficulty when attempting to extract ancient DNA from fossils and prepare it for analysis. DNA is continuously being split up. While the organism is alive these splits are repaired; however, once an organism has died, the DNA will begin to deteriorate without repair. This results in samples having strands of DNA measuring around 100 base pairs in length. Contamination is another significant challenge at multiple steps throughout the process. Often other DNA, such as bacterial DNA, will be present in the original sample. To avoid contamination, it is necessary to take many precautions such as separate ventilation systems and workspaces for ancient DNA extraction work. The best samples to use are fresh fossils as uncareful washing can lead to mold growth. DNA coming from fossils also occasionally contains a compound that inhibits DNA replication. Coming to a consensus on which methods are best at mitigating challenges is also difficult due to the lack of repeatability caused by the uniqueness of specimens.

Silica-based DNA extraction is a method used as a purification step to extract DNA from archaeological bone artifacts and yield DNA that can be amplified using polymerase chain reaction (PCR) techniques. This process works by using silica as a means to bind DNA and separate it from other components of the fossil process that inhibit PCR amplification. However, silica itself is also a strong PCR inhibitor, so careful measures must be taken to ensure that silica is removed from the DNA after extraction. The general process for extracting DNA using the silica-based method is outlined by the following:

1. Bone specimen is cleaned and the outer layer is scraped off
2. Sample is collected from preferably compact section
3. Sample is ground to fine powder and added to an extraction solution to release DNA
4. Silica solution is added and centrifuged to facilitate DNA binding
5. Binding solution is removed and a buffer is added to the solution to release the DNA from the silica

One of the main advantages of silica-based DNA extraction is that it is relatively quick and efficient, requiring only a basic laboratory setup and chemicals. It is also independent of sample size, as the process can be scaled to accommodate larger or smaller quantities. Another benefit is that the process can be executed at room temperature. However, this method does contain some drawbacks. Mainly, silica-based DNA extraction can only be applied to bone and teeth samples; they cannot be used on soft tissue. While they work well with a variety of different fossils, they may be less effective in fossils that are not fresh (e.g. treated fossils for museums). Also, contamination poses a risk for all DNA replication in general, and this method may result in misleading results if applied to contaminated material.

Polymerase chain reaction is a process that can amplify segments of DNA and is often used on extracted ancient DNA. It has three main steps: denaturation, annealing, and extension. Denaturation splits the DNA into two single strands at high temperatures. Annealing involves attaching primer strands of DNA to the single strands that allow Taq polymerase to attach to the DNA. Extension occurs when Taq polymerase is added to the sample and matches base pairs to turn the two single strands into two complete double strands. This process is repeated many times, and is usually repeated a higher number of times when used with ancient DNA. Some issues with PCR is that it requires overlapping primer pairs for ancient DNA due to the short sequences. There can also be "jumping PCR" which causes recombination during the PCR process which can make analyzing the DNA more difficult in inhomogeneous samples.

Methods of DNA Analysis

DNA extracted from fossil remains is primarily sequenced using massive parallel sequencing, which allows simultaneous amplification and sequencing of all DNA segments in a sample, even when it is highly fragmented and of low concentration. It involves attaching a generic sequence to every single strand that generic primers can bond to, and thus all of the DNA present is amplified. This is generally more costly and time intensive than PCR but due to the difficulties involved in ancient DNA amplification it is cheaper and more efficient. One method of massive parallel sequencing, developed by Margulies et al., employs bead-based emulsion PCR and pyro sequencing, and was found to be powerful in analyses of aDNA because it avoids potential loss of sample, substrate competition for templates, and error propagation in replication.

The most common way to analyze aDNA sequence is to compare it with a known sequence from other sources, and this could be done in different ways for different purposes.

The identity of the fossil remain can be uncovered by comparing its DNA sequence with those of known species using software such as BLASTN. This archaeogenetic approach is especially helpful when the morphology of the fossil is ambiguous. Apart from that, species identification can also be done by finding specific genetic markers in an aDNA sequence. For example, the American indigenous population is characterized by specific mitochondrial RFLPs and deletions defined by Wallace et al.

aDNA comparison study can also reveal the evolutionary relationship between two species. The number of base differences between DNA of an ancient species and that of a closely related extant species can be used to estimate the divergence time of those two species from their last common ancestor. The phylogeny of some extinct species, such as Australian marsupial wolves and American ground sloths, has been constructed by this method. Mitochondrial DNA in animals and chloroplast DNA in plants are usually used for this purpose because they have hundreds of copies per cell and thus, are more easily accessible in ancient fossils.

Another method to investigate relationship between two species is through DNA hybridization. Single-stranded DNA segments of both species are allowed to form complementary pair bonding with each other. More closely related species have a more similar genetic makeup, and thus a stronger hybridization signal. Southern blot hybridization was conducted on Neanderthal aDNA (extracted from fossil remain W-NW and Krapina). The results showed weak ancient human-Neanderthal hybridization and strong ancient human-modern human hybridization. The human-chimpanzee and Neanderthal-chimpanzee hybridization are of similarly weak strength. This suggests that humans and Neanderthals are not as closely related as two individuals of the same species are, but they are more related to each other than to chimpanzees.

There have also been some attempts to decipher aDNA to provide valuable phenotypic information of ancient species. This is always done by mapping aDNA sequence onto the karyotype of a well-studied closely related species, which share a lot of similar phenotypic traits. For example, Green et al. compared the aDNA sequence from Neanderthal Vi-80 fossil with modern human X and Y chromosome sequence, and they found a similarity in 2.18 and 1.62 bases per 10,000 respectively, suggesting Vi-80 sample was from a male individual. Other similar studies include finding of a mutation associated with dwarfism in *Arabidopsis* in ancient Nubian cotton, and investigation on the bitter taste perception locus in Neanderthals.

Gene-centered View of Evolution

In the gene centered view there are assumed to be indivisible elementary units of the genome (thought of as individual genes) that are preserved from generation to generation. Different versions of the gene (alleles) compete and mutate rather than the organism as a whole. Thus the subject of evolution is the allele, and, in effect, the selection is of alleles rather than organisms.

Correlations between genes arise when the presence of one allele in one place in the genome affects the probability of another allele appearing in another place in the genome. One of the confusing points about the gene centered theory is that there are two stages in which the dynamic introduction of correlations must be considered: selection and sexual reproduction (gene mixing). Correlations occur in selection when the probability of survival favors certain combinations of alleles, rather than being determined by a product of terms given by each allele separately. Correlations occur in reproduction when parents are more likely to mate if they have certain combinations of alleles. If correlations only occur in selection and not in reproduction, the mean field approximation continues to be at least partially valid. However, if there are correlations in both selection and sexual reproduction then the mean field approximation and the gene centered view break down. Indeed, there are cases for which it is sufficient for there to be very weak correlations in sexual reproduction for the breakdown to occur. For example, populations of organisms distributed over space and an assumption that reproductive coupling is biased toward organisms that are born closer to each other can self-consistently generate allelic correlations in sexual reproduction by symmetry breaking. This is thus particularly relevant to considering trait divergence of subpopulations. Simulations of models that illustrate trait divergence through symmetry breaking can be found elsewhere.

Formalizing the Gene Centered View

A standard first model of sexual reproduction assumes that recombination of the genes during sexual reproduction results in a complete mixing of the possible alleles not just in each pair of mating organisms but rather throughout the species—the group of organisms that is mating and reproducing. Offspring are assumed to be selected from the ensemble which represents all possible combinations of the genomes from reproducing organisms.

If we further simplify the model by assuming that each gene controls a particular phenomic trait for which selection occurs independent of other gene-related traits, then each gene would evolve independently; a selected allele reproduces itself and its presence within an organism is irrelevant. Without this further assumption, selection should be considered to operate on the genome of organism, which may induce correlations in the allele populations in the surviving (reproducing) organisms. As the presence of one allele in the population changes in the population due to evolution over generations, the fitness of another allele at a different gene will be affected. However, due to the assumption of complete mixing in sexual reproduction, the correlations disappear in the offspring and only the average effect (mean field) of one gene on another is relevant. From the point of view of a particular allele at a particular gene, the complete mixing means that at all other genes alleles will be present in the same proportion that they appear in the population. Thus the assumption of complete mixing in sexual reproduction is equivalent to a gene based mean field approximation.

The mean field approximation is widely used in statistical physics as a "zeroth" order approximation to understanding the properties of systems. There are many cases where it provides important insight to some aspects of a system (e.g. the Ising model of magnets) and others where is essentially valid (conventional BCS superconductivity). The application of the mean field approximation to a problem involves assuming an element (or small part of the system) can be treated in the average environment that it finds throughout the system. This is equivalent to assuming that the probability distribution of the states of the elements factor.

This qualitative discussion of standard models of evolution and their relationship to the mean field approximation can be shown formally. In the mean field approximation, the probability of appearance of a particular state of the system s (e.g. a particular genome) is considered as the product of probabilities of the components a_i (e.g. its alleles):

$$P(s) = P(a_1,...,a_n) = \prod_i p_i(a_i)$$

In the usual application of this approximation, it can be shown to be equivalent to allowing each of the components to be placed in an environment which is an average over the possible environments formed by the other components of the system, hence the term "mean field approximation."

The key to applying this in the context of evolution is to consider carefully the effect of the reproduction step, not just the selection step. The two steps of reproduction and selection can be written quite generally as:

$$\{N(s, t+1\} = R\left[\{N'(s, t)\}\right]$$

$$\{N'(s, t)\} = D\left[\{N(s, t)\}\right]$$

The first equation describes reproduction. The number of offspring $N(s, t+1$ having a particular genome s is written as a function of the reproducing organisms $N'(s, t)$ from the previous generation. The second equation describes selection. The reproducing population $N'(s, t)$ is written as a function of the same generation at birth $N(s, t)$. The brackets on the left indicate that each equation represents a set of equations for each value of the genome. The brackets within the functions indicate, for example, that each of the offspring populations depends on the entire parent population.

The proportion of alleles can be written as the number of organisms, which have a particular allele a_i at gene *i* divided by the total number of organisms:

$$P_i'(a_i, t) = \frac{1}{N_0'(t)} \sum_{a_j,\, j\neq i} N'(s,t)$$

Where $s = (a_1,...,a_n)$ represents the genome in terms of alleles a_i. The sum is over all alleles of genes j except gene i that is fixed to allele a_i. $N_0(t)$ is the total reproducing population at time t. Using the assumption of complete allelic mixing by sexual reproduction, the frequency of allele a_i in the offspring is determined by only the proportion of a_i in the parent population. Then, the same offspring would be achieved by an 'averaged' population with a number of reproducing organisms given by

$$\tilde{N}'(s, t) = N_0'(t)\prod_i p_i'(a_i, t)$$

Since this $\tilde{N}'(s, t)$ has the same allelic proportions as $N'(s, t)$ in $P_i'(a_i, t) = \frac{1}{N_0'(t)} \sum_{a_j, j \neq i} N'(s,t)$. Thus, complete reproductive mixing assumes that:

$$R\left[\left\{\tilde{N}'(s, t)\right\}\right] \approx R\left[\left\{N'(s, t)\right\}\right]$$

The form of $\tilde{N}'(s, t) = N_0'(t) \prod p_i'(a_i, t)$ indicates that the effective probability of a particular genome can be considered as a product of the probabilities of the individual genes— as if they were independent. It follows that a complete step including both reproduction and selection can also be written in terms of the allele probabilities in the whole population. Given the above equations the update of an allele probability is:

$$P_i'(a_i, t+1) \approx \frac{1}{N_0'(t+1)} \sum_{a_j, j \neq i} D_s\left[R\left[\left\{\tilde{N}'(s, t)\right\}\right]\right]$$

where D_s is a function, which satisfies, $N'(s, t) = D_s R\left[\left\{N(s, t)\right\}\right]$ Given the form of $\tilde{N}'(s, t) = N_0'(t) \prod_i p_i'(a_i, t)$ and the additional assumption that the relative dynamics of change of genome proportions is not affected by the absolute population size N_0', we could write this as an effective one-step update

$$P_i'(a_i, t+1) \approx \tilde{D}\left[\left\{p_i'(a_i, t)\right\}\right].$$

Which describes the allele population change from one generation to the next of offspring. Since this equation describes the behavior of a single allele it corresponds to the gene centered view.

There is still a difficulty pointed out by Sober and Lewontin. The effective fitness of each allele depends on the distribution of alleles in the population. Thus, the fitness of an allele is coupled to the evolution of other alleles. This is apparent in $P_i'(a_i, t+1) = \tilde{D}\left[\left\{p_i'(a_i, t)\right\}\right]$ which, as indicated by the brackets, is a function of all the allele populations. It corresponds, as in other mean field approximations, to placing an allele in an average environment formed from the other alleles. This problem with fitness assignment would not be present if each allele separately coded for an organism trait. While this is a partial violation of the simplest conceptual view of evolution, however, the applicability of a gene centered view can still be justified, as long as the contextual assignment of fitness is included. When the fitness of organism phenotype is dependent on the relative frequency of phenotypes in a population of organisms it is known as frequency dependent selection, which is a concept that is being applied to genes in this context. A more serious breakdown of the mean field approximation occurs when the assumption of complete mixing during reproduction does not hold. This corresponds to symmetry breaking.

Breakdown of the Gene Centered View

We can provide a specific example of breakdown of the mean field approximation using a simple example. We start by using a simple model for population growth, where an organism that reproduces at a rate of λ offspring per individual per generation has a population growth described by an iterative equation:

$$N(t+1) = \lambda N(t)$$

We obtain a standard model for fitness and selection by taking two equations of the form $N(t+1) = \lambda N(t)$ for two populations $N_1(t)$ and $N_2(t)$ with λ_1 and λ_2 respectively, and normalize the population at every step so that the total number of organisms remains fixed at N_o. We have that:

$$N_1(t+1) = \frac{\lambda_1 N_1(t)}{\lambda_1 N_1(t) + \lambda l N_2(t)} No$$
$$N_2(t+1) = \frac{\lambda_2 N_2(t)}{\lambda_1 N_1(t) + \lambda_2 N_2(t)} No$$

The normalization does not change the relative dynamics of the two populations, thus the faster-growing population will dominate the slower-growing one according to their relative reproduction rates. If we call λ_i the fitness of the *ith* organism we see that according to this model the organism populations grow at a rate that is determined by the ratio of their fitness to the average fitness of the population

Consider now sexual reproduction where we have multiple genes. In particular, consider two non homologue genes with selection in favor of a particular combination of alleles on genes. Specifically, after selection, when allele A_1 appears in one gene, allele B_1 must appear on the second gene, and when allele A_{-1} appears on the first gene allele B_{-1} must appear on the second gene. We can write these high fitness organisms with the notation (1, 1) and (−1, −1), and the organisms with lower fitness (for simplicity, $\lambda = 0$) as (1, −1) and (−1, 1). When correlations in reproduction are neglected there are two stable states of the population with all organisms (1, 1) or all organisms (−1, −1). If we start with exactly 50% of each allele, then there is an unstable steady state in which 50% of the organisms reproduce and 50% do not in every generation. Any small bias in the proportion of one or the other will cause there to be progressively more of one type over the other, and the population will eventually have only one set of alleles.

We can solve this example explicitly for the change in population in each generation when correlations in reproduction are neglected. It simplifies matters to realize that the reproducing parent population (either (1, 1) or (−1, −1)) must contain the same proportion of the correlated alleles (A_1 and B_1) so that:

$$P_{1,1}(t) + P_{1,-1}(t) = P_{1,1}(t) + P_{-1,1}(t) = p(t)$$
$$P_{-1,1}(t) + P_{-1,-1}(t) = P_{1,-1}(t) + P_{-1,-1}(t) = 1 - p(t)$$

Where p is a proportion of allele A_1 or B_1. The reproduction equations are:

$$P_{1,1}(t+1) = p(t)^2$$
$$P_{1,-1}(t+1) = P_{-1,1}(t+1) = p(t)(1-p(t))$$
$$P_{-1,-1}(t+1) = (1-p(t))^2$$

The proportion of the alleles in the generation *t* is given by the selected organisms:

$$p(t) = P'_{1,1}(t) + P'_{1,-1}(t) = P'_{1,1}(t) + P'_{-1,1}(t)$$

Since the less fit organisms (1, −1) and (−1, 1) do not reproduce this is described by:

$$p(t) = P'_{1,1}(t) = \frac{P_{1,1}(t)}{P_{1,1}(t) + P_{-1,-1}(t)}$$

This gives the update equation

$$p(t+1) = \frac{p(t)^2}{p(t)^2 + (1 - p(t))^2}$$

which has the behavior described above and shown in figure. This problem is reminiscent of an using ferromagnet at very low temperature. Starting from a nearly random state with a slight bias in the number of up and down spins, the spins align becoming either all up or all down.

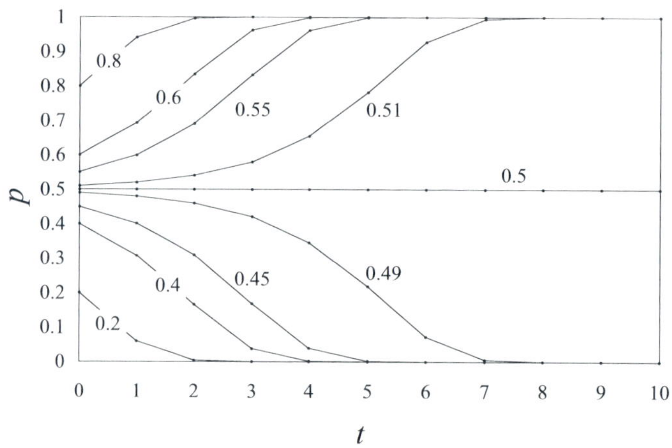

Figure: Behavior of *p* in $p(t+1) = \frac{p(t)^2}{p(t)^2 + (1 - p(t))^2}$ with several different initial values.

Since, we can define the proportion of a gene in generation *t* and in generation *t + 1* we can always write an expression for allele evolution in the form:

$$P_i(a_i, t+1) = \frac{\lambda_{a_i}}{\sum_{a_i} \lambda_{a_i} P_i(a_i, t)} P_i(a_i, t)$$

So that we have evolution that can be described in terms of gene rather than organism behavior. The fitness coefficient λ_1 for allele A_1 or B_1 is seen from $p(t+1) = \frac{p(t)^2}{p(t)^2 + (1 - p(t))^2}$ to be:

$$\lambda_1(t) = p(t)$$

With the corresponding $\lambda_{-1} = 1 - \lambda_1$. The assignment of a fitness to an allele reflects the gene centered view. The explicit dependence on the population composition has been objected to on grounds of biological appropriateness. For our purposes, we recognize this dependence as the natural outcome of a mean field approximation.

It is interesting to consider when this picture breaks down more severely due to a breakdown in the assumption of complete reproductive mixing. In this example, if there is a spatial distribution in the organism population with mating correlated by spatial location and fluctuations so that the starting population has more of the alleles represented by 1 in one region and more of the alleles represented by −1 in another region, then, patches of organisms that have predominantly (1, 1) or (−1, −1) will form after several generations. This symmetry breaking, like in a ferromagnet, is the usual breakdown of the mean field approximation. Here it creates correlations in the genetic makeup of the population. When the correlations become significant then the whole population becomes to contain a number of types. The formation of organism types depends on the existence of correlations in reproduction that are, in effect, a partial form of speciation. For an example of such symmetry breaking and pattern formation.

Thus, we see that the most dramatic breakdown of the mean field approximation / gene centered view occurs when multiple organism types form. This is consistent with our understanding of ergodicity breaking, phase transitions and the mean field approximation. Interdependence at the genetic level is echoed in the population through the development of subpopulations. We should emphasize again that this symmetry breaking required both selection and reproduction to be coupled to gene correlations.

The gene-centered view of evolution is a model for the evolution of social characteristics such as selfishness and altruism.

Acquired Characteristics

The formulation of the central dogma of molecular biology was summarized by Maynard Smith:

If the central dogma is true, and if it is also true that nucleic acids are the only means whereby information is transmitted between generations, this has crucial implications for evolution. It would imply that all evolutionary novelty requires changes in nucleic acids, and that these changes – mutations – are essentially accidental and non-adaptive in nature. Changes elsewhere – in the egg cytoplasm, in materials transmitted through the placenta, in the mother's milk – might alter the development of the child, but, unless the changes were in nucleic acids, they would have no long-term evolutionary effects.

The rejection of the inheritance of acquired characters, combined with Ronald Fisher the statistician, giving the subject a mathematical footing, and showing how Mendelian genetics was compatible with natural selection in his 1930 book. J. B. S. Haldane, and Sewall Wright, paved the way to the formulation of the selfish-gene theory.

Gene as the Unit of Selection

The view of the gene as the unit of selection was developed mainly in the works of Richard Dawkins, W. D. Hamilton, Colin Pittendrigh and George C. Williams It was mainly popularized and expanded by Dawkins in his book.

According to Williams' 1966 book,

The essence of the genetical theory of natural selection is a statistical bias in the relative rates of survival of alternatives (genes, individuals, etc.). The effectiveness of such bias in producing adaptation is contingent on the maintenance of certain quantitative relationships among the operative factors. One necessary condition is that the selected entity must have a high degree of permanence and a low rate of endogenous change, relative to the degree of bias (differences in selection coefficients).

Williams argued that "the natural selection of phenotypes cannot in itself produce cumulative change, because phenotypes are extremely temporary manifestations." Each phenotype is the unique product of the interaction between genome and environment. It does not matter how fit and fertile a phenotype is, it will eventually be destroyed and will never be duplicated.

Since 1954, it has been known that DNA is the main physical substrate to genetic information, and it is capable of high-fidelity replication through many generations. So, a particular gene coded in a nucleobase sequence of a lineage of replicated DNA molecules can have a high permanence and a low rate of endogenous change.

In normal sexual reproduction, an entire genome is the unique combination of father's and mother's chromosomes produced at the moment of fertilization. It is generally destroyed with its organism, because "meiosis and recombination destroy genotypes as surely as death." Only half of it is transmitted to each descendant due to independent segregation.

And the high prevalence of horizontal gene transfer in bacteria and archaea means that genomic combinations of these asexually reproducing groups are also transient in evolutionary time: "The traditional view, that prokaryotic evolution can be understood primarily in terms of clonal divergence and periodic selection, must be augmented to embrace gene exchange as a creative force."

The gene as an informational entity persists for an evolutionarily significant span of time through a lineage of many physical copies.

In his book, Dawkins coins the phrase *God's utility function* to explain his view on genes as units of selection. He uses this phrase as a synonym of the "meaning of life" or the "purpose of life". By rephrasing the word *purpose* in terms of what economists call a utility function, meaning "that which is maximized", Dawkins attempts to reverse-engineer the purpose in the mind of the Divine Engineer of Nature, or the *utility function of god*. Finally, Dawkins argues that it is a mistake to assume that an ecosystem or a species as a whole exists for a purpose. He writes that it is incorrect to suppose that individual organisms lead a meaningful life either; in nature, only genes have a utility function – to perpetuate their own existence with indifference to great sufferings inflicted upon the organisms they build, exploit and discard.

Organisms as Vehicles

Genes are usually packed together inside a genome, which is itself contained inside an organism. Genes group together into genomes because "genetic replication makes use of energy and substrates that are supplied by the metabolic economy in much greater quantities than would be possible without a genetic division of labor." They build vehicles to promote their mutual interests of jumping into the next generation of vehicles. As Dawkins puts it, organisms are the "survival machines" of genes.

The phenotypic effect of a particular gene is contingent on its environment, including the fellow genes constituting with it the total genome. A gene never has a fixed effect, so how is it possible to speak of a gene for long legs? It is because of the phenotypic differences between alleles. One may say that one allele, all other things being equal or varying within certain limits, causes greater legs than its alternative. This difference enables the scrutiny of natural selection.

"A gene can have multiple phenotypic effects, each of which may be of positive, negative or neutral value. It is the net selective value of a gene's phenotypic effect that determines the fate of the gene." For instance, a gene can cause its bearer to have greater reproductive success at a young age, but also cause a greater likelihood of death at a later age. If the benefit outweighs the harm, averaged out over the individuals and environments in which the gene happens to occur, then phenotypes containing the gene will generally be positively selected and thus the abundance of that gene in the population will increase.

Even so, it becomes necessary to model the genes in combination with their vehicle as well as in combination with the vehicle's environment.

Selfish-gene Theory

The selfish-gene theory of natural selection can be restated as follows:

Genes do not present themselves naked to the scrutiny of natural selection, instead they present their phenotypic effects. Differences in genes give rise to differences in these phenotypic effects. Natural selection acts on the phenotypic differences and thereby on genes. Thus genes come to be represented in successive generations in proportion to the selective value of their phenotypic effects.

The result is that "the prevalent genes in a sexual population must be those that, as a mean condition, through a large number of genotypes in a large number of situations, have had the most favorable phenotypic effects for their own replication." In other words, we expect selfish genes ("selfish" meaning that it promotes its own survival without necessarily promoting the survival of the organism, group or even species). This theory implies that adaptations are the phenotypic effects of genes to maximize their representation in future generations. An adaptation is maintained by selection if it promotes genetic survival directly, or else some subordinate goal that ultimately contributes to successful reproduction.

Individual Altruism and Genetic Egoism

The gene is a unit of hereditary information that exists in many physical copies in the world, and which particular physical copy will be replicated and originate new copies does not matter from the gene's point of view. A selfish gene could be favored by selection by producing altruism among organisms containing it. The idea is summarized as follows:

If a gene copy confers a benefit *B* on another vehicle at cost *C* to its own vehicle, its costly action is strategically beneficial if $p_{B>C}$, where *p* is the probability that a copy of the gene is present in the vehicle that benefits. Actions with substantial costs therefore require significant values of *p*. Two kinds of factors ensure high values of p: relatedness (kinship) and recognition (green beards).

A gene in a somatic cell of an individual may forego replication to promote the transmission of its copies in the germ line cells. It ensures the high value of $p = 1$ due to their constant contact and their common origin from the zygote.

The kin selection theory predicts that a gene may promote the recognition of kinship by historical continuity: a mammalian mother learns to identify her own offspring in the act of giving birth; a male preferentially directs resources to the offspring of mothers with whom he has copulated; the other chicks in a nest are siblings; and so on. The expected altruism between kin is calibrated by the value of p, also known as the coefficient of relatedness. For instance, an individual has a $p = 1/2$ in relation to his brother, and $p = 1/8$ to his cousin, so we would expect, *ceteris paribus*, greater altruism among brothers than among cousins. In this vein, geneticist J. B. S. Haldane famously joked, "Would I lay down my life to save my brother? No, but I would to save two brothers or eight cousins." However, examining the human propensity for altruism, kin selection theory seems incapable of explaining cross-familiar, cross-racial and even cross-species acts of kindness.

Green-beard Effect

Green-beard effects gained their name from a thought-experiment of Richard Dawkins, who considered the possibility of a gene that caused its possessors to develop a green beard and to be nice to other green-bearded individuals. Since then, "green-beard effect" has come to refer to forms of genetic self-recognition in which a gene in one individual might direct benefits to other individuals that possess the gene. Such genes would be especially selfish, benefiting themselves regardless of the fates of their vehicles. After Dawklns predicted them, green-beard genes have been discovered in nature, such as *Gp-9* in fire ants (*Solenopsis invicta*), *csA* in social amoeba (*Dictyostelium discoideum*), and FLO1 in budding yeast (*Saccharomyces cerevisiae*).

Kinds of Altruism

Kindness

On the other hand, a single trait, group reciprocal kindness, is capable of explaining the vast majority of altruism that is generally accepted as "good" by modern societies. Imagine a green-bearding behavioral trait whose recognition does not depend on the recognition of some external feature such as beard color, but relies on recognition of the behavior itself. Imagine now that the behavior is altruistic. The success of such a trait in sufficiently intelligent and undeceived organisms is implicit. Moreover, the existence of such a trait predicts a tendency for kindness to unrelated organisms that are apparently kind, even if the organisms are of a completely different species. Moreover, the gene need not be exactly the same, so long as the effect is similar. Multiple versions of the gene—or even meme—would have virtually the same effect in a sort of symbiotic green-bearding cycle of altruism.

Deceit

Whenever recognition plays a role in evolution, so does deception. Just like the harmless lizard that has evolved a pattern that mimics its poisonous cousin and therefore tricks predators, the selfish creature may pretend to be kind by "growing a green beard" (whatever that green beard may be). Thus green-bearding and the selfish-gene theory also give rise to an explanation for the evolution of lies and deceit, characteristics that do not benefit the population as a whole.

Adaptive Evolution in the Human Genome

Positive natural selection, or the tendency of beneficial traits to increase in prevalence (frequency) in a population, is the driving force behind adaptive evolution. For a trait to undergo positive selection, it must have two characteristics. First, the trait must be beneficial; in other words, it must increase the organism's probability of surviving and reproducing. Second, the trait must be heritable so that it can be passed to an organism's offspring. Beneficial traits are extremely varied and may include anything from protective coloration, to the ability to utilize a new food source, to a change in size or shape that might be useful in a particular environment. If a trait results in more offspring who share the trait, then that trait is more likely to become common in the population than a trait that arises randomly. At the molecular level, selection occurs when a particular DNA variant becomes more common because of its effect on the organisms that carry it.

Charles Darwin and Alfred Wallace famously proposed that positive selection could explain the many marvelous adaptations that suit organisms to their environments and lifestyles, and this simple process remains the central explanation for all evolutionary adaptation yet today. Positive selection is by no means the only component of evolution, however. In humans, at least, the great majority of mutations are thought to be selectively neutral, conferring neither benefit nor cost on their bearers. The frequency of some of these neutral genetic variants (alleles) increases simply by chance, and the resulting "genetic drift" is thought to be the most common process in human evolution. Moreover, when selection does occur, it is most often in the form of negative, or purifying, selection, which removes new deleterious mutations as they arise, rather than promoting the spread of new traits.

Advantageous Alleles and Selective Sweep

As advantageous alleles that are under positive selection increase in prevalence, these alleles leave distinctive signatures, or patterns of genetic variation, in the DNA sequence. Consider a population of individuals for which, before selection, there are hundreds of thousands of varied chromosomes in the population, all with different combinations of genetic variants. Now, say that an advantageous allele arises as a mutation on one copy of a chromosome. Through succeeding generations, the descendants of this copy, including the selected allele and nearby "hitchhiking" alleles, become more and more common through a process called a "selective sweep". Note that the entire chromosome is not passed down as a unit, however; rather, because of recombination, segments of the chromosome are inherited. Thus, while the selected allele and hitchhiking alleles increase in prevalence in a selective sweep, at the same time, the segment that includes the selected allele is slowly reduced in size by recombination. Investigators are interested in the types of signals that can be detected in a selective sweep, as well as their properties and technical challenges.

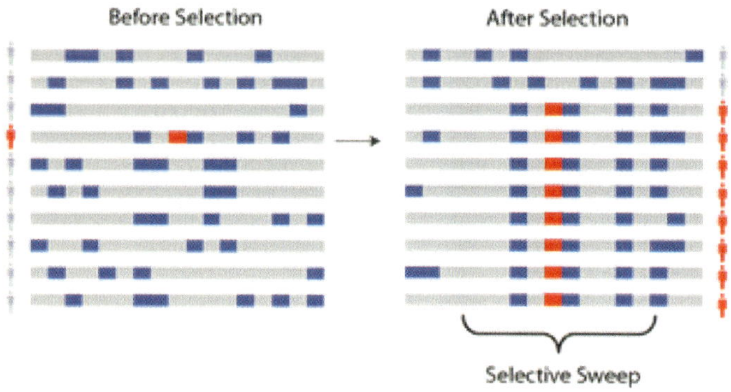

Figure: A selective sweep

Under natural selection, a new beneficial mutation will rise in frequency (prevalence) in a population. A schematic shows polymorphisms along a chromosome, including the selected allele, before and after selection. Ancestral alleles are shown in grey and derived (non-ancestral) alleles are shown in blue. As a new positively-selected allele (red) rises to high frequency, nearby linked alleles on the chromosome 'hitchhike' along with it to high frequency, creating a 'selective sweep.

Evidence of Positive Selection in Humans

Within the last decade, our ability to probe our own species for evidence of selection has increased dramatically due to the flood of genetic data that have been generated. Starting with the complete sequence of the human genome, which provides a framework and standard reference for all human genetics, key data sets include the completed or near-completed genomes of several related species (e.g., chimpanzee, macaque, gorilla, and orangutan), a public database of known genetic variants in humans, and surveys of genetic variation in hundreds of individuals in multiple populations. With these new data, it is now possible to scan the entire human genome in search of signals of natural selection.

Although the study of natural selection in humans is still in an early stage, the new data, building on decades of earlier work, are beginning to reveal some of the landscape of selection in our species. In fact, researchers have identified many genetic loci at which selection has likely occurred, and some of the selective pressures involved have been elucidated. Three significant forces that have been identified thus far include changes in diet, changes in climate, and infectious disease.

Lactose Tolerance

The domestication of plants and animals roughly 10,000 years ago profoundly changed human diets, and it gave those individuals who could best digest the new foods a selective advantage. The best understood of these adaptations is lactose tolerance. The ability to digest lactose, a sugar found in milk, usually disappears before adulthood in mammals, and the same is true in most human populations. However, for some people, including a large fraction of individuals of European descent, the ability to break down lactose persists because of a mutation in the lactase gene (*LCT*). This suggests that the allele became common in Europe because of increased nutrition from cow's milk, which became available after the domestication of cattle. This hypothesis was eventually confirmed by Todd Bersaglieri and his colleagues, who demonstrated that the lactase persistence allele is common in Europeans (nearly 80% of people of European descent carry this allele), and it has evidence of a selective sweep spanning roughly 1 million base pairs (1 megabase). Indeed, lactose tolerance is one of the strongest signals of selection seen anywhere in the genome. Sarah Tishkoff and colleagues subsequently found a distinct LCT mutation also conferring lactose tolerance, in this case in African pastoralist populations, suggesting the action of convergent evolution.

Malaria Resistance

The development of agriculture also changed the selective pressures on humans in another way: Increased population density made the transmission of infectious diseases easier, and it probably expanded the already substantial role of pathogens as agents of natural selection. That role is reflected in the traces left by selection in human genetic diversity; multiple loci associated with disease resistance have been identified as probable sites of selection. In most cases, the resistance is to the same disease—malaria.

Malaria's power to drive selection is not surprising, as it is one of the human population's oldest diseases and remains one of the greatest causes of morbidity and mortality in the world today, infecting hundreds of millions of people and killing 1 to 2 million children in Africa each year. In fact, malaria was responsible for the first case of positive selection demonstrated genetically in humans. In the 1940s and 1950s, J. B. S. Haldane and A. C. Allison demonstrated that the geographical distribution of the sickle-cell mutation (Glu6Val) in the beta hemoglobin gene (*HBB*) was limited to Africa and correlated with malaria endemicity, and that individuals who carry the sickle-cell trait are resistant to malaria. Since then, many more alleles for malaria resistance have shown evidence of selection, including more mutations in *HBB*, as well as mutations causing other red blood cell disorders (e.g., a-thalassemia, G6PD deficiency, and ovalocytosis).

Malaria also drove one of the most striking genetic differences between populations. This difference involves the Duffy antigen gene (*FY*), which encodes a membrane protein used by the *Plasmodium vivax* malaria parasite to enter red blood cells, a critical first step in its life cycle. A mutation in *FY* that disrupts the protein, thus conferring protection against *P. vivax* malaria, is at a frequency of 100% throughout most of sub-Saharan Africa and virtually absent elsewhere; such an extreme difference in allele frequency is very rare for humans.

Pigmentation

As proto-Europeans and Asians moved northward out of Africa, they experienced less sunlight and colder temperature, new environmental forces that exerted selective pressure on the migrants. Exactly why reduced sunlight should be a potent selective force is still debated, but it has become clear that humans have experienced positive selection at numerous genes to finely tune the amount of skin pigment they produce, depending on the amount of sunlight exposure.

The role of selection in controlling human pigmentation is not a new idea; in fact, it was first advanced by William Wells in 1813, long before Darwin's formulation of natural selection. In recent years, signals of positive selection have been identified in many genes, with some signals solely in Europeans, some solely in Asians, and some shared across both continents. Evidence for purifying selection has also been found to maintain dark skin color in Africa, where sunlight exposure is great.

A good example of selection for lighter pigmentation is the gene *SLC24A5*, which was one of the first to be characterized. Rebecca Lamason and her colleagues identified a mutation in the zebrafish homologue of this gene that is responsible for pigmentation phenotype. The investigators then demonstrated that a human variant in the gene explains roughly one-third of the variation in pigmentation between Europeans and West Africans, and that the European variant had likely been a target of selection. In related work, Angela Hancock and her colleagues examined many genes involved in metabolism, and they showed that alleles of these genes show evidence of positive selection and correlate strongly with climate, suggesting that humans adapted to cooler climates by changing their metabolic rates.

Signals for Positive Selection

Signal of selection	Target, variant	Normal biological function	Phenotypic variation	Other variations in the pathway	Evolutionary adaptation
chr11, Africa	HBB, coding	oxygen carrying capacity in RBCs	sickle cell, malaria resistance	CD36, HBA, SLC4A1, DARC, HBB*	protection from malaria
chr2, Europe, Africa	LCT, regulatory	enzyme lactase to digest milk	prolonged expression of LCT in adulthood	LCT*	milk digestion as adults after cattle domestication
chr15, Europe	SLC24A5, coding	pigment formation	decreased pigment formation	KITLG, SLC42A2, OCA2	lighter skin as humans moved to less sunlight
chr2, Asia	EDAR, coding	development of hair, sweat, and teeth	linked to thicker hair on head, largely unknown	EDA2R	
chr16, Asia	ABCC11, coding	ear wax type determinant	dry ear wax		
chr18, Europe and Asia	DTNA, unknown				

Figure: Characterizing signals of selection in humans

While these instances of selection illustrate the power this line of research has to answer important biological and historical questions, in most cases, little or nothing of the underlying story is understood. For the great majority of selective sweeps, the pressure that drove selection, the trait selected for, and even the specific gene involved is unknown. Understanding these will require case-by-case study, identifying the possible causal mutations within each region based on strength of signal and function (e.g., mutations that alter amino acids or gene regulatory regions), and then finding the biological effects of each.

Such detailed investigations are underway, and they are intriguing. For example, a strong signal of selection in Asia localizes to amino acid substitution in the gene *EDAR*. Mutations in *EDAR* cause defects in the development of hair, teeth, and exocrine glands in both mice and humans. Meanwhile, there is also evidence for selection at other genes in the same pathway in humans, as well as in stickleback fish, where the pathway regulates scale development. The phenotypic variation for this mutation is only just being elucidated, but it has already been linked to thicker head hair in Asia and has been shown to affect gene activity in the molecular pathway, although what trait was actually under selection is not yet clear. In another case, Scott Williamson and his colleagues found the strongest signal of selection in Europe and Asia at the gene *DTNA*, a component of the dystrophin complex. While the target polymorphism and genetic variation have yet to be elucidated, the dystrophin complex is known to be important in the architecture of muscle tissue, as well as in the pathogenesis of many infectious agents, including arenaviruses and mycobacterium leprae. Another candidate gene for selection, *LARGE*, is also important for dystrophin function, and it has been shown to be critical for entry of various arenaviruses, including Lassa virus.

Understanding the biology behind these cases, and the many others like them, will not be easy, and it will require contributions from diverse fields, including genetics, molecular biology, developmental biology, and the study of model organisms. Nevertheless, the potential rewards are high. Through the study of natural selection in humans, researchers hope to learn more about how our species has changed over time, about the challenges the species has faced and how it has overcome them, and about past and present causes of disease.

Methods

The methods used to identify adaptive evolution are generally devised to test the null hypothesis of neutral evolution, which, if rejected, provides evidence of adaptive evolution. These tests can be broadly divided into two categories.

Firstly, there are methods that use a comparative approach to search for evidence of function altering mutations. The dN/dS rates-ratio test estimates ω, the rates at which nonsynonymous ('dN') and synonymous ('dS') nucleotide substitutions occur ('synonymous' nucleotide substitutions do not lead to a change in the coding amino acid, while 'nonsynonymous' ones do). In this model, neutral evolution is considered the null hypothesis, in which dN and dS approximately balance so that $\omega \approx 1$. The two alternative hypotheses are a relative absence of nonsynonymous substitutions ($dN < dS$; $\omega < 1$), suggesting the effect on fitness ('fitness effect', or 'selection pressure') of such mutations is negative (purifying selection has operated over time; or a *relative excess* of nonsynonymous substitutions ($dN > dS$; $\omega > 1$), indicating positive effect on fitness, i.e. diversifying selection.

The McDonald-Kreitman (MK) test quantifies the amount of adaptive evolution occurring by estimating the proportion of nonsynonymous substitutions which are adaptive, referred to as α. α is calculated as: $\alpha = 1-(dspn/dnps)$, where dn and ds are as above, and pn and ps are the number of nonsynonymous (fitness effect assumed neutral or deleterious) and synonymous (fitness effect assumed neutral) polymorphisms respectively.

Note, both these tests are presented here in basic forms, and these tests are normally modified considerably to account for other factors, such as the effect of slightly deleterious mutations.

The other methods for detecting adaptive evolution use genome wide approaches, often to look for evidence of selective sweeps. Evidence of complete selective sweeps is shown by a decrease in genetic diversity, and can be inferred from comparing the patterns of the site frequency spectrum (SFS, i.e. the allele frequency distribution) obtained with the SFS expected under a neutral model.

Partial selective sweeps provide evidence of the most recent adaptive evolution, and the methods identify adaptive evolution by searching for regions with a high proportion of derived allele.

Examining patterns of linkage disequilibrium (LD) can locate signatures of adaptive evolution. LD tests work on the basic principle that, assuming equal recombination rates, LD will rise with increasing natural selection. These genomic methods can also be applied to search for adaptive evolution in non-coding DNA, where putatively neutral sites are hard to identify.

Another recent method used to detect selection in non-coding sequences examines insertions and deletions (indels), rather than point mutations, although the method has only been applied to examine patterns of negative selection.

Amount of Adaptive Evolution

Coding DNA

Many different studies have attempted to quantify the amount of adaptive evolution in the human genome, the vast majority using the comparative approaches outlined above. Although there are discrepancies between studies, generally there is relatively little evidence of adaptive evolution in protein coding DNA, with estimates of adaptive evolution often near 0%. The most obvious exception to this is the 35% estimate of α. This comparatively early study used relatively few loci (fewer than 200) for their estimate, and the polymorphism and divergence data used was obtained from different genes, both of which may have led to an overestimate of α. The next highest estimate is the 20% value of α. However, the MK test used in this study was sufficiently weak that the authors state that this value of α is not statistically significantly different from 0%.Nielsen estimate that 9.8% of genes have undergone adaptive evolution also has a large margin of error associated with it, and their estimate shrinks dramatically to 0.4% when they stipulate that the degree of certainty that there has been adaptive evolution must be 95% or more. This raises an important issue, which is that many of these tests for adaptive evolution are very weak. Therefore the fact that many estimates are at (or very near to) 0% does not rule out the occurrence of any adaptive evolution in the human genome, but simply shows that positive selection is not frequent enough to be detected by the tests. The generally low estimates of adaptive evolution in human coding DNA can be contrasted with other species.Bakewell found more evidence of adaptive evolution in chimpanzees than humans, with 1.7% of chimpanzee genes showing evidence of adaptive evolution (compared with the 1.1% estimate for humans). Comparing humans with more distantly related animals, an early estimate for α in Drosophila species was 45%, and later estimates largely agree with this. Bacteria and viruses generally show even more evidence of adaptive evolution; research shows values of α in a range of 50-85%, depending on the species examined. Generally, there does appear to be a positive correlation between (effective) population size of the species, and amount of adaptive evolution occurring in the coding DNA regions. This may be because random genetic drift becomes less powerful at altering allele frequencies, compared to natural selection, as population size increases.

Non-coding DNA

Estimates of the amount of adaptive evolution in non-coding DNA are generally very low, although fewer studies have been done on non-coding DNA. As with the coding DNA however, the methods currently used are relatively weak. Ponting and Lunter speculate that underestimates may be even more severe in non-coding DNA, because non-coding DNA may undergo periods of functionality (and adaptive evolution), followed by periods of neutrality. If this is true, current methods for detecting adaptive evolution are inadequate to account for such patterns. Additionally, even if low estimates of the amount of adaptive evolution are correct, this can still equate to a large amount of adaptively evolving non-coding DNA, since non-coding DNA makes up approximately 98% of the DNA in the human genome. For example, Ponting and Lunter detect a modest 0.03% of non-coding DNA showing evidence of adaptive evolution, but this still equates to approximately 1 Mb of adaptively evolving DNA. Where there is evidence of adaptive evolution (which implies functionality) in non-coding DNA, these regions are generally thought to be involved in the regulation of protein coding sequences. As with humans, fewer studies have searched for adaptive evolution in non-coding regions of other organisms. However, where research has been done on Drosophila, there appears to be large amounts of adaptively evolving non-coding DNA. Andolfatto estimated that adaptive evolution has occurred in 60% of untranslated mature portions of mRNAs, and in 20% of intronic and intergenic regions. If this is true, this would imply that much non-coding DNA could be of more functional importance than coding DNA, dramatically altering the consensus view. However, this would still leave unanswered what function all this non-coding DNA performs, as the regulatory activity observed thus far is in just a tiny proportion of the total amount of non-coding DNA. Ultimately, significantly more evidence needs to be gathered to substantiate this viewpoint.

Variation between Human Populations

Several recent studies have compared the amounts of adaptive evolution occurring between different populations within the human species. Williamson found more evidence of adaptive evolution in European and Asian populations than African American populations. Assuming African Americans are representative of Africans, these results makes sense intuitively, because humans spread out of Africa approximately 50,000 years ago (according to the consensus Out-of-Africa hypothesis of human origins), and these humans would have adapted to the new environments they encountered. By contrast, African populations remained in a similar environment for the following tens of thousands of years, and were therefore probably nearer their adaptive peak for the environment. However, Voight found evidence of more adaptive evolution in Africans, than in Non-Africans (East Asian and European populations examined), and Boyko found no significant difference in the amount of adaptive evolution occurring between different human populations. Therefore, the evidence obtained so far is inconclusive as to what extent different human populations have undergone different amounts of adaptive evolution.

Rate of Adaptive Evolution

The rate of adaptive evolution in the human genome has often assumed to be constant over time. For example, the 35% estimate for α calculated by Fay led them to conclude that there was one adaptive substitution in the human lineage every 200 years since human divergence from Old World monkeys. However, even if the original value of α is accurate for a particular time period, this extrapolation is still invalid. This is because there has been a large acceleration in the amount of positive selection in the human lineage over the last 40,000 years, in terms of the number of genes that have undergone adaptive evolution. This agrees with simple theoretical predictions, because the human population size has expanded massively in the last 40,000 years, and with more people, there should be more adaptive substitutions. Hawks argue that demographic changes (particularly population expansion) may greatly facilitate adaptive evolution, an argument that corroborates somewhat with the positive correlation inferred between population size and amount of adaptive evolution occurring mentioned previously. It has been suggested that cultural evolution may have replaced genetic evolution, and hence slowed the rate of adaptive evolution over the past 10,000 years. However, it is possible that cultural evolution could actually increase genetic adaption. Cultural evolution has vastly increased communication and contact between different populations, and this provides much greater opportunities for genetic admixture between the different populations. However, recent cultural phenomena, such as modern medicine and the smaller variation in modern family sizes, may reduce genetic adaption as natural selection is relaxed, overriding the increased potential for adaptation due to greater genetic admixture.

Regions of the Genome which Show Evidence of Adaptive Evolution

A considerable number of studies have used genomic methods to identify specific human genes that show evidence of adaptive evolution. Below are listed some of the types of gene which show strong evidence of adaptive evolution in the human genome.

- Disease genes

Bakewell found that a relatively large proportion (9.7%) of positively selected genes were associated with diseases. This may be because diseases can be adaptive in some contexts. For example, schizophrenia has been linked with increased creativity, perhaps a useful trait for obtaining food or attracting mates in Palaeolithic times. Alternatively, the adaptive mutations may be the ones which reduce the chance of disease arising due to other mutations. However, this second explanation seems unlikely, because the mutation rate in the human genome is fairly low, so selection would be relatively weak.

- Immune genes

417 genes involved in the immune system showed strong evidence of adaptive evolution in the study of Nielsen. This is probably because the immune genes may become involved in a evolutionary arms race with bacteria and viruses. These pathogens evolve very rapidly, so selection pressures change quickly, giving more opportunity for adaptive evolution.

- Testes genes

247 genes in the testes showed evidence of adaptive evolution in the study of Nielsen. This could be partially due to sexual antagonism. Male-female competition could facilitate an arms race of adaptive evolution. However, in this situation you would expect to find evidence of adaptive evolution in the female sexual organs also, but there is less evidence of this. Sperm competition is another possible explanation. Sperm competition is strong, and sperm can improve their chances of fertilizing the female egg in a variety of ways, including increasing their speed, stamina or response to chemo attractants.

- Olfactory genes

Genes involved in detecting smell show strong evidence of adaptive evolution, probably due to the fact that the smells encountered by humans have changed recently in their evolutionary history. Humans' sense of smell has played an important role in determining the safety of food sources.

- Nutrition genes

Genes involved in lactose metabolism show particularly strong evidence of adaptive evolution amongst the genes involved in nutrition. A mutation linked to lactase persistence shows very strong evidence of adaptive evolution in European and American populations, populations where pastoral farming for milk has been historically important.

- Pigmentation genes

Pigmentation genes show particularly strong evidence of adaptive evolution in non-African populations. This is likely to be because those humans that left Africa approximately 50,000 years ago, entered less sunny climates, and so were under new selection pressures to obtain enough Vitamin D from the weakened sunlight.

- Brain genes

There is some evidence of adaptive evolution in genes linked to brain development, but some of these genes are often associated with diseases, e.g. microcephaly. However, there is a particular interest in the search for adaptive evolution in brain genes, despite the ethical issues surrounding such research. If more adaptive evolution was discovered in brain genes in one human population than another, then this information could be interpreted as showing greater intelligence in the more adaptively evolved population.

- Other

Other gene types showing considerable evidence of adaptive evolution (but generally less evidence than the types discussed) include: genes on the X chromosome, nervous system genes, genes involved in apoptosis, genes coding for skeletal traits, and possibly genes associated with speech.

Difficulties in Identifying Positive Selection

Many of the tests used to detect adaptive evolution have very large degrees of uncertainty surrounding their estimates. It is beyond the purview of this topic to look at all the modifications applied to individual tests to overcome the associated problems. However, it will briefly discuss in general terms two of what may be the most important confounding variables that may hinder accurate detection of adaptive evolution. Demographic changes are particularly problematic and may severely bias estimates of adaptive evolution. The human lineage has undergone both rapid population size contractions and expansions over its evolutionary history, and these events will change many of the signatures thought to be characteristic of adaptive evolution. Some genomic methods have been shown through simulations to be relatively robust to demographic changes. However, no tests are completely robust to demographic changes, and new genetic phenomena linked to demographic changes have recently been discovered. This includes the concept of "surfing mutations", where new mutations can be propagated with a population expansion. A phenomenon which could severely alter the way we look for signatures of adaptive evolution is biased gene conversion (BGC). Meiotic recombination between homologous chromosomes that are heterozygous at a particular locus can produce a DNA mismatch. DNA repair mechanisms are biased towards repairing a mismatch to the CG base pair. This will lead allele frequencies to change, leaving a signature of non-neutral evolution. The excess of AT to GC mutations in human genomic regions with high substitution rates (human accelerated regions, HARs) implies that BGC has occurred frequently in the human genome. Initially, it was postulated that BGC could have been adaptive, but more recent observations have made this seem unlikely. Firstly, some HARs show no substantial signs of selective sweeps around them. Secondly, HARs tend to be present in regions with high recombination rates. In fact, BGC could lead to HARs containing a high frequency of deleterious mutations. However, it is unlikely that HARs are generally maladaptive, because DNA repair mechanisms themselves would be subject to strong selection if they propagated deleterious mutations. Either way, BGC should be further investigated, because it may force radical alteration of the methods which test for the presence of adaptive evolution.

Robustness

Robustness is the persistence of an organismal trait under perturbations. Many different organismal features could qualify as *traits* in this definition of robustness. A trait could be the proper fold or activity of a protein, a gene expression pattern produced by a regulatory gene network, the regular progression of a cell division cycle, the communication of a molecular signal from cell surface to nucleus, a cell interaction necessary for embryogenesis or the proper formation of a viable organism or organ.

Robustness is important in ensuring the stability of phenotypic traits that are constantly exposed to genetic and non-genetic variation. To better understand robustness is of paramount importance for understanding organismal evolution, because robustness permits cryptic genetic variation to accumulate. Such variation may serve as a source of new adaptations and evolutionary innovations.

Detection of Robustness

Robustness is not an all-or-nothing property. It is a matter of degree. For a quantitative trait, lack of robustness can be expressed using the coefficient of variation (square root of the variance over the mean) for the trait or, when comparing two conditions, the unsigned difference in the means. For a complex qualitative trait such as a protein sequence or the vulval cell fate pattern, robustness (or a lack thereof) can be expressed using the proportion of deviant phenotypes produced in response to perturbations. For example, a given environmental condition or mutation may produce a deviant phenotype for a large (e.g., 10^{-2}) or small (10^{-10}) fraction of organisms. In addition, the types of deviation ('errors') that a system produces – an amino-acid misincorporation in a protein sequence during translation, a deviant cell fate pattern or the shape of an organ – and their consequence on the organism's fitness influence crucially how natural selection acts on a system, yet they are often not investigated.

Robustness of a trait to noise is best detected by assaying individuals of an isogenic strain in a given constant environment. The use of isogenic strains eliminates confounding effects from genetic variation between individuals in assessing the effect of stochastic noise. For organisms that have a prominent haploid life cycle stage (many fungi, bacteria) or are commonly selfing (such as *C. elegans*), isogenic strains are easy to obtain. Vulva development of *C. elegans* has been mostly studied using the isogenic N2 reference strain in one standard culture condition. In these conditions, vulva cell fate patterning errors are found at a low frequency (on the order of 10^{-3} or less, for deviations that disrupt the cell fate pattern, but do not necessarily prevent egg-laying), implying that this aspect of vulva development is precise and robust to stochastic noise. The degree of robustness and the types of error can be compared between different isogenic backgrounds. A second way to eliminate confounding effects from genetic variation in measuring robustness to noise is to quantify the developmental variation between the right and left sides of an animal (fluctuating asymmetry).

Robustness of a trait to environmental variation is detected by subjecting organisms to a given environmental change or an array of environmental changes that may mimic ecologically relevant environments, possibly including some 'stressful' environments. In the vulva example, under starvation conditions in the second larval stage (one test environment), *C. elegans* N2 individuals are prone to miscenter their vulva on P5.p instead of P6.p. This centering variation of the cell fate pattern results in a quasi-normal vulva because P4.p is competent to form vulval tissue and adopts a 2° fate in these animals. Furthermore, the incidence and patterning of vulva variants vary with environmental conditions. They also vary with the wild-type genetic background, which means that they are subject to evolutionary change, possibly via the action of natural selection.

Robustness to a given mutation is detected by comparing the mutant to the reference wild-type genotype, and asking whether the mutation is silent or neutral, that is, whether it lacks an effect on the trait. The question whether a mutation is truly neutral is surprisingly difficult to answer. For instance, a mutation might have an effect at one developmental stage, but not on the final phenotype, or vice versa. In addition, a mutation's effect may critically depend on the genotype at other loci. For instance, in *C. elegans* vulva development, null mutations in the gene coding for the Ras GTPase activating protein, or for an activator of EGF receptor degradation (SLI-1), are silent with respect to the final cell fate pattern. The system is robust to these mutations. In contrast, the double mutant displays an excess of vulval fates, showing that these two molecules indeed modulate Ras pathway activity and are thus not silent at this level.

This test of robustness to a given mutation can be extended to a statistical measure (e.g. the mutational variance for quantitative traits; Lynch) of the effect of thousands of random mutations that are produced either spontaneously or through systematic mutagenesis studies. Systematic gene inactivation libraries are becoming available in several organisms. However, many of these 'inactivations' may be partial and result in a reduction of a gene's function. They, thus, only represent a narrow band within a broader spectrum of mutational effects in the wild. More 'natural' mutational patterns are best reconstituted using spontaneous mutation accumulation lines. These lines are obtained by propagating multiple populations (lines) of organisms by only retaining one or two randomly chosen individuals per line for reproduction at each generation. The resulting severe bottleneck reduces the efficacy of natural selection and allows the accumulation of deleterious mutations over many generations. The phenotypic effect of random mutation on the vulva system was probed using a set of mutation accumulation lines derived from the N2 genotype over the course of 400 generations: 'errors' in cell fate patterning and centering increased in most of the lines compared to the N2 control.

Another indirect approach to inferring robustness to genetic change uses genetic variation that occurs in natural populations. In this comparative approach, one considers genetic variation among individuals of the same or different species. These species share an invariant trait that may be produced by a varying developmental process. For example, in several species related to *C. elegans* the final vulval cell fate pattern is invariant, but the developmental route to this final pattern varies strikingly among them. This qualitative approach is powerful because it allows the comparison of organisms and genotypes that are only remotely related. Such organisms have accumulated much greater genetic change than can be produced in laboratory evolution experiments. However, the approach does not provide a quantitative measure of robustness to random genetic change. It also has the disadvantage that the adaptive significance of the existing variation (truly neutral, beneficial, or slightly deleterious) is often not known.

Finally, a generic approach in estimating robustness applies to traits whose mechanistic basis is experimentally well studied. For such traits, one can build quantitative models of the developmental process producing a trait. Such models permit estimation of the trait's sensitivity to changes in model parameters. Changes in parameters (e.g., the affinity of a transcription factor for its target site, or the degradation rate of a protein) may result either from environmental variation or from mutational change. To systematically perturb model parameters thus allows one to assay a system's robustness to multiple types of change. One challenge for this approach is to provide a quantitative framework to integrate information about mutational variation and population structure on the one hand, and environmental variation on the other. In addition, experimental data for model building and validation are sorely needed.

Proximate (Mechanistic) Causes of Robustness

Different categorizations of mechanistic causes of robustness are conceivable. We here emphasize a simple yet very fundamental one: redundancy versus distributed robustness.

Both panels of the figure show a hypothetical signal transduction or metabolic pathway in which information about an upstream signal (upper white circles, e.g., the presence of a growth factor ligand) is communicated via a number of intermediate pathway components (black circles) to a downstream effector (lower white circles, e.g., a transcription factor). If a pathway like this shows distributed robustness (left), it is robust because the flow of information is distributed among several alternative paths, with no two parts performing the same function. In contrast, if robustness is achieved through redundancy (right), several components perform the same function.

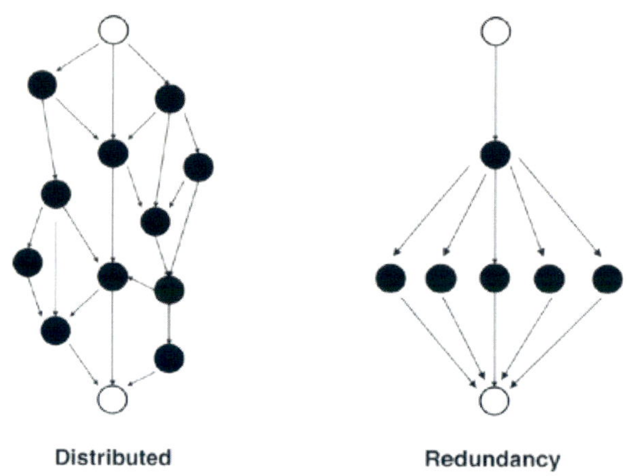

Figure: An illustration of distributed robustness versus redundancy.

In a system with redundant parts, multiple components of a system have the same function.

Redundancy is generally an important cause of robustness in systems whose parts are genes. The reason is that genomes are littered with duplicate genes, and gene duplication is a process that produces genes with redundant functions. Redundancy may also be found at other levels, for example between cells. An example is the redundancy between cells of the vulval competence group, where one cell can replace another (defective) one in making vulval tissue.

Distributed robustness, in contrast, can exist even in systems where no two parts exert the same function. Prominent candidate examples of distributed robustness can be found in metabolic systems. For example, many metabolic functions have long feedback loops, where the end-product of a long chain of chemical reaction allosterically inhibits the enzyme catalyzing the first reaction, thus providing homeostatic regulation. Similarly, in complex metabolic reaction networks, blockage of one metabolic pathway may have little consequence if an important metabolite can be produced through an alternative pathway, even though the two pathways may not share a single enzyme with identical (redundant) functions.

Which of these causes of robustness, redundancy versus distributed robustness, is prevalent in biological systems is a matter of some debate. However, the often rapid divergence in both sequence and function of gene duplicates suggests that gene redundancy may be less important in providing robustness than one might think. Although a systematic study of the robustness of altered vulva signaling networks is still missing, the available evidence indicates that distributed robustness is important in vulva development. Specifically, the vulva system appears to have several mechanistic features that involve distributed robustness.

First, the dynamic behavior of core components of the Ras pathway results in nonlinearities and may thus contribute to robustness to a broad range of variation in EGF signaling. For example, the multiple phosphorylations of mitogen-activated protein (MAP) kinase and the positive feedback loop from the activated MAP kinase to the EGF receptor are likely to create a switch between at least two activity plateaus, a high Ras pathway activity triggering a 1° fate, a low Ras activity a 3° fate.

Second, the Ras pathway has many additional inputs of silent positive and negative regulators that can buffer genetic (or non-genetic) variation. As mentioned above with the SLI-1/GAP-1 example, the knockout of one regulator is silent, but the inactivation of two of these regulators may have an effect. The affected regulators are not redundant, in the sense that they usually do not perform the same molecular activity, nor do they act at the same step in the pathway. One exception is the gene duplication of the positive regulator KSR.

Third, the cross-talk between the Ras and Notch pathways is a typical case of distributed robustness contributing to the specification of three cell fates. A high Ras activity triggers Notch degradation in the 1° cell and thus ensures that the cell does not adopt a 2° fate. A high Ras activity also activates the expression of several Delta-like molecules (the Notch ligands) by the 1° cell. The Delta-like molecules activate Notch in neighboring cells, which in turn inhibits Ras pathway activity in those cells. This interaction probably helps form a robust switch between the 2 and 1° fates.

Fourth, at least in some experimental conditions, the 2° vulval fate can be specified either through morphogen action of the EGF inducer at intermediate doses, or through lateral activation of the

Notch pathway by the 1° cell, which itself acts downstream of EGF/Ras signaling in the 1° cell. If developmental perturbations inhibit one mechanism, the alternative mechanism may guarantee a stable output. Again, these two mechanisms may be said to act redundantly in a wide sense, but they do not perform equivalent activities in the vulva signaling network: one is directly downstream of the EGF inducer, whereas the other is downstream of lateral signaling through Notch. Overall network topology thus contributes to the robustness of the vulva system.

Ultimate (Evolutionary) Causes of Robustness

The robustness of a trait to perturbations can have two evolutionary causes. One such cause – you might call it 'robustness for free' – is rooted in the observation that most biological processes (from enzymatic catalysis to organismal development) have an astronomical number of alternative yet equivalent solutions. These solutions can be thought of existing in a neutral space, in which individual solutions can often be connected through a series of neutral genetic changes. We use the term 'neutral' in the sense that the change has no effect on the final phenotype because it is very difficult to assess whether any change is neutral for 'fitness'. In other words, the robustness of a trait may simply derive from the existence of many alternative ways of building it. A second possibility is that robustness is an evolutionary adaptation *to* perturbations. Where robustness of a trait is advantageous, natural selection can favor genotypes that render the trait robust. For developmental traits, such evolved robustness is called canalization.

A sizable theoretical literature has arisen around the question under what conditions natural selection will lead to a trait's increased robustness. A general insight that has emerged from this theoretical literature is that high robustness can only readily evolve to perturbations that are abundant. Except under high mutation rates, noise and environmental change are likely to be more important driving forces for the evolution of robustness. However, it is likely that the effect of mutation and of non-genetic change on a system are partially correlated, because both affect the same underlying biological processes. For example, an environmental change that results in a higher degradation rate of a protein may have effects similar to that of a reduction-of-function mutation causing a reduced gene expression level or reduced protein activity. In this case, robustness to the environmental change may result in robustness to the genetic change. Obviously, exceptions to this correlation are possible: a given environmental variation and a given mutation may have different effects on a system. Unfortunately, a systematic experimental test of the relationship between environmental and genetic robustness of a trait is still lacking.

Despite an abundance of theoretical work, it is currently not clear which of the two potential causes – robustness for free or natural selection – is prevalent. For example, in the vulva system, robustness to stochastic and environmental variations may be an adaptation, the simple result of a selective process eliminating genetic variants that are less robust and thus deleterious in ecologically relevant environments. The comparison of vulva phenotypes in mutation accumulation lines with those of natural wild strains indeed suggests that several vulva phenotypes are under selection pressure (directly or indirectly), as they are easy to change through mutations yet very rare in the natural wild strains. Some robust features of the vulva network are thus likely to have evolved under selection, rather than merely as an accidental byproduct of the system's architecture. On the other hand, nonlinear effects that contribute to robustness may be unavoidable consequences of system properties that were not subject to direct selection on robustness. For example, enzymatic reactions are often relatively insensitive to enzyme concentrations. (Developmental signal transduction pathways involve many enzymes such as protein kinases and GTP-ases.) Such insensitivity implies a large fraction of neutral mutations among all mutations that affect enzyme concentration, which can thus evolve by neutral drift. Because robustness is not controlled independently from the core components of a system, it is not straightforward to disentangle buffering mechanisms that have been subject to natural selection from those that have not. This is a major challenge for future work.

Another open question is the extent to which trade-offs between different functions of a biological system influence the evolution of robustness. One might think, for example, that a gene regulatory network that needs to function in many different biological processes is more constrained in its evolution than a network deployed in only one process. For example, components of the Ras/MAP kinase pathway that are important in vulval fate induction also play a role in several other developmental decisions in *C. elegans*, as well as in olfaction and in response to pathogens. A key question here is how the different selection pressures affecting pleiotropic mutations shape the evolution of robustness.

Evolutionary Consequences of Robustness

Mutational robustness causes an organism to tolerate changes. One immediate consequence is that for a robust trait, little genetic variation will be expressed as phenotypic variation. Natural selection, in turn, will be less effective in acting on the trait, at least in the short run, because the extent of phenotypic change that natural selection can cause strongly depends on phenotypically expressed genetic variation. Yet another immediate consequence is that cryptic genetic variation can accumulate, because neutral genetic variation accumulates faster than deleterious variation. The system can drift in neutral genotype space, and the larger the available neutral space, the more the system can drift. In other words, variation in an intermediate trait can accumulate without change in the robust final trait. In the face of environmental stressors that drive a system to the limit of its buffered range, such variation can become expressed at the level of the final phenotype. The vast majority of such expressed variation may be deleterious in these new conditions. However, a tiny fraction of it can harbor the seeds of new adaptations, which can change the evolutionary trajectory of an organism. Cryptic genetic variation may thus play two roles in phenotypic variation: allowing variation in intermediate phenotypes in the short term, and potential future phenotypic evolution in the final phenotype in the long term. Present controversies that remain to be experimentally addressed are twofold: (i) assessing whether such cryptic genetic variation evolves neutrally or under some kind of selection in the short term and (ii) determining whether it may have a role in adaptation to new conditions in the long term.

Cryptic genetic variation is by definition, difficult to detect. One way to uncover it is to experimentally drive the system out of its buffered range, using either environmental challenges such as heat shock or ether exposure as in the classical experiments by Waddington, or mutations. In the latter case, the same mutation is introduced (usually by repeated crosses leading to introgression) into different wild genetic backgrounds. Cryptic variation in these wild genetic backgrounds can be detected as variation in mutational effects among the different backgrounds. For example, robustness properties of the vulva network ensure that three precursor cells adopt vulval fates in all wild isolates of *C. elegans*. However, cryptic variation between these wild isolates can be unmasked by displacing the system from the plateau of three induced cells. This is done by strongly reducing or increasing Ras pathway activity through mutations. Preliminary results suggest that the effect of Ras, Notch and Wnt pathway mutations does indeed vary significantly among different *C. elegans* wild genetic backgrounds. The robust vulva system thus accumulates cryptic variation, much like the robust cell fate patterning system of the *Drosophila* eye. In the latter case, the genetic architecture of the cryptic variation is complex, involving variation at many loci and epistatic effects among them. Molecular variation at the EGF receptor locus contributes to a small but significant part of this variation. Understanding the genetic structure of cryptic genetic variation and the patterns of molecular evolution at the corresponding loci is an important current challenge.

An alternative way to detect cryptic variation is to turn to an 'intermediate' phenotype, which may show variation between the tested conditions. One needs to clearly distinguish between the final output of the system, which is robust and invariant, and intermediate phenotypes that may be plastic in response to environmental variations and accumulate genetic variation (which is 'cryptic' when referring to the final phenotype). For example, the level of Ras pathway activity may vary between different wild *C. elegans* isolates without effect on the final cell fate pattern, either because the change is small and does not displace the population from the robust plateau, or because it is compensated by a change at another level (e.g. downstream in the same pathway). Using such an 'intermediate' developmental phenotype, one can, in principle, reveal not only cryptic genetic variation, but also environmental or stochastic variation between individuals. Unraveling such variation remains an experimental challenge in robust developmental model systems.

References

- Williams, G. C. (1992). Natural Selection: Domains, Levels and Challenges. Oxford University Press, Oxford, UK. ISBN 0-19-506932-3
- What-is-the-difference-between-paleogenetics-and-archaeogenetics: quora.com, Retrieved 28 June 2018
- Krieger, Michael J. B.; Ross, Kenneth G. (2002-01-11). "Identification of a Major Gene Regulating Complex Social Behavior". Science. 295 (5553): 328–332. doi:10.1126/science.1065247. ISSN 0036-8075. PMID 11711637
- Evolutionary-adaptation-in-the-human-lineage-12397: nature.com, Retrieved 11 July 2018
- Zimmer, Carl (21 September 2017). "Clues to Africa's Mysterious Past Found in Ancient Skeletons". The New York Times. Retrieved 21 September 2017
- Hamilton, W. D. (1964). "The genetical evolution of social behaviour I". Journal of Theoretical Biology. 7 (1): 1–16. doi:10.1016/0022-5193(64)90038-4. PMID 5875341

Chapter 3

DNA: Sequencing, Damage and Repair

A molecule that is made of two strands of nucleotides forming a double helix carrying the genetic information vital for the functioning, growth and development of organisms is called the DNA. The aim of this chapter is to provide an insight into DNA sequencing, damage and repair.

DNA, or deoxyribonucleic acid, is the hereditary material in humans and almost all other organisms. Nearly every cell in a person's body has the same DNA. Most DNA is located in the cell nucleus (where it is called nuclear DNA), but a small amount of DNA can also be found in the mitochondria (where it is called mitochondrial DNA or mtDNA). Mitochondria are structures within cells that convert the energy from food into a form that cells can use.

The information in DNA is stored as a code made up of four chemical bases: adenine (A), guanine (G), cytosine (C), and thymine (T). Human DNA consists of about 3 billion bases, and more than 99 percent of those bases are the same in all people. The order, or sequence, of these bases determines the information available for building and maintaining an organism, similar to the way in which letters of the alphabet appear in a certain order to form words and sentences.

DNA bases pair up with each other, A with T and C with G, to form units called base pairs. Each base is also attached to a sugar molecule and a phosphate molecule. Together, a base, sugar, and phosphate are called a nucleotide. Nucleotides are arranged in two long strands that form a spiral called a double helix. The structure of the double helix is somewhat like a ladder, with the base pairs forming the ladder's rungs and the sugar and phosphate molecules forming the vertical side-pieces of the ladder.

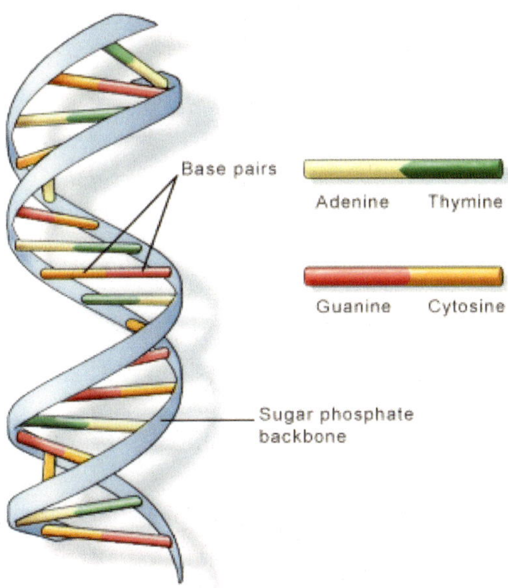

Figure: Structure of DNA.

An important property of DNA is that it can replicate, or make copies of itself. Each strand of DNA in the double helix can serve as a pattern for duplicating the sequence of bases. This is critical when cells divide because each new cell needs to have an exact copy of the DNA present in the old cell.

DNA is a double helix formed by base pairs attached to a sugar-phosphate backbone.

Packaging DNA: Chromatin and Chromosomes

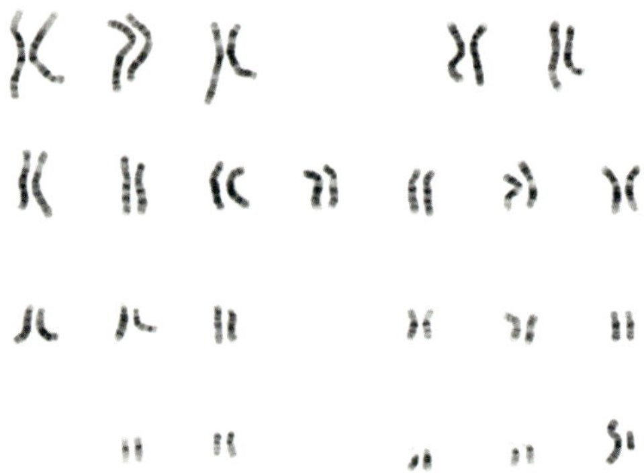

Figure: The complete set of chromosomes in a human male.

Most DNA lives in the nuclei of cells and some is found in mitochondria, which are the powerhouses of the cells.

Because we have so much DNA (2 meters in each cell) and our nuclei are so small, DNA has to be packaged incredibly neatly.

Strands of DNA are looped, coiled and wrapped around proteins called histones. In this coiled state, it is called chromatin.

Chromatin is further condensed, through a process called supercoiling, and it is then packaged into structures called chromosomes. These chromosomes form the familiar "X" shape as seen in the image above.

Each chromosome contains one DNA molecule. Humans have 23 pairs of chromosomes or 46 chromosomes in total. Interestingly, fruit flies have 8 chromosomes, and pigeons have 80.

Chromosome 1 is the largest and contains around 8,000 genes. The smallest is chromosome 21 with around 3,000 genes.

Properties

DNA is a long polymer made from repeating units called nucleotides. The structure of DNA is dynamic along its length, being capable of coiling into tight loops, and other shapes. In all species it is composed of two helical chains, bound to each other by hydrogen bonds. Both chains are coiled round the same axis, and have the same pitch of 34 ångströms (3.4 nanometres). The pair of chains has a radius of 10 ångströms (1.0 nanometre). According to another study, when measured in a different solution, the DNA chain measured 22 to 26 ångströms wide (2.2 to 2.6 nanometres), and one nucleotide unit measured 3.3 Å (0.33 nm) long. Although each individual nucleotide repeating unit is very small, DNA polymers can be very large molecules containing millions to hundreds of millions of nucleotides. For instance, the DNA in the largest human chromosome, chromosome number 1, consists of approximately 220 million base pairs and would be 85 mm long if straightened.

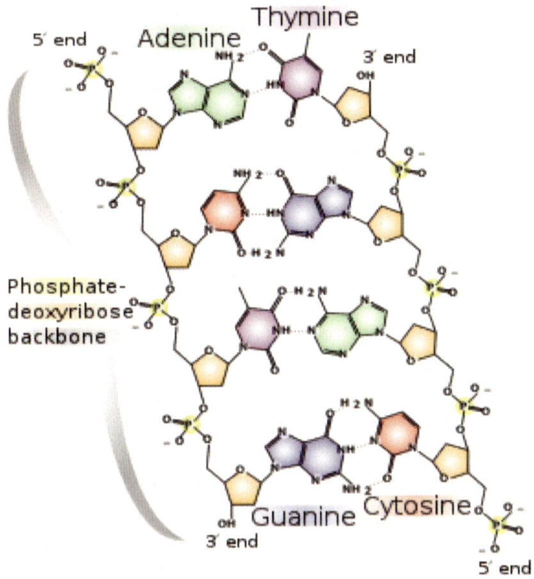

Figure: Chemical structure of DNA; hydrogen bonds shown as dotted lines

In living organisms, DNA does not usually exist as a single molecule, but instead as a pair of molecules that are held tightly together. These two long strands entwine like vines, in the shape of a double helix. The nucleotide contains both a segment of the backbone of the molecule (which holds the chain together) and a nucleobase (which interacts with the other DNA strand in the helix). A nucleobase linked to a sugar is called a nucleoside and a base linked to a sugar and one or more phosphate groups is called a nucleotide. A polymer comprising multiple linked nucleotides (as in DNA) is called a polynucleotide.

The backbone of the DNA strand is made from alternating phosphate and sugar residues. The sugar in DNA is 2-deoxyribose, which is a pentose (five-carbon) sugar. The sugars are joined together by phosphate groups that form phosphodiester bonds between the third and fifth carbon atoms of adjacent sugar rings, which are known as the 3′and 5′ carbons, the prime symbol being used to distinguish these carbon atoms from those of the base to which the deoxyribose forms a glycosidic bond. When imagining DNA, each phosphoryl is normally considered to “belong” to the nucleotide whose 5′ carbon forms a bond therewith. Any DNA strand therefore normally has one end at which there is a phosphoryl attached to the 5′ carbon of a ribose (the 5′ phosphoryl) and another end at which there is a free hydroxyl attached to the 3′ carbon of a ribose (the 3′ hydroxyl). The orientation of the 3′ and 5′ carbons along the sugar-phosphate backbone confers directionality (sometimes called polarity) to each DNA strand. In a double helix, the direction of the nucleotides in one strand is opposite to their direction in the other strand: the strands are antiparallel. The asymmetric ends of DNA strands are said to have a directionality of five prime (5′) and three prime (3′), with the 5′ end having a terminal phosphate group and the 3′ end a terminal hydroxyl group. One major difference between DNA and RNA is the sugar, with the 2-deoxyribose in DNA being replaced by the alternative pentose sugar ribose in RNA.

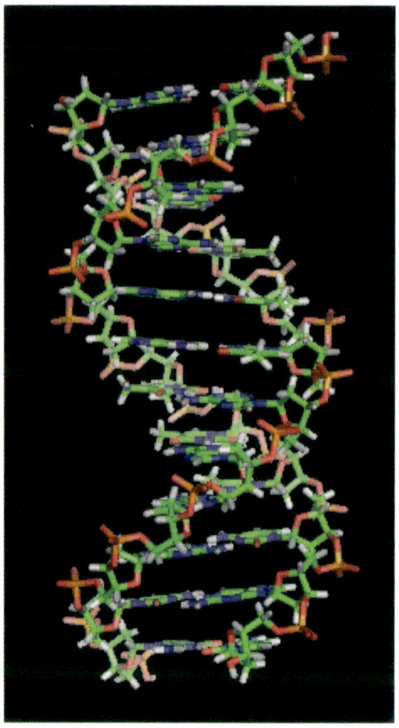

Figure: A section of DNA

The DNA double helix is stabilized primarily by two forces: hydrogen bonds between nucleotides and base-stacking interactions among aromatic nucleobases. In the aqueous environment of the cell, the conjugated π bonds of nucleotide bases align perpendicular to the axis of the DNA molecule, minimizing their interaction with the solvation shell. The four bases found in DNA are adenine (A), cytosine (C), guanine (G) and thymine (T). These four bases are attached to the sugar-phosphate to form the complete nucleotide, as shown for adenosine monophosphate. Adenine pairs with thymine and guanine pairs with cytosine. It was represented by A-T base pairs and G-C base pairs.

Nucleobase Classification

The nucleobases are classified into two types: the purines, A and G, being fused five- and six-membered heterocyclic compounds, and the pyrimidines, the six-membered rings C and T. A fifth pyrimidine nucleobase, uracil (U), usually takes the place of thymine in RNA and differs from thymine by lacking a methyl group on its ring. In addition to RNA and DNA, many artificial nucleic acid analogues have been created to study the properties of nucleic acids, or for use in biotechnology.

Non Canonical Bases

Uracil is not usually found in DNA, occurring only as a breakdown product of cytosine. However, in several bacteriophages, such as *Bacillus subtilis* bacteriophages PBS1 and PBS2 and *Yersinia* bacteriophage piR1-37, thymine has been replaced by uracil. Another phage - Staphylococcal phage S6 - has been identified with a genome where thymine has been replaced by uracil.

5-hydroxymethyldeoxyuridine,(hm5dU) is also known to replace thymidine in several genomes including the *Bacillus* phages SPO1, ϕe, SP8, H1, 2C and SP82. Another modified uracil - 5-dihydroxypentauracil – has also been described.

Base J (beta-d-glucopyranosyloxymethyluracil), a modified form of uracil, is also found in several organisms: the flagellates *Diplonema* and *Euglena*, and all the kinetoplastid genera. Biosynthesis of J occurs in two steps: in the first step, a specific thymidine in DNA is converted into hydroxymethyldeoxyuridine; in the second, HOMedU is glycosylated to form J. Proteins that bind specifically to this base have been identified. These proteins appear to be distant relatives of the Tet1 oncogene that is involved in the pathogenesis of acute myeloid leukemia. J appears to act as a termination signal for RNA polymerase II.

In 1976, the S-2La bacteriophage, which infects species of the genus *Synechocystis*, was found to have all the adenosine bases within its genome replaced by 2,6-diaminopurine. In 2016 deoxyarchaeosine was found to be present in the genomes of several bacteria and the *Escherichia* phage 9g.

Modified bases also occur in DNA. The first of these recognised was 5-methylcytosine, which was found in the genome of *Mycobacterium tuberculosis* in 1925. The complete replacement of cytosine by 5-glycosylhydroxymethylcytosine in T even phages (T2, T4 and T6) was observed in 1953. In the genomes of Xanthomonas oryzae bacteriophage Xp12 and halovirus FH the full complement of cystosine has been replaced by 5-methylcytosine. 6N-methyadenine was discovered to be present in DNA in 1955. N6-carbamoyl-methyladenine was described in 1975. 7-methylguanine was described in 1976. N4-methylcytosine in DNA was described in 1983. In 1985 5-hydroxycytosine was found in the genomes of the Rhizobium phages RL38JI and N17. α-putrescinylthymine occurs in both the genomes of the Delftia phage ΦW-14 and the Bacillus phage SP10. α-glutamylthymidine is found in the Bacillus phage SP01 and 5-dihydroxypentyluracil is found in the Bacillus phage SP15.

The reason for the presence of these non canonical bases in DNA is not known. It seems likely that at least part of the reason for their presence in bacterial viruses (phages) is to avoid the restriction enzymes present in bacteria. This enzyme system acts at least in part as a molecular immune system protecting bacteria from infection by viruses.

This does not appear to be the entire story. Four modifications to the cytosine residues in human DNA have been reported. These modifications are the addition of methyl (CH_3)-, hydroxymethyl (CH_2OH)-, formyl (CHO)- and carboxyl (COOH)- groups. These modifications are thought to have regulatory functions.

Uracil is found in the centromeric regions of at least two human chromosomes (6 and 11).

Listing of Non Canonical bases found in DNA

Seventeen non canonical bases are known to occur in DNA. Most of these are modifications of the canonical bases plus uracil.

- Modified Adenosine
    - N6-carbamoyl-methyladenine
    - N6-methyadenine
- Modified Guanine
    - 7-Methylguanine

- Modified Cytosine
    - N4-Methylcytosine
    - 5-Carboxylcytosine
    - 5-Formylcytosine
    - 5-Glycosylhydroxymethylcytosine
    - 5-Hydroxycytosine
    - 5-Methylcytosine
- Modified Thymidine
    - α-Glutamythymidine
    - α-Putrescinylthymine
- Uracil and modifications
    - Base J
    - Uracil
    - 5-Dihydroxypentauracil
    - 5-Hydroxymethyldeoxyuracil
- Others
    - Deoxyarchaeosine
    - 2,6-Diaminopurine

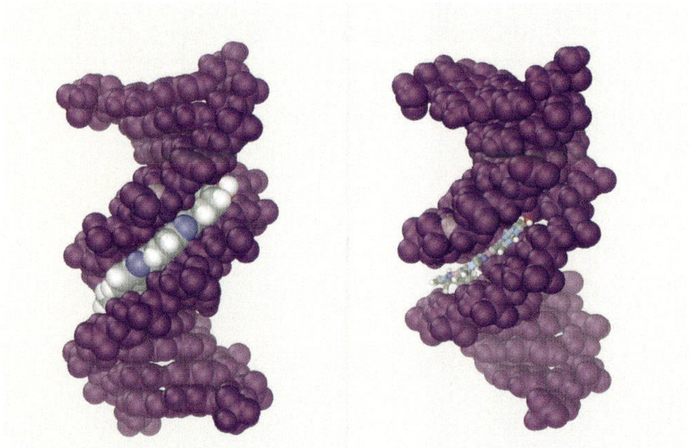

Figure: DNA major and minor grooves. The latter is a binding site for the Hoechst stain dye 33258.

Grooves

Twin helical strands form the DNA backbone. Another double helix may be found tracing the spaces, or grooves, between the strands. These voids are adjacent to the base pairs and may provide a binding site. As the strands are not symmetrically located with respect to each other, the grooves are unequally sized. One groove, the major groove, is 22 Å wide and the other, the minor groove, is 12 Å wide. The width of the major groove means that the edges of the bases are more accessible in the major groove than in the minor groove. As a result, proteins such as transcription factors that can bind to specific sequences in double-stranded DNA usually make contact with the sides of the bases exposed in the major groove. This situation varies in unusual conformations of DNA within the cell, but the major and minor grooves are always named to reflect the differences in size that would be seen if the DNA is twisted back into the ordinary B form.

Base Pairing

In a DNA double helix, each type of nucleobase on one strand bonds with just one type of nucleobase on the other strand. This is called complementary base pairing. Here, purines form hydrogen bonds to pyrimidines, with adenine bonding only to thymine in two hydrogen bonds, and cytosine bonding only to guanine in three hydrogen bonds. This arrangement of two nucleotides binding together across the double helix is called a Watson-Crick base pair. Another type of base pairing is Hoogsteen base pairing where two hydrogen bonds form between guanine and cytosine. As hydrogen bonds are not covalent, they can be broken and rejoined relatively easily. The two strands of DNA in a double helix can thus be pulled apart like a zipper, either by a mechanical force or high temperature. As a result of this base pair complementarity, all the information in the double-stranded sequence of a DNA helix is duplicated on each strand, which is vital in DNA replication. This reversible and specific interaction between complementary base pairs is critical for all the functions of DNA in living organisms.

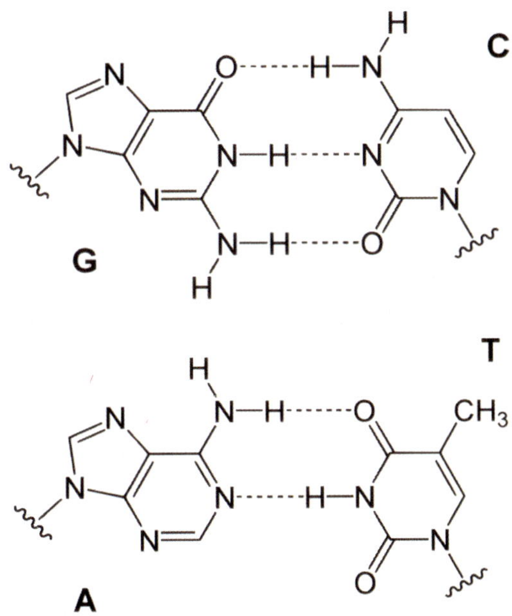

Figure: Top, a GC base pair with three hydrogen bonds. Bottom, an AT base pair with two hydrogen bonds. Non-covalent hydrogen bonds between the pairs are shown as dashed lines.

The two types of base pairs form different numbers of hydrogen bonds, AT forming two hydrogen bonds, and GC forming three hydrogen bonds. DNA with high GC-content is more stable than DNA with low GC-content.

As noted above, most DNA molecules are actually two polymer strands, bound together in a helical fashion by noncovalent bonds; this double-stranded (dsDNA) structure is maintained largely by the intrastrand base stacking interactions, which are strongest for G,C stacks. The two strands can come apart – a process known as melting – to form two single-stranded DNA (ssDNA) molecules. Melting occurs at high temperature, low salt and high pH (low pH also melts DNA, but since DNA is unstable due to acid depurination, low pH is rarely used).

The stability of the dsDNA form depends not only on the GC-content (% G,C basepairs) but also on sequence (since stacking is sequence specific) and also length (longer molecules are more stable). The stability can be measured in various ways; a common way is the "melting temperature", which is the temperature at which 50% of the ds molecules are converted to ss molecules; melting temperature is dependent on ionic strength and the concentration of DNA. As a result, it is both the percentage of GC base pairs and the overall length of a DNA double helix that determines the strength of the association between the two strands of DNA. Long DNA helices with a high GC-content have stronger-interacting strands, while short helices with high AT content have weaker-interacting strands. In biology, parts of the DNA double helix that need to separate easily, such as the TATAAT Pribnow box in some promoters, tend to have a high AT content, making the strands easier to pull apart.

In the laboratory, the strength of this interaction can be measured by finding the temperature necessary to break the hydrogen bonds, their melting temperature (also called T_m value). When all the base pairs in a DNA double helix melt, the strands separate and exist in solution as two entirely independent molecules. These single-stranded DNA molecules have no single common shape, but some conformations are more stable than others.

Sense and Antisense

A DNA sequence is called "sense" if its sequence is the same as that of a messenger RNA copy that is translated into protein. The sequence on the opposite strand is called the "antisense" sequence. Both sense and antisense sequences can exist on different parts of the same strand of DNA (i.e. both strands can contain both sense and antisense sequences). In both prokaryotes and eukaryotes, antisense RNA sequences are produced, but the functions of these RNAs are not entirely clear. One proposal is that antisense RNAs are involved in regulating gene expression through RNA-RNA base pairing.

A few DNA sequences in prokaryotes and eukaryotes, and more in plasmids and viruses, blur the distinction between sense and antisense strands by having overlapping genes. In these cases, some DNA sequences do double duty, encoding one protein when read along one strand, and a second protein when read in the opposite direction along the other strand. In bacteria, this overlap may be involved in the regulation of gene transcription, while in viruses, overlapping genes increase the amount of information that can be encoded within the small viral genome.

Supercoiling

DNA can be twisted like a rope in a process called DNA supercoiling. With DNA in its "relaxed" state, a strand usually circles the axis of the double helix once every 10.4 base pairs, but if the DNA is twisted the strands become more tightly or more loosely wound. If the DNA is twisted in the direction of the helix, this is positive supercoiling, and the bases are held more tightly together. If they are twisted in the opposite direction, this is negative supercoiling, and the bases come apart more easily. In nature, most DNA has slight negative supercoiling that is introduced by enzymes called topoisomerases. These enzymes are also needed to relieve the twisting stresses introduced into DNA strands during processes such as transcription and DNA replication.

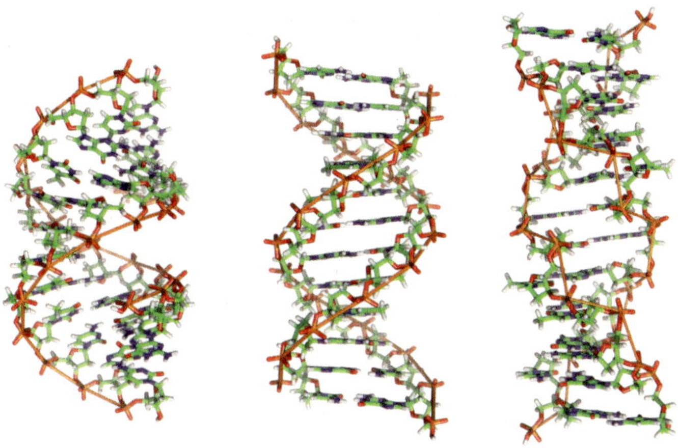

Figure: From left to right, the structures of A, B and Z DNA

Alternative DNA Structures

DNA exists in many possible conformations that include A-DNA, B-DNA, and Z-DNA forms, although, only B-DNA and Z-DNA have been directly observed in functional organisms. The conformation that DNA adopts depends on the hydration level, DNA sequence, the amount and direction of supercoiling, chemical modifications of the bases, the type and concentration of metal ions, and the presence of polyamines in solution.

The first published reports of A-DNA X-ray diffraction patterns—and also B-DNA—used analyses based on Patterson transforms that provided only a limited amount of structural information for oriented fibers of DNA. An alternative analysis was then proposed by Wilkins *et al.*, in 1953, for the *in vivo* B-DNA X-ray diffraction-scattering patterns of highly hydrated DNA fibers in terms of squares of Bessel functions. James Watson and Francis Crick presented their molecular modeling analysis of the DNA X-ray diffraction patterns to suggest that the structure was a double-helix.

Although the *B-DNA form* is most common under the conditions found in cells, it is not a well-defined conformation but a family of related DNA conformations that occur at the high hydration levels present in living cells. Their corresponding X-ray diffraction and scattering patterns are characteristic of molecular paracrystals with a significant degree of disorder.

Compared to B-DNA, the A-DNA form is a wider right-handed spiral, with a shallow, wide minor groove and a narrower, deeper major groove. The A form occurs under non-physiological conditions in partly dehydrated samples of DNA, while in the cell it may be produced in hybrid pairings of DNA and RNA strands, and in enzyme-DNA complexes. Segments of DNA where the bases have been chemically modified by methylation may undergo a larger change in conformation and adopt the Z form. Here, the strands turn about the helical axis in a left-handed spiral, the opposite of the more common B form. These unusual structures can be recognized by specific Z-DNA binding proteins and may be involved in the regulation of transcription.

Alternative DNA Chemistry

For many years, exobiologists have proposed the existence of a shadow biosphere, a postulated microbial biosphere of Earth that uses radically different biochemical and molecular processes than currently known life. One of the proposals was the existence of lifeforms that use arsenic instead of phosphorus in DNA. A report in 2010 of the possibility in the bacterium GFAJ-1, was announced, though the research was disputed, and evidence suggests the bacterium actively prevents the incorporation of arsenic into the DNA backbone and other biomolecules.

Quadruplex Structures

At the ends of the linear chromosomes are specialized regions of DNA called telomeres. The main function of these regions is to allow the cell to replicate chromosome ends using the enzyme telomerase, as the enzymes that normally replicate DNA cannot copy the extreme 3′ ends of chromosomes. These specialized chromosome caps also help protect the DNA ends, and stop the DNA repair systems in the cell from treating them as damage to be corrected. In human cells, telomeres are usually lengths of single-stranded DNA containing several thousand repeats of a simple TTAGGG sequence.

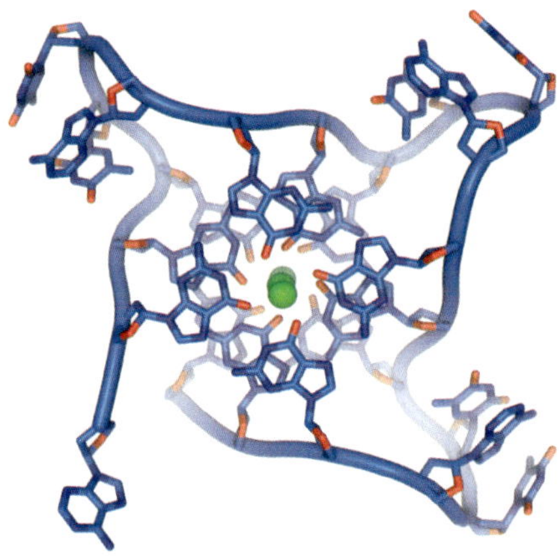

Figure: DNA quadruplex formed by telomere repeats.

The looped conformation of the DNA backbone is very different from the typical DNA helix. The green spheres in the center represent potassium ions.

These guanine-rich sequences may stabilize chromosome ends by forming structures of stacked sets of four-base units, rather than the usual base pairs found in other DNA molecules. Here, four guanine bases form a flat plate and these flat four-base units then stack on top of each other, to form a stable G-quadruplex structure. These structures are stabilized by hydrogen bonding between the edges of the bases and chelation of a metal ion in the center of each four-base unit. Other structures can also be formed, with the central set of four bases coming from either a single strand folded around the bases, or several different parallel strands, each contributing one base to the central structure.

In addition to these stacked structures, telomeres also form large loop structures called telomere loops, or T-loops. Here, the single-stranded DNA curls around in a long circle stabilized by telomere-binding proteins. At the very end of the T-loop, the single-stranded telomere DNA is held onto a region of double-stranded DNA by the telomere strand disrupting the double-helical DNA and base pairing to one of the two strands. This triple-stranded structure is called a displacement loop or D-loop.

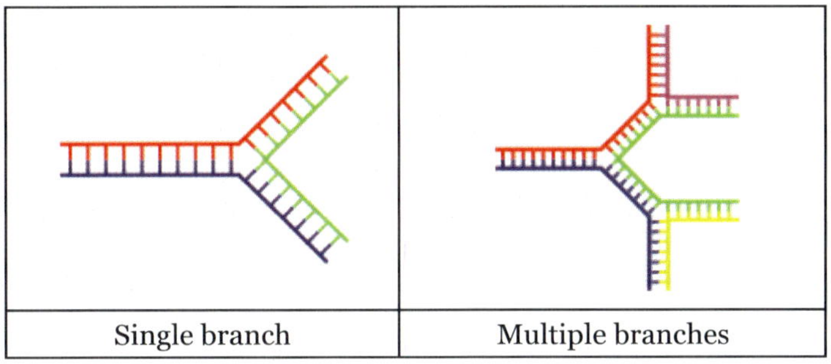

Figure: Branched DNA can form networks containing multiple branches.

Branched DNA

In DNA, fraying occurs when non-complementary regions exist at the end of an otherwise complementary double-strand of DNA. However, branched DNA can occur if a third strand of DNA is introduced and contains adjoining regions able to hybridize with the frayed regions of the pre-existing double-strand. Although the simplest example of branched DNA involves only three strands of DNA, complexes involving additional strands and multiple branches are also possible. Branched DNA can be used in nanotechnology to construct geometric shapes.

Biological Functions

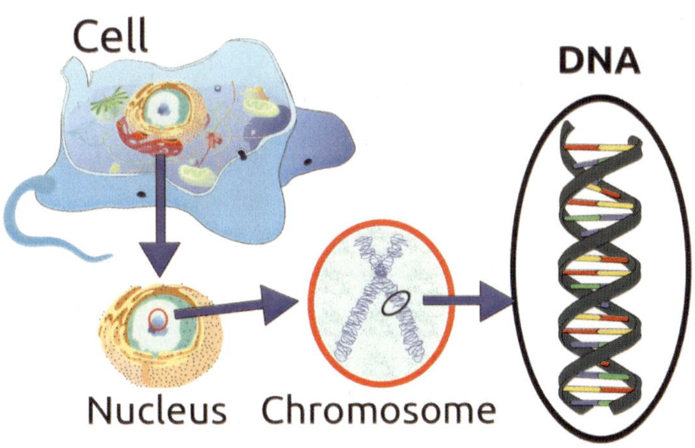

Figure: Location of eukaryote nuclear DNA within the chromosomes

DNA usually occurs as linear chromosomes in eukaryotes, and circular chromosomes in prokaryotes. The set of chromosomes in a cell makes up its genome; the human genome has approximately 3 billion base pairs of DNA arranged into 46 chromosomes. The information carried by DNA is held in the sequence of pieces of DNA called genes. Transmission of genetic information in genes is achieved via complementary base pairing. For example, in transcription, when a cell uses the information in a gene, the DNA sequence is copied into a complementary RNA sequence through the attraction between the DNA and the correct RNA nucleotides. Usually, this RNA copy is then used to make a matching protein sequence in a process called translation, which depends on the same interaction between RNA nucleotides. In alternative fashion, a cell may simply copy its genetic information in a process called DNA replication. The details of these functions are covered in other articles; here the focus is on the interactions between DNA and other molecules that mediate the function of the genome.

Genes and Genomes

Genomic DNA is tightly and orderly packed in the process called DNA condensation, to fit the small available volumes of the cell. In eukaryotes, DNA is located in the cell nucleus, with small amounts in mitochondria and chloroplasts. In prokaryotes, the DNA is held within an irregularly shaped body in the cytoplasm called the nucleoid. The genetic information in a genome is held within genes, and the complete set of this information in an organism is called its genotype. A gene is a unit of heredity and is a region of DNA that influences a particular characteristic in an organism. Genes contain an open reading frame that can be transcribed, and regulatory sequences such as promoters and enhancers, which control transcription of the open reading frame.

In many species, only a small fraction of the total sequence of the genome encodes protein. For example, only about 1.5% of the human genome consists of protein-coding exons, with over 50% of human DNA consisting of non-coding repetitive sequences. The reasons for the presence of so much noncoding DNA in eukaryotic genomes and the extraordinary differences in genome size, or *C-value*, among species, represent a long-standing puzzle known as the "C-value enigma". However, some DNA sequences that do not code protein may still encode functional non-coding RNA molecules, which are involved in the regulation of gene expression.

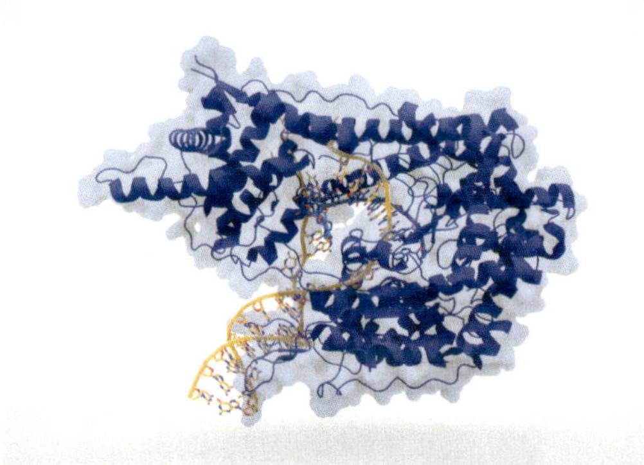

Figure: T7 RNA polymerase (blue) producing an mRNA (green) from a DNA template (orange)

Some noncoding DNA sequences play structural roles in chromosomes. Telomeres and centromeres typically contain few genes but are important for the function and stability of chromosomes. An abundant form of noncoding DNA in humans is pseudogenes, which are copies of genes that have been disabled by mutation. These sequences are usually just molecular fossils, although they can occasionally serve as raw genetic material for the creation of new genes through the process of gene duplication and divergence.

Transcription and Translation

A gene is a sequence of DNA that contains genetic information and can influence the phenotype of an organism. Within a gene, the sequence of bases along a DNA strand defines a messenger RNA sequence, which then defines one or more protein sequences. The relationship between the nucleotide sequences of genes and the amino-acid sequences of proteins is determined by the rules of translation, known collectively as the genetic code. The genetic code consists of three-letter 'words' called *codons* formed from a sequence of three nucleotides (e.g. ACT, CAG, TTT).

In transcription, the codons of a gene are copied into messenger RNA by RNA polymerase. This RNA copy is then decoded by a ribosome that reads the RNA sequence by base-pairing the messenger RNA to transfer RNA, which carries amino acids. Since there are 4 bases in 3-letter combinations, there are 64 possible codons (4^3 combinations). These encode the twenty standard amino acids, giving most amino acids more than one possible codon. There are also three 'stop' or 'nonsense' codons signifying the end of the coding region; these are the TAA, TGA, and TAG codons.

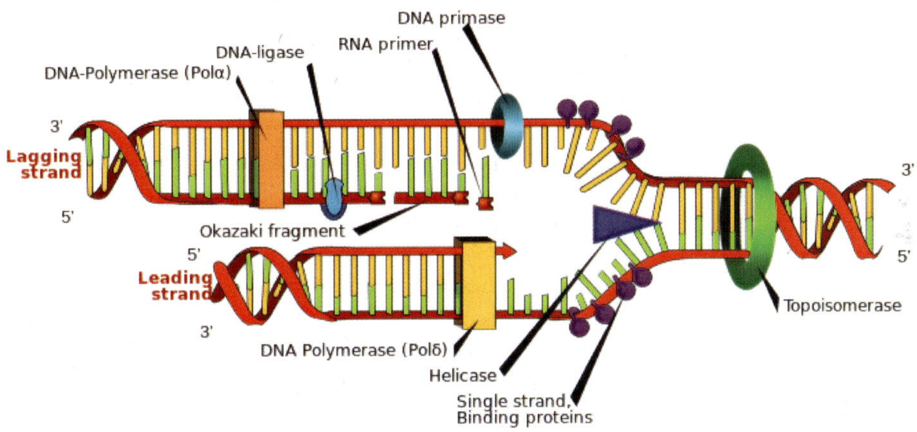

Figure: DNA replication.

The double helix is unwound by a helicase and topoisomerase. Next, one DNA polymerase produces the leading strand copy. Another DNA polymerase binds to the lagging strand. This enzyme makes discontinuous segments (called Okazaki fragments) before DNA ligase joins them together.

Replication

Cell division is essential for an organism to grow, but, when a cell divides, it must replicate the DNA in its genome so that the two daughter cells have the same genetic information as their parent. The double-stranded structure of DNA provides a simple mechanism for DNA replication. Here, the two strands are separated and then each strand's complementary DNA sequence is recreated by an enzyme called DNA polymerase. This enzyme makes the complementary strand by finding the correct base through complementary base pairing and bonding it onto the original strand. As DNA polymerases can only extend a DNA strand in a 5′ to 3′ direction, different mechanisms are used to copy the antiparallel strands of the double helix. In this way, the base on the old strand dictates which base appears on the new strand, and the cell ends up with a perfect copy of its DNA.

Extracellular Nucleic Acids

Naked extracellular DNA (eDNA), most of it released by cell death, is nearly ubiquitous in the environment. Its concentration in soil may be as high as 2 μg/L, and its concentration in natural aquatic environments may be as high at 88 μg/L. Various possible functions have been proposed for eDNA: it may be involved in horizontal gene transfer; it may provide nutrients; and it may act as a buffer to recruit or titrate ions or antibiotics. Extracellular DNA acts as a functional extracellular matrix component in the biofilms of several bacterial species. It may act as a recognition factor to regulate the attachment and dispersal of specific cell types in the biofilm; it may contribute to biofilm formation; and it may contribute to the biofilm's physical strength and resistance to biological stress.

Cell-free fetal DNA is found in the blood of the mother, and can be sequenced to determine a great deal of information about the developing fetus.

Interactions with Proteins

All the functions of DNA depend on interactions with proteins. These protein interactions can be non-specific, or the protein can bind specifically to a single DNA sequence. Enzymes can also bind to DNA and of these, the polymerases that copy the DNA base sequence in transcription and DNA replication are particularly important.

DNA-binding Proteins

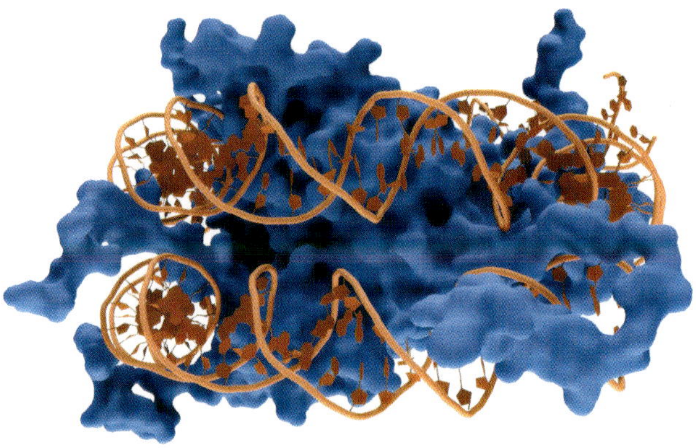

Figure: Interaction of DNA (in orange) with histones (in blue)

Structural proteins that bind DNA are well-understood examples of non-specific DNA-protein interactions. Within chromosomes, DNA is held in complexes with structural proteins. These proteins organize the DNA into a compact structure called chromatin. In eukaryotes, this structure involves DNA binding to a complex of small basic proteins called histones, while in prokaryotes multiple types of proteins are involved. The histones form a disk-shaped complex called a nucleosome, which contains two complete turns of double-stranded DNA wrapped around its surface. These non-specific interactions are formed through basic residues in the histones, making ionic bonds to the acidic sugar-phosphate backbone of the DNA, and are thus largely independent of the base sequence. Chemical modifications of these basic amino acid residues include methylation, phosphorylation, and acetylation. These chemical changes alter the strength of the interaction between the DNA and the histones, making the DNA more or less accessible to transcription factors and changing the rate of transcription. Other non-specific DNA-binding proteins in chromatin include the high-mobility group proteins, which bind to bent or distorted DNA. These proteins are important in bending arrays of nucleosomes and arranging them into the larger structures that make up chromosomes.

A distinct group of DNA-binding proteins is the DNA-binding proteins that specifically bind single-stranded DNA. In humans, replication protein A is the best-understood member of this family and is used in processes where the double helix is separated, including DNA replication, recombination, and DNA repair. These binding proteins seem to stabilize single-stranded DNA and protect it from forming stem-loops or being degraded by nucleases.

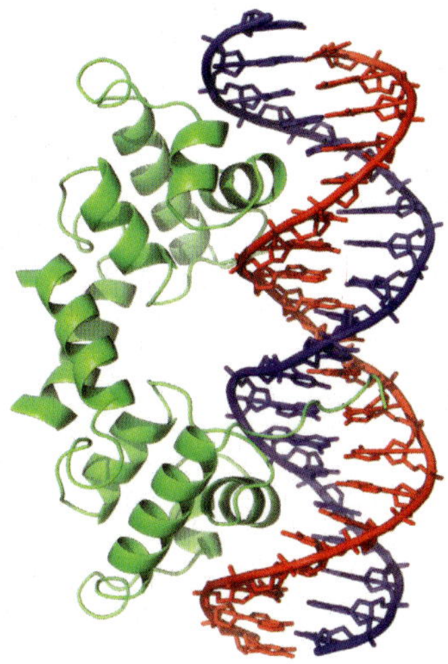

Figure: The lambda repressor helix-turn-helix transcription factor bound to its DNA target.

In contrast, other proteins have evolved to bind to particular DNA sequences. The most intensively studied of these are the various transcription factors, which are proteins that regulate transcription. Each transcription factor binds to one particular set of DNA sequences and activates or inhibits the transcription of genes that have these sequences close to their promoters. The transcription factors do this in two ways. Firstly, they can bind the RNA polymerase responsible for transcription, either directly or through other mediator proteins; this locates the polymerase at the promoter and allows it to begin transcription. Alternatively, transcription factors can bind enzymes that modify the histones at the promoter. This changes the accessibility of the DNA template to the polymerase.

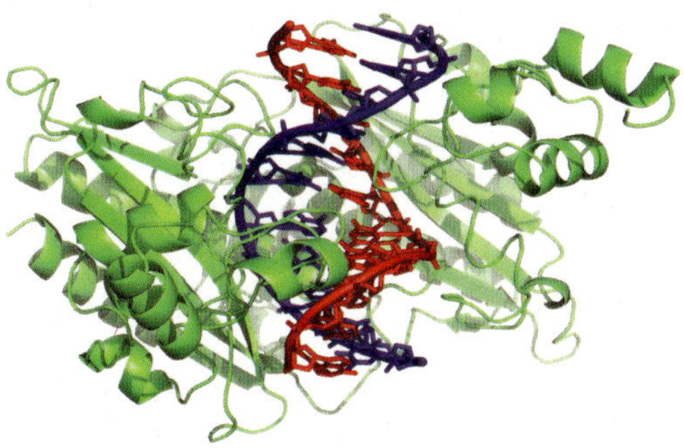

Figure: The restriction enzyme EcoRV (green) in a complex with its substrate DNA

As these DNA targets can occur throughout an organism's genome, changes in the activity of one type of transcription factor can affect thousands of genes. Consequently, these proteins are often the targets of the signal transduction processes that control responses to environmental changes or cellular differentiation and development. The specificity of these transcription factors' interactions with DNA come from the proteins making multiple contacts to the edges of the DNA bases, allowing them to "read" the DNA sequence. Most of these base-interactions are made in the major groove, where the bases are most accessible.

DNA-modifying Enzymes

Nucleases and Ligases

Nucleases are enzymes that cut DNA strands by catalyzing the hydrolysis of the phosphodiester bonds. Nucleases that hydrolyze nucleotides from the ends of DNA strands are called exonucleases, while endonucleases cut within strands. The most frequently used nucleases in molecular biology are the restriction endonucleases, which cut DNA at specific sequences. For instance, the EcoRV enzyme shown to the left recognizes the 6-base sequence 5′-GATATC-3′ and makes a cut at the horizontal line. In nature, these enzymes protect bacteria against phage infection by digesting the phage DNA when it enters the bacterial cell, acting as part of the restriction modification system. In technology, these sequence-specific nucleases are used in molecular cloning and DNA fingerprinting.

Enzymes called DNA ligases can rejoin cut or broken DNA strands. Ligases are particularly important in lagging strand DNA replication, as they join together the short segments of DNA produced at the replication fork into a complete copy of the DNA template. They are also used in DNA repair and genetic recombination.

Topoisomerases and Helicases

Topoisomerases are enzymes with both nuclease and ligase activity. These proteins change the amount of supercoiling in DNA. Some of these enzymes work by cutting the DNA helix and allowing one section to rotate, thereby reducing its level of supercoiling; the enzyme then seals the DNA break. Other types of these enzymes are capable of cutting one DNA helix and then passing a second strand of DNA through this break, before rejoining the helix. Topoisomerases are required for many processes involving DNA, such as DNA replication and transcription.

Helicases are proteins that are a type of molecular motor. They use the chemical energy in nucleoside triphosphates, predominantly adenosine triphosphate (ATP), to break hydrogen bonds between bases and unwind the DNA double helix into single strands. These enzymes are essential for most processes where enzymes need to access the DNA bases.

Polymerases

Polymerases are enzymes that synthesize polynucleotide chains from nucleoside triphosphates. The sequence of their products is created based on existing polynucleotide chains—which are called templates. These enzymes function by repeatedly adding a nucleotide to the 3′ hydroxyl group at the end of the growing polynucleotide chain. As a consequence, all polymerases work in a 5′ to 3′ direction. In the active site of these enzymes, the incoming nucleoside triphosphate base pairs to the template: this allows polymerases to accurately synthesize the complementary strand of their template. Polymerases are classified according to the type of template that they use.

In DNA replication, DNA-dependent DNA polymerases make copies of DNA polynucleotide chains. To preserve biological information, it is essential that the sequence of bases in each copy is precisely complementary to the sequence of bases in the template strand. Many DNA polymerases have a proofreading activity. Here, the polymerase recognizes the occasional mistakes in the synthesis reaction by the lack of base pairing between the mismatched nucleotides. If a mismatch is detected, a 3′ to 5′ exonuclease activity is activated and the incorrect base removed. In most organisms, DNA polymerases function in a large complex called the replisome that contains multiple accessory subunits, such as the DNA clamp or helicases.

RNA-dependent DNA polymerases are a specialized class of polymerases that copy the sequence of an RNA strand into DNA. They include reverse transcriptase, which is a viral enzyme involved in the infection of cells by retroviruses, and telomerase, which is required for the replication of telomeres. For example, HIV reverse transcriptase is an enzyme for AIDS virus replication. Telomerase is an unusual polymerase because it contains its own RNA template as part of its structure. It synthesizes telomeres at the ends of chromosomes. Telomeres prevent fusion of the ends of neighboring chromosomes and protect chromosome ends from damage.

Transcription is carried out by a DNA-dependent RNA polymerase that copies the sequence of a DNA strand into RNA. To begin transcribing a gene, the RNA polymerase binds to a sequence of DNA called a promoter and separates the DNA strands. It then copies the gene sequence into a messenger RNA transcript until it reaches a region of DNA called the terminator, where it halts and detaches from the DNA. As with human DNA-dependent DNA polymerases, RNA polymerase II, the enzyme that transcribes most of the genes in the human genome, operates as part of a large protein complex with multiple regulatory and accessory subunits.

Genetic Recombination

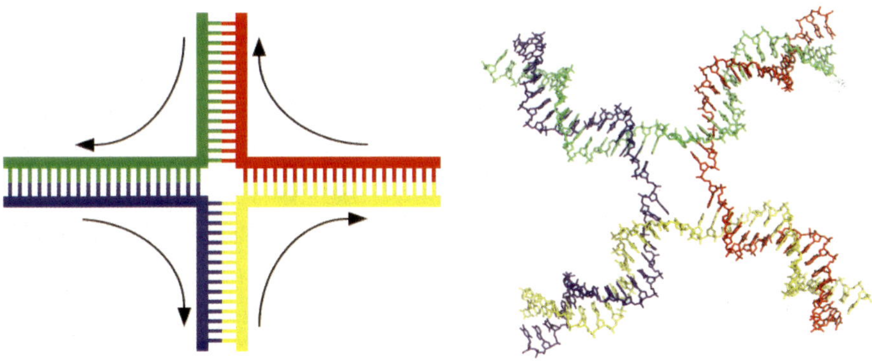

Figure: Structure of the Holliday junction intermediate in genetic recombination. The four separate DNA strands are colored red, blue, green and yellow.

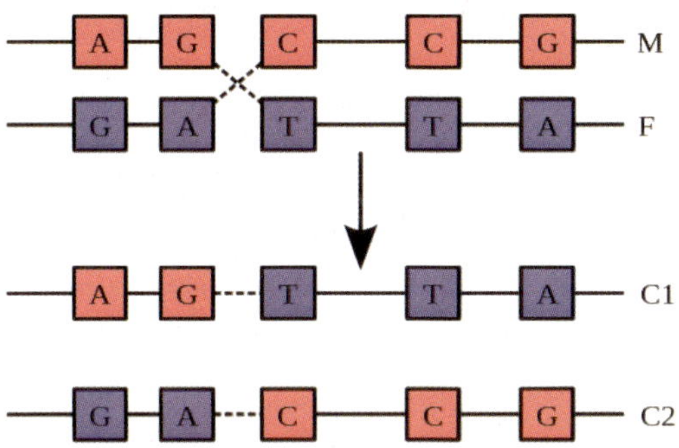

Figure: Recombination involves the breaking and rejoining of two chromosomes (M and F) to produce two rearranged chromosomes (C1 and C2).

A DNA helix usually does not interact with other segments of DNA, and in human cells, the different chromosomes even occupy separate areas in the nucleus called "chromosome territories". This physical separation of different chromosomes is important for the ability of DNA to function as a stable repository for information, as one of the few times chromosomes interact is in chromosomal crossover which occurs during sexual reproduction, when genetic recombination occurs. Chromosomal crossover is when two DNA helices break, swap a section and then rejoin.

Recombination allows chromosomes to exchange genetic information and produces new combinations of genes, which increases the efficiency of natural selection and can be important in the rapid evolution of new proteins. Genetic recombination can also be involved in DNA repair, particularly in the cell's response to double-strand breaks.

The most common form of chromosomal crossover is homologous recombination, where the two chromosomes involved share very similar sequences. Non-homologous recombination can be damaging to cells, as it can produce chromosomal translocations and genetic abnormalities. The recombination reaction is catalyzed by enzymes known as recombinases, such as RAD51. The first step in recombination is a double-stranded break caused by either an endonuclease or damage to the DNA. A series of steps catalyzed in part by the recombinase then leads to joining of the two helices by at least one Holliday junction, in which a segment of a single strand in each helix is annealed to the complementary strand in the other helix. The Holliday junction is a tetrahedral junction structure that can be moved along the pair of chromosomes, swapping one strand for another. The recombination reaction is then halted by cleavage of the junction and re-ligation of the released DNA. Only strands of like polarity exchange DNA during recombination. There are two types of cleavage: east-west cleavage and north-south cleavage. The north-south cleavage nicks both strands of DNA, while the east-west cleavage has one strand of DNA intact. The formation of a Holliday junction during recombination makes it possible for genetic diversity, genes to exchange on chromosomes, and expression of wild-type viral genomes.

Evolution

DNA contains the genetic information that allows all modern living things to function, grow and reproduce. However, it is unclear how long in the 4-billion-year history of life DNA has performed this function, as it has been proposed that the earliest forms of life may have used RNA as their genetic material. RNA may have acted as the central part of early cell metabolism as it can both transmit genetic information and carry out catalysis as part of ribozymes. This ancient RNA world where nucleic acid would have been used for both catalysis and genetics may have influenced the evolution of the current genetic code based on four nucleotide bases. This would occur, since the number of different bases in such an organism is a trade-off between a small number of bases increasing replication accuracy and a large number of bases increasing the catalytic efficiency of ribozymes. However, there is no direct evidence of ancient genetic systems, as recovery of DNA from most fossils is impossible because DNA survives in the environment for less than one million years, and slowly degrades into short fragments in solution. Claims for older DNA have been made, most notably a report of the isolation of a viable bacterium from a salt crystal 250 million years old, but these claims are controversial.

Building blocks of DNA (adenine, guanine, and related organic molecules) may have been formed extra terrestrially in outer space. Complex DNA and RNA organic compounds of life, including uracil, cytosine, and thymine, have also been formed in the laboratory under conditions mimicking those found in outer space, using starting chemicals, such as pyrimidine, found in meteorites. Pyrimidine, like polycyclic aromatic hydrocarbons (PAHs), the most carbon-rich chemical found in the universe, may have been formed in red giants or in interstellar cosmic dust and gas clouds.

Uses in Technology

Genetic Engineering

Methods have been developed to purify DNA from organisms, such as phenol-chloroform extraction, and to manipulate it in the laboratory, such as restriction digests and the polymerase chain reaction. Modern biology and biochemistry make intensive use of these techniques in recombinant DNA technology. Recombinant DNA is a man-made DNA sequence that has been assembled from other DNA sequences. They can be transformed into organisms in the form of plasmids or in the appropriate format, by using a viral vector. The genetically modified organisms produced can be used to produce products such as recombinant proteins, used in medical research, or be grown in agriculture.

DNA Profiling

Forensic scientists can use DNA in blood, semen, skin, saliva or hair found at a crime scene to identify a matching DNA of an individual, such as a perpetrator. This process is formally termed DNA profiling, but may also be called "genetic fingerprinting". In DNA profiling, the lengths of variable sections of repetitive DNA, such as short tandem repeats and minisatellites, are compared between people. This method is usually an extremely reliable technique for identifying a matching DNA. However, identification can be complicated if the scene is contaminated with DNA from several people. DNA profiling was developed in 1984 by British geneticist Sir Alec Jeffreys, and first used in forensic science to convict Colin Pitchfork in the 1988 Enderby murders case.

The development of forensic science and the ability to now obtain genetic matching on minute samples of blood, skin, saliva, or hair has led to re-examining many cases. Evidence can now be uncovered that was scientifically impossible at the time of the original examination. Combined with the removal of the double jeopardy law in some places, this can allow cases to be reopened where prior trials have failed to produce sufficient evidence to convince a jury. People charged with serious crimes may be required to provide a sample of DNA for matching purposes. The most obvious defense to DNA matches obtained forensically is to claim that cross-contamination of evidence has occurred. This has resulted in meticulous strict handling procedures with new cases of serious crime.

DNA profiling is also used successfully to positively identify victims of mass casualty incidents, bodies or body parts in serious accidents, and individual victims in mass war graves, via matching to family members.

DNA profiling is also used in DNA paternity testing to determine if someone is the biological parent or grandparent of a child with the probability of parentage is typically 99.99% when the alleged parent is biologically related to the child. Normal DNA sequencing methods happen after birth, but there are new methods to test paternity while a mother is still pregnant.

DNA Enzymes or Catalytic DNA

Deoxyribozymes, also called DNAzymes or catalytic DNA, are first discovered in 1994. They are mostly single stranded DNA sequences isolated from a large pool of random DNA sequences through a combinatorial approach called in vitro selection or systematic evolution of ligands by exponential enrichment (SELEX). DNAzymes catalyze variety of chemical reactions including RNA-DNA cleavage, RNA-DNA ligation, amino acids phosphorylation-dephosphorylation, carbon-carbon bond formation, and etc. DNAzymes can enhance catalytic rate of chemical reactions up to 100,000,000,000-fold over the uncatalyzed reaction. The most extensively studied class of DNAzymes is RNA-cleaving types which have been used to detect different metal ions and designing therapeutic agents. Several metal-specific DNAzymes have been reported including the GR-5 DNAzyme (lead-specific), the CA1-3 DNAzymes (copper-specific), the 39E DNAzyme (uranyl-specific) and the NaA43 DNAzyme (sodium-specific). The NaA43 DNAzyme, which is reported to be more than 10,000-fold selective for sodium over other metal ions, was used to make a real-time sodium sensor in living cells.

Bioinformatics

Bioinformatics involves the development of techniques to store, data mine, search and manipulate biological data, including DNA nucleic acid sequence data. These have led to widely applied advances in computer science, especially string searching algorithms, machine learning, and database theory. String searching or matching algorithms, which find an occurrence of a sequence of letters inside a larger sequence of letters, were developed to search for specific sequences of nucleotides. The DNA sequence may be aligned with other DNA sequences to identify homologous sequences and locate the specific mutations that make them distinct. These techniques, especially multiple sequence alignment, are used in studying phylogenetic relationships and protein function. Data sets representing entire genomes' worth of DNA sequences, such as those produced by the Human Genome Project, are difficult to use without the annotations that identify the locations of genes and regulatory elements on each chromosome. Regions of DNA sequence that have the characteristic patterns associated with protein- or RNA-coding genes can be identified by gene finding algorithms, which allow researchers to predict the presence of particular gene products and their possible functions in an organism even before they have been isolated experimentally. Entire genomes may also be compared, which can shed light on the evolutionary history of particular organism and permit the examination of complex evolutionary events.

DNA Nanotechnology

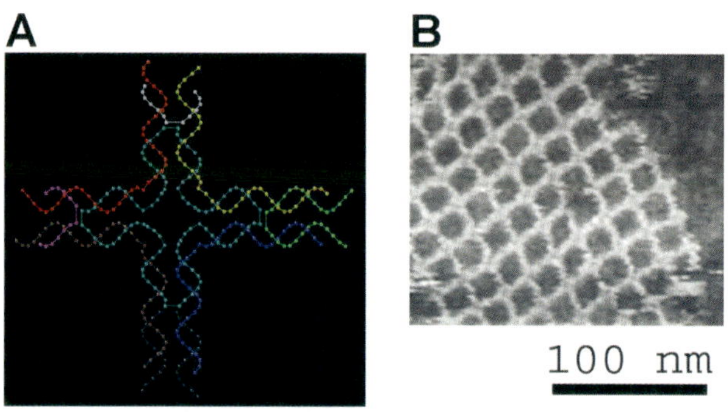

Figure: The DNA structure at left (schematic shown) will self-assemble into the structure visualized by atomic force microscopy at right.

DNA nanotechnology is the field that seeks to design nanoscale structures using the molecular recognition properties of DNA molecules.

DNA nanotechnology uses the unique molecular recognition properties of DNA and other nucleic acids to create self-assembling branched DNA complexes with useful properties. DNA is thus used as a structural material rather than as a carrier of biological information. This has led to the creation of two-dimensional periodic lattices (both tile-based and using the *DNA origami* method) and three-dimensional structures in the shapes of polyhedra. Nanomechanical devices and algorithmic self-assembly have also been demonstrated, and these DNA structures have been used to template the arrangement of other molecules such as gold nanoparticles and streptavidin proteins.

History and Anthropology

Because DNA collects mutations over time, which are then inherited, it contains historical information, and, by comparing DNA sequences, geneticists can infer the evolutionary history of organisms, their phylogeny. This field of phylogenetics is a powerful tool in evolutionary biology. If DNA sequences within a species are compared, population geneticists can learn the history of particular populations. This can be used in studies ranging from ecological genetics to anthropology.

Information Storage

In a paper in January 2013, scientists from the European Bioinformatics Institute and Agilent Technologies proposed a mechanism to use DNA's ability to code information as a means of digital data storage. The group was able to encode 739 kilobytes of data into DNA code, synthesize the actual DNA, then sequence the DNA and decode the information back to its original form, with a reported 100% accuracy. The encoded information consisted of text files and audio files. A prior experiment was published in August 2012. It was conducted by researchers at Harvard University, where the text of a 54,000-word book was encoded in DNA.

Moreover, in living cells, the storage can be turned active by enzymes. Light-gated protein domains fused to DNA processing enzymes are suitable for that task *in vitro*. Fluorescent exonucleases can transmit the output according to the nucleotide they have read.

Complementary DNA

The term cDNA refers to complementary DNA. cDNA is known to be synthesized, or manufactured from an mRNA or messenger RNA template. It is synthesized in a reaction that is catalyzed by the reverse transcriptase and DNA polymerase enzymes. Essential to note is that cDNA is usually used to clone eukaryotic genes in prokaryotes.

Scientists usually use cDNA when they want to express s certain protein in a cell that does not normally express such a protein. This process is referred to as heterologous expression. The expression of such a protein will be done by transferring the cDNA that codes for that protein to the recipient cell. Also, essential to note is that cDNA can also be produced by retroviruses like Simian Immunodeficiency Virus, HIV-1 and HIV-2 among others. Once the cDNA is created from such viruses, it is integrated into the genome of the host, where it goes on to create a provirus.

Research shows that when a protein is being synthesized, a gene's DNA is transcribed into an mRNA, which is then translated into a protein. Genes are usually divided into eukaryotic and prokaryotic genes. The only difference between these genes is that the eukaryotic genes contain introns instead of extrons that are contained in the prokaryotic genes.

Introns are not coding sequences, while extrons are DNA coding systems. During the transcription of the proteins, all intron RNA are cut from the primary RNA and the remaining piece is sliced back to become an mRNA. In other words, the mRNA is formed after all introns are removed from the primary RNA. Once formed, the mRNA is then translated into an amino acid and comprises a newly formed protein. From the above, it is noted that prokaryotic genes do not contain any introns, so their RNAs are not subject to cutting or splicing.

In some instances, it might be necessary to make a prokaryotic cell express the genes of a eukaryotic cell. One of the ways of making this possible is by adding eukaryotic DNA directly into the prokaryotic cell, so as to allow it to make a protein of its own. As has been noted, the eukaryotic DNA has introns, while the prokaryotic DNA does not have the machinery for removing introns from the RNA that has been transcribed. As a result, all intron sequences need to be removed from the eukaryotic DNA before it is transferred to the prokaryotic cell. This way, the cell will not be placed with the burden of having to remove introns. The intron-free DNA that is created is as a result of intron-free mRNA, which is why it is referred as a complementary copy of the mRNA. It is for this reason that is referred to as a complementary DNA or cDNA.

Though there are numerous processes for synthesizing cDNA, the best way of doing so is by using mature or fully spliced mRNA. This is usually done by using the enzyme reverse transcriptase. This enzyme is used because it mainly operates as a single strand of mRNA. It generates its cDNA by pairing RNA base pairs to their DNA complements.

Synthesis

Although there are several methods for doing so, cDNA is most often synthesized from mature (fully spliced) mRNA using the enzyme reverse transcriptase. This enzyme, which naturally occurs in retroviruses, operates on a single strand of mRNA, generating its complementary DNA based on the pairing of RNA base pairs (A, U, G and C) to their DNA complements (T, A, C and G, respectively).

To obtain eukaryotic cDNA whose introns have been removed:

1. A eukaryotic cell transcribes the DNA (from genes) into RNA (pre-mRNA).
2. The same cell processes the pre-mRNA strands by removing introns, and adding a poly-A tail and 5' Methyl-Guanine cap (this is known as post-transcriptional modification)
3. This mixture of mature mRNA strands is extracted from the cell. The poly-A tail of the post-transcriptional mRNA can be taken advantage of with oligo(dT) beads in an affinity chromatography assay.
4. A poly-T oligonucleotide primer is hybridized onto the poly-A tail of the mature mRNA template, or random hexamer primers can be added which contain every possible 6 base single strand of DNA and can therefore hybridize anywhere on the RNA (Reverse transcriptase requires this double-stranded segment as a primer to start its operation.)

5. Reverse transcriptase is added, along with deoxynucleotide triphosphates (A, T, G, C). This synthesizes one complementary strand of DNA hybridized to the original mRNA strand.
6. To synthesize an additional DNA strand, traditionally one would digest the RNA of the hybrid strand, using an enzyme like RNase H, or through alkali digestion method.
7. After digestion of the RNA, a single stranded DNA (ssDNA) is left and because single stranded nucleic acids are hydrophobic, it tends to loop around itself. It is likely that the ssDNA forms a hairpin loop at the 3' end.
8. From the hairpin loop, a DNA polymerase can then use it as a primer to transcribe a complementary sequence for the ss cDNA.
9. Now, you should be left with a double stranded cDNA with identical sequence as the mRNA of interest.

Applications

Complementary DNA is often used in gene cloning or as gene probes or in the creation of a cDNA library. When scientists transfer a gene from one cell into another cell in order to express the new genetic material as a protein in the recipient cell, the cDNA will be added to the recipient (rather than the entire gene), because the DNA for an entire gene may include DNA that does not code for the protein or that interrupts the coding sequence of the protein (e.g., introns). Partial sequences of cDNAs are often obtained as expressed sequence tags (EST).

With amplification of DNA sequences via polymerase chain reaction (PCR) now commonplace, one will typically conduct reverse transcription as an initial step, followed by PCR to obtain an exact sequence of cDNA for intra-cellular expression. This is achieved by designing sequence-specific DNA primers that hybridize to the 5' and 3' ends of a cDNA region coding for a protein. Once amplified, the sequence can be cut at each end with nucleases and inserted into one of many small circular DNA sequences known as expression vectors. Such vectors allow for self-replication, inside the cells, and potentially integration in the host DNA. They typically also contain a strong promoter to drive transcription of the target cDNA into mRNA, which is then translated into protein

FANTOM

FANTOM is an international research consortium established by Dr. Hayashizaki and his colleagues in 2000 to assign functional annotations to the full-length cDNAs that were collected during the Mouse Encyclopedia Project at RIKEN. FANTOM has since developed and expanded over time to encompass the fields of transcriptome analysis. The object of the project is moving steadily up the layers in the system of life, progressing thus from an understanding of the 'elements' - the transcripts - to an understanding of the 'system' - the transcriptional regulatory network, in other words the 'system' of an individual life form.

Viruses

Some viruses also use cDNA to turn their viral RNA into mRNA (viral RNA → cDNA → mRNA). The mRNA is used to make viral proteins to take over the host cell.

An example of this first step from viral DNA to cDNA can be seen in the HIV cycle of infection. Here, the host cell membrane becomes attached to the virus' lipid envelope which allows the viral capsid with two copies of viral genome RNA to enter the host. The cDNA copy is then made though reverse transcription of the viral RNA, a process facilitated by the chaperone CypA and a viral capsid associated reverse transcriptase.

Nuclear DNA

Virtually all the cells in our body contain genetic material, with the exception of red blood cells and platelets. These have no nucleus and consequently no DNA. Otherwise every cell in the body is made up of a cell wall (cell membrane), cell fluid (cytoplasm) and a nucleus.

DNA is actually nuclear DNA: DNA in the nucleus. In resting cells this looks like an unraveled tangle but it becomes arranged into a more compact structure during cell division. Then the DNA organizes itself into smaller, well-defined sections and can be visualized as the 23 pairs of chromosomes that we call our 'karyotype'.

Nuclear DNA is made up of genetic material from your father and mother, and the nucleus therefore contains pairs of chromosomes: you have two copies of each chromosome, one from each parent. Distributed between these 46 chromosomes we find more than 20,000 pairs of genes.

Structure

Nuclear DNA is a nucleic acid, a polymeric biomolecule or biopolymer, found in the nucleus of eukaryotic organisms. Its structure is a double helix, with two strands wound around each other. This double helix structure was first described by Francis Crick and James D. Watson using data collected by Rosalind Franklin. Each strand is a long polymer chain of repeating nucleotides. Each nucleotide is composed of a five-carbon sugar, a phosphate group, and an organic base. Nucleotides are distinguished by their bases. There are the purines, large bases which include adenine and guanine, and pyrimidines, small bases which include thymine and cytosine. Chargaff's rules state that adenine will always pair with thymine and guanine will always pair with cytosine. The phosphate groups are held together by a phosphodiester bond and the bases are held together by hydrogen bonds.

Forensics

Nuclear DNA is known as the molecule of life and contains the genetic instructions for the development of all living organisms. It is found in almost every cell in the human body, with exceptions such as red blood cells. Everyone has a unique genetic blueprint, even identical twins. Forensic departments such as the Bureau of Criminal Apprehension (BCA) and Federal Bureau of Investigation (FBI) are able to use techniques involving nuclear DNA to compare samples in a case. Techniques used include polymerase chain reaction (PCR), which allows one to utilize very small amounts of DNA by making copies of targeted regions on the molecule, also known as short tandem repeats (STRs).

Cell Division

Like mitosis, meiosis is a form of eukaryotic cell division. Meiosis gives rise to four unique daughter cells, each of which has half the number of chromosomes as the parent cell. Because meiosis creates cells that are destined to become gametes (or reproductive cells), this reduction in chromosome number is critical — without it, the union of two gametes during fertilization would result in offspring with twice the normal number of chromosomes.

Meiosis creates new combinations of genetic material in each of the four daughter cells. These new combinations result from the exchange of DNA between paired chromosomes. Such exchange means that the gametes produced through meiosis often exhibit considerable genetic variation.

Meiosis involves two rounds of nuclear division, not just one. Prior to undergoing meiosis, a cell goes through an interphase period in which it grows, replicates its chromosomes, and checks all of its systems to ensure that it is ready to divide.

Like mitosis, meiosis also has distinct stages called prophase, metaphase, anaphase, and telophase. A key difference, however, is that during meiosis, each of these phases occurs twice — once during the first round of division, called meiosis I, and again during the second round of division, called meiosis II.

Replication

Prior to cell division, the DNA material in the original cell must be duplicated so that after cell division, each new cell contains the full amount of DNA material. The process of DNA duplication is usually called replication. The replication is termed semiconservative since each new cell contains one strand of original DNA and one newly synthesized strand of DNA. The original polynucleotide strand of DNA serves as a template to guide the synthesis of the new complementary polynucleotide of DNA. The DNA single strand template serves to guide the synthesis of a complementary strand of DNA.

DNA replication begins at a specific site in the DNA molecule called the origin of replication. The enzyme helicase unwinds and separates a portion of the DNA molecule after which single-strand binding proteins react with and stabilize the separated, single-stranded sections of the DNA molecule. The enzyme complex DNA polymerase engages the separated portion of the molecule and initiates the process of replication. DNA polymerase can only connect new DNA nucleotides to a pre-existing chain of nucleotides. Therefore, replication begins as an enzyme called primase assembles an RNA primer at the origin of replication. The RNA primer consists of a short sequence of RNA nucleotides, complementary to a small, initial section of the DNA strand being prepared for replication. DNA polymerase is then able to add DNA nucleotides to the RNA primer and thus begin the process of constructing a new complementary strand of DNA. Later the RNA primer is enzymatically removed and replaced with the appropriate sequence of DNA nucleotides. Because the two complementary strands of the DNA molecule are oriented in opposite directions and the DNA polymerase can only accommodate replication in one direction, two different mechanisms for copying the strands of DNA are employed. One strand is replicated continuously towards the unwinding, separating portion of the original DNA molecule; while the other strand is replicated discontinuously in the opposite direction with the formation of a series of short DNA segments called Okazaki fragments. Each Okazaki fragment requires a separate RNA primer. As the Okazaki fragments are synthesized, the RNA primers are replaced with DNA nucleotides and the fragments are bonded together in a continuous complementary strand.

DNA Damage and Repair

Damage of nuclear DNA is a persistent problem arising from a variety of disruptive endogenous and exogenous sources. Eukaryotes have evolved a diverse set of DNA repair processes that remove nuclear DNA damages. These repair processes include base excision repair, nucleotide excision repair, homologous recombinational repair, non-homologous end joining and microhomology-mediated end joining. Such repair processes are essential for maintaining nuclear DNA stability. Failure of repair activity to keep up with the occurrence of damages has various negative consequences. Nuclear DNA damages, as well as the mutations and epigenetic alterations that such damages cause, are considered to be a major cause of cancer. Nuclear DNA damages are also implicated in aging and neurodegenerative diseases.

Mutation

Nuclear DNA is subject to mutation. A major cause of mutation is inaccurate DNA replication, often by specialized DNA polymerases that synthesize past DNA damages in the template strand (error-prone trans-lesion synthesis). Mutations also arise by inaccurate DNA repair. The microhomology-mediated end joining pathway for repair of double-strand breaks is particularly prone to mutation. Mutations arising in the nuclear DNA of the germline are most often neutral or adaptively disadvantageous. However, the small proportions of mutations that prove to be advantageous provide the genetic variation upon which natural selection operates to generate new adaptations.

Mitochondrial DNA

Mitochondrial DNA (mtDNA) is a type of DNA located outside the nucleus in the liquid portion of the cell (cytoplasm) and inside cellular organelles called mitochondria. Mitochondria are found in all complex or eukaryotic cells, including plant, animal, fungi, and single celled protists, which contain their own mtDNA genome. In animals with a backbone, or vertebrates, mtDNA is a double stranded molecule that forms a circular genome, which ranges in size from sixteen to eighteen kilo-base pairs, depending on species. Each mitochondrion in a cell can have multiple copies of the mtDNA genome. In humans, the mature egg cell, or oocyte, contains the highest number of mitochondria among human cells, ranging from 100,000 to 600,000 mitochondria per cell, but each mitochondrion contains only one copy of mtDNA. In human embryonic development, the number of mitochondria, the content of mtDNA in each mitochondrion, and the subsequent mtDNA activity affects the production of the oocytes, fertilization of the oocytes, and early embryonic growth and development.

Mitochondria were once free-living bacteria that took up residence inside a primitive eukaryotic cell in the process called endosymbiosis. Much of the evidence for the claim is in the mtDNA genome and the nuclear genome. The genomes co-evolved, and control of mitochondria involves exchange of information between the nucleus and the many copies of the mtDNA. In the developing embryo, ninety-nine percent of the mitochondria, and therefore the mtDNA, come from the mother. Point mutations and deletions in the mtDNA can lead to serious developmental mitochondrial diseases.

In 1890, Richard Altmann, who studied diseases in Germany, observed the occurrence of mitochondria in many different animal cell types and noted they were similar to bacteria. Altmann proposed that mitochondria were the fundamental particles of life, or the living part of the cell, and in 1896 he called them bioblasts. In 1901, Carl Benda, a physician from Germany, named the organelles mitochondria from Greek *mitos*, meaning thread, and *chondros*, meaning grains.

In 1963 Margit M. K. Nass and Sylvan Nass published describing DNA in mitochondria from chick embryos. Nass and Nass, who worked at the Wenner-Gren Institute for Experimental Biology at the University of Stockholm, Sweden, used an electron microscope to detect DNA in chick embryos. The paper provided early evidence that mitochondria contained DNA (mtDNA), supporting the hypothesis that mitochondria were direct descendants of bacteria. In her 1967 paper, Lynn (Sagan) Margulis proposed the theory of endosymbiosis, which claimed that organelles, including mitochondria, were once free-living bacteria that came to reside inside of complex cells about two billion years ago.

In the 1960s and 1970s, researchers investigated mtDNA by using yeast mitochondria. In 1975, Peter L. Molloy, Anthony W. Limmane, and H. B. Lukins published a yeast (*Saccharomyces cerevisiae*) mtDNA genome sequence map. The sequence was a rough draft of the entire yeast mtDNA genome. From 1974 to 1976, several laboratories began using enzymes to break DNA at specific places, a method called restriction enzyme analysis. The use of restriction enzyme analysis resulted in mtDNA maps of yeast and several other species including humans (*Homo sapiens*). In 1981, Fredrick Sanger's group in Cambridge, England, reported a complete sequence of the human mtDNA genome.

Sanger found that circular mtDNA in vertebrates consists of a light strand and a heavy strand. Both strands are coding sequences, and the process of DNA replication proceeds in both strands simultaneously in opposite directions. Sanger's team also found that vertebrate mtDNA is extremely compact and conserved through evolution, as most animals have the similar sets of mitochondrial genes. In vertebrates, mtDNA codes for thirty-seven gene products, and thirteen of the mtDNA genes code for proteins. Twenty-two mtDNA genes code for molecules that carry the building blocks of proteins (amino acids), called transfer RNAs (tRNA), and two genes code for the structures where cells assemble proteins, called ribosomal RNAs (rRNA).

In animal DNA, some of the protein and rRNA genes are located next to a tRNA gene. Justin C. St. John at Monash University in Melbourne, Australia, reported in 2010 that some coding regions overlap, meaning that a sequence of mtDNA codes for more than one product. The only region that does not code for a protein is the displacement loop (D-loop), which is organized as a triple-stranded structure that contains the main regulatory region involved with mtDNA replication. In contrast, the yeast mtDNA has non-coding sequences between protein and mtRNA gene coding sequences.

The thirteen proteins coded for in mtDNA are all involved with the production of what the cell uses as energy, a molecule called adenosine-5'-triphosphate (ATP). Mitochondria generate ATP in a process called oxidative phosphorylation (OXPHOS). The large OXPHOS protein complexes requires hundreds of proteins, thirteen of which are coded in mtDNA. The DNA from the cell's nucleus (nDNA) encodes the remaining proteins. Specific systems transport the proteins into the mitochondria from the cytoplasm. Transcription of mtDNA is under the control of both the nuclear and mitochondrial genomes. The mtDNA genome and the nuclear genome work together to regulate the energy production, otherwise several problems can occur in the cell that can affect the entire organism and may lead to disease.

Researchers first reported a patient suffering from a mitochondrial disease in 1959, a few years before they discovered mtDNA. The patient was a woman from Sweden who had the highest human metabolic rate then recorded. Researchers stated that the problem she had related to a defect in mitochondria. Her mitochondria produced energy in the form of ATP and heat; even when the woman was at rest, she would sweat. The mitochondrial defect, called Luft disease after the endocrinologist Rolf Luft, who first described it in 1962, is one of the rarest of all mitochondrial disorders.

In 1988, scientists began to describe pathogenic mutations in mtDNA. Researchers had studied mtDNA since 1963, but clinical scientists paid little attention to it. In 1988, Ian Holt's group at the Institute of Neurology in London, United Kingdom, identified large-scale deletions of base pairs of mtDNA in patients with mitochondrial muscle disease (myopathies). In the same year Douglas Wallace's group at Emory University School of Medicine in Atlanta, Georgia, described mutations in mtDNA in a human family whose members had Leber hereditary optic neuropathy (LHON). LHON results in optic nerve degeneration and blindness. Mutations within the mtDNA link to a number of primary neurological disorders. With a prevalence of ten in one-hundred thousand people, the disorders are one of the most common inherited neurological disorders. Mitochondrial diseases result from substitutions of a single mtDNA base, deletions of one or several bases, rearrangement of gene sequences, and duplication of genes. There are hundreds of mitochondrial diseases.

Humans inherit mitochondria from their mothers and mtDNA through the oocyte. In a human female embryo, the first primary oocytes develop from the primordial germ cells from two to three weeks into the process of embryo development. As reported by various scientists, the number of mitochondria in the primary oocyte ranges from fewer than ten to two hundred. Robert P.S. Jansen in his 2000 article reports fewer than ten mitochondria per human primordial germ cell. However, by the time the female infant is born, each primary oocyte has approximately 10,000 mitochondria per cell. There is another tenfold increase in mitochondrial number during adult growth and development. For most female mammals, the mature oocyte has from 100,000 to 600,000 mitochondria. The amount of mtDNA in each mitochondria in the female germ-line is slightly more mtDNA than the number of mitochondria. Ovarian insufficiency is associated with major depletion of mtDNA in the oocyte.

In the late 1990s, Jacques Cohen at Saint Barnabas Medical Center in Livingston, New Jersey, and his colleagues investigated the phenomenon of ovarian insufficiency. They transferred a small amount of cytoplasm from a cell of a donor who was fertile into the oocytes of a woman who had undergone several rounds of IVF without success. The procedure used by Cohen and his colleagues became called ooplasmic transfer or cytoplasmic transfer. Over the course of four years, at least thirty infants were born using this technique. One problem with ooplasmic transfer, which researchers noted, was that the offspring can retain mtDNA from the mother as well as from the donor. The mixture of mtDNA, called heteroplasmy, can lead to mitochondrial diseases. For example, scientists showed how mice experience problems if their normal mtDNA mixes with dissimilar mtDNA. In 2012, Mark S. Sharpley at the University of Pennsylvania in Philadelphia, Pennsylvania, and his group published a study on mice in which they generated mice with mixtures of different strains of mtDNA. The mice with mixtures had abnormal behavior and cognition.

Scientists correlated mtDNA mutations with a increasing number of diseases, and into the first decades of the twentieth century there were few treatments to alleviate the symptoms. Nuclear transfer is an alternate technique for preventing mitochondrial disease. There are several nuclear transfer techniques. These techniques use a donor oocyte with healthy mtDNA that has its nucleus removed. In 2010, Helen Tuppen's group in the UK at Newcastle University transferred fertilized oocytes to a donor oocyte that had its nucleus removed. A group led by Shoukhrat Mitalipov at Oregon Health and Science University in Beaverton, Oregon, used an unfertilized oocyte, removed the nucleus, transferred it to an unfertilized oocyte of a healthy donor, and then fertilized the oocyte with sperm.

Mitochondrial Inheritance

In most multicellular organisms, mtDNA is inherited from the mother (maternally inherited). Mechanisms for this include simple dilution (an egg contains on average 200,000 mtDNA molecules, whereas a healthy human sperm was reported to contain on average 5 molecules), degradation of sperm mtDNA in the male genital tract, in the fertilized egg, and, at least in a few organisms, failure of sperm mtDNA to enter the egg. Whatever the mechanism, this single parent (uniparental inheritance) pattern of mtDNA inheritance is found in most animals, most plants and in fungi as well.

Female Inheritance

In sexual reproduction, mitochondria are normally inherited exclusively from the mother; the mitochondria in mammalian sperm are usually destroyed by the egg cell after fertilization. Also, most mitochondria are present at the base of the sperm's tail, which is used for propelling the sperm cells; sometimes the tail is lost during fertilization. In 1999 it was reported that paternal sperm mitochondria (containing mtDNA) are marked with ubiquitin to select them for later destruction inside the embryo. Some *in vitro* fertilization techniques, particularly injecting a sperm into an oocyte, may interfere with this.

The fact that mitochondrial DNA is maternally inherited enables genealogical researchers to trace maternal lineage far back in time. (Y-chromosomal DNA, paternally inherited, is used in an analogous way to determine the patrilineal history.) This is usually accomplished on human mitochondrial DNA by sequencing the hypervariable control regions (HVR1 or HVR2), and sometimes the complete molecule of the mitochondrial DNA, as a genealogical DNA test. HVR1, for example, consists of about 440 base pairs. These 440 base pairs are then compared to the control regions of other individuals (either specific people or subjects in a database) to determine maternal lineage. Most often, the comparison is made to the revised Cambridge Reference Sequence. Vila have published studies tracing the matrilineal descent of domestic dogs to wolves. The concept of the Mitochondrial Eve is based on the same type of analysis, attempting to discover the origin of humanity by tracking the lineage back in time.

mtDNA is highly conserved, and its relatively slow mutation rates (compared to other DNA regions such as microsatellites) make it useful for studying the evolutionary relationships—phylogeny—of organisms. Biologists can determine and then compare mtDNA sequences among different species and use the comparisons to build an evolutionary tree for the species examined. However, due to the slow mutation rates it experiences, it is often hard to distinguish between closely related species to any large degree, so other methods of analysis must be used.

Mitochondrial Bottleneck

Entities undergoing uniparental inheritance and with little to no recombination may be expected to be subject to Muller's ratchet, the accumulation of deleterious mutations until functionality is lost. Animal populations of mitochondria avoid this buildup through a developmental process known as the mtDNA bottleneck. The bottleneck exploits stochastic processes in the cell to increase in the cell-to-cell variability in mutant load as an organism develops: a single egg cell with some proportion of mutant mtDNA thus produces an embryo where different cells have different mutant loads. Cell-level selection may then act to remove those cells with more mutant mtDNA, leading to a stabilisation or reduction in mutant load between generations. The mechanism underlying the bottleneck is debated, with a recent mathematical and experimental metastudy providing evidence for a combination of random partitioning of mtDNAs at cell divisions and random turnover of mtDNA molecules within the cell.

Male Inheritance

Doubly uniparental inheritance of mtDNA is observed in bivalve mollusks. In those species, females have only one type of mtDNA (F), whereas males have F type mtDNA in their somatic cells, but M type of mtDNA (which can be as much as 30% divergent) in germline cells. Paternally inherited mitochondria have additionally been reported in some insects such as fruit flies, honeybees, and periodical cicadas.

Male mitochondrial inheritance was recently discovered in Plymouth Rock chickens. Evidence supports rare instances of male mitochondrial inheritance in some mammals as well. Specifically, documented occurrences exist for mice, where the male-inherited mitochondria were subsequently rejected. It has also been found in sheep, and in cloned cattle. It has been found in a single case in a human male.

Although many of these cases involve cloned embryos or subsequent rejection of the paternal mitochondria, others document *in vivo* inheritance and persistence under lab conditions.

Mitochondrial Donation

An IVF technique known as mitochondrial donation or mitochondrial replacement therapy (MRT) results in offspring containing mtDNA from a donor female, and nuclear DNA from the mother and father. In the spindle transfer procedure, the nucleus of an egg is inserted into the cytoplasm of an egg from a donor female which has had its nucleus removed, but still contains the donor female's mtDNA. The composite egg is then fertilized with the male's sperm. The procedure is used when a woman with genetically defective mitochondria wishes to procreate and produce offspring with healthy mitochondria. The first known child to be born as a result of mitochondrial donation was a boy born to a Jordanian couple in Mexico on 6 April 2016.

Mutations and Diseases

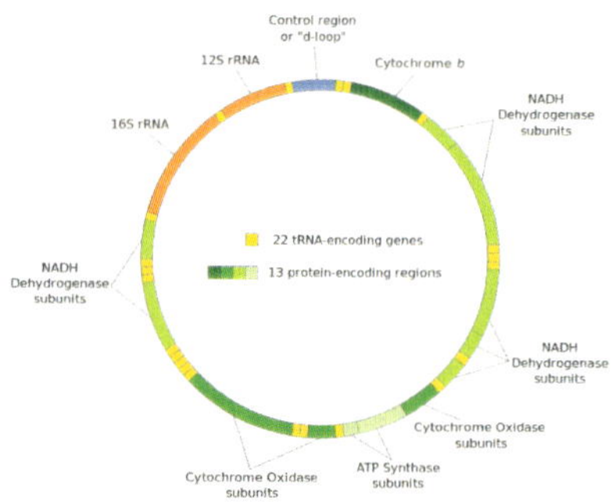

Figure: Human mitochondrial DNA with groups of protein-, rRNA- and tRNA-encoding genes

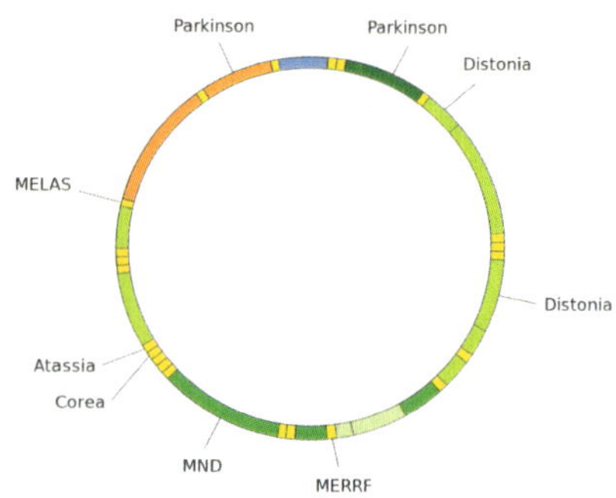

Figure: Involvement of mitochondrial DNA in several human diseases

Susceptibility

The concept that mtDNA is particularly susceptible to reactive oxygen species generated by the respiratory chain due to its proximity remains controversial· mtDNA does not accumulate any more oxidative base damage than nuclear DNA. It has been reported that at least some types of oxidative DNA damage are repaired more efficiently in mitochondria than they are in the nucleus. mtDNA is packaged with proteins which appear to be as protective as proteins of the nuclear chromatin. Moreover, mitochondria evolved a unique mechanism which maintains mtDNA integrity through degradation of excessively damaged genomes followed by replication of intact/repaired mtDNA. This mechanism is not present in the nucleus and is enabled by multiple copies of mtDNA present in mitochondria. The outcome of mutation in mtDNA may be an alteration in the coding instructions for some proteins, which may have an effect on organism metabolism and/or fitness.

Genetic Illnes

Mutations of mitochondrial DNA can lead to a number of illnesses including exercise intolerance and Kearns–Sayre syndrome (KSS), which causes a person to lose full function of heart, eye, and muscle movements. Some evidence suggests that they might be major contributors to the aging process and age-associated pathologies. Particularly in the context of disease, the proportion of mutant mtDNA molecules in a cell is termed heteroplasmy. The within-cell and between-cell distributions of heteroplasmy dictate the onset and severity of disease and are influenced by complicated processes within the cell and during development·

Mutations in mitochondrial tRNAs can be responsible for severe diseases like the MELAS and MERRF syndromes.

Mutations in nuclear genes that encode proteins that mitochondria use can also contribute to mitochondrial diseases. These diseases do not follow mitochondrial inheritance patterns, but instead follow Mendelian inheritance patterns.

Use in Disease Diagnosis

Recently, a mutation in mtDNA has been used to help diagnose prostate cancer in patients with negative prostate biopsy.

Relationship with Aging

Though the idea is controversial, some evidence suggests a link between aging and mitochondrial genome dysfunction. In essence, mutations in mtDNA upset a careful balance of reactive oxygen species (ROS) production and enzymatic ROS scavenging (by enzymes like superoxide dismutase, catalase, glutathione peroxidase and others). However, some mutations that increase ROS production (e.g., by reducing antioxidant defenses) in worms increase, rather than decrease, their longevity. Also, naked mole rats, rodents about the size of mice, live about eight times longer than mice despite having reduced, compared to mice, antioxidant defenses and increased oxidative damage to biomolecules. Once, there was thought to be a positive feedback loop at work (a 'Vicious Cycle'); as mitochondrial DNA accumulates genetic damage caused by free radicals, the mitochondria lose function and leak free radicals into the cytosol. A decrease in mitochondrial function reduces overall metabolic efficiency. However, this concept was conclusively disproved when it was demonstrated that mice, which were genetically altered to accumulate mtDNA mutations at accelerated rate do age prematurely, but their tissues do not produce more ROS as predicted by the 'Vicious Cycle' hypothesis. Supporting a link between longevity and mitochondrial DNA, some studies have found correlations between biochemical properties of the mitochondrial DNA and the longevity of species. Extensive research is being conducted to further investigate this link and methods to combat aging. Presently, gene therapy and nutraceutical supplementation are popular areas of ongoing research. Bjelakovic analyzed the results of 78 studies between 1977 and 2012, involving a total of 296,707 participants, and concluded that antioxidant supplements do not reduce all-cause mortality nor extend lifespan, while some of them, such as beta carotene, vitamin E, and higher doses of vitamin A, may actually increase mortality.

Neurodegenerative Diseases

Increased mtDNA damage is a feature of several neurodegenerative diseases.

The brains of individuals with Alzheimer's disease have elevated levels of oxidative DNA damage in both nuclear DNA and mtDNA, but the mtDNA has approximately 10-fold higher levels than nuclear DNA. It has been proposed that aged mitochondria is the critical factor in the origin of neurodegeneration in Alzheimer's disease.

In Huntington's disease, mutant huntingtin protein causes mitochondria dysfunction involving inhibition of mitochondrial electron transport, higher levels of reactive oxygen species and increased stress. Mutant huntingtin protein promotes oxidative damage to mtDNA, as well as nuclear DNA, that may contribute to Huntington's disease pathology.

The DNA oxidation product 8-oxoguanine (8-oxoG) is a well-established marker of oxidative DNA damage. In persons with amyotrophic lateral sclerosis (ALS), the enzymes that normally repair 8-oxoG DNA damages in the mtDNA of spinal motor neurons are impaired. Thus, oxidative damage to mtDNA of motor neurons may be a significant factor in the etiology of ALS.

Correlation of the mtDNA Base Composition with Animals Lifespan

Animal species mtDNA base composition was retrieved from the MitoAge database and compared to their maximum life span from An Age database.

Over the past decade, an Israeli research group led by Professor Vadim Fraifeld has shown that extraordinarily strong and significant correlations exist between the mtDNA base composition and animal species-specific maximum life spans. As demonstrated in their work, higher mtDNA guanine + cytosine content (GC%) strongly associates with longer maximum life spans across animal species. An additional astonishing observation is that the mtDNA GC% correlation with the maximum life spans is independent of the well-known correlation between animal species metabolic rate and maximum life spans. The mtDNA GC% and resting metabolic rate explain the differences in animal species maximum life spans in a multiplicative manner (i.e., species maximum life span = their mtDNA GC% * metabolic rate) To support the scientific community in carrying out comparative analyses between mtDNA features and longevity across animals, a dedicated database was built named MitoAge.

Relationship with Non-B (Non-canonical) DNA Structures

Deletion breakpoints frequently occur within or near regions showing non-canonical (non-B) conformations, namely hairpins, cruciform and cloverleaf-like elements. Moreover, there is data supporting the involvement of helix-distorting intrinsically curved regions and long G-tetrads in eliciting instability events. In addition, higher breakpoint densities were consistently observed within GC-skewed regions and in the close vicinity of the degenerate sequence motif YMMYMN-NMMHM. Recently was found that all mitochondrial genomes sequenced so far contain many of inverted repeats necessary for cruciform DNA formation and these loci are particularly enriched in replication origin sites, D-loops and stem loops.

SiDNA

Signal interfering DNA (siDNA) is a class of short modified double stranded DNA molecules, 8–64 base pairs in length. siDNA molecules are capable of inhibiting DNA repair activities by interfering with multiple repair pathways. In general, these molecules act by mimicking DNA breaks and interfering with recognition and repair of DNA damage induced on chromosomes by irradiation or genotoxic products.

Mechanism of Action

The siDNA family, led by *Dbait* consists of 32 base pairs deoxyribonucleotide forming an intramolecular double helix, which mimicks DNA double-strand break lesions. Dbait binds to and hyperactivate DNA-PK, an enzyme involved in DNA breaks signaling and repair. DNA-PK hyperactivation induces pan-nuclear phosphorylation of histone H_2AX among all the chromatin. H_2AX phosphorylation is the signal, which allows double-strand break repair proteins (from NHEJ and homologous recombination pathways) to form DNA repair complexes selectively on DNA double-strand breaks. Dbait-dependent unspecific phosphorylation of H_2AX results in inefficient double strand break recognition and repair.

Possible Therapeutic Application

Most of the anticancer therapies act by induction of DNA damage (chemotherapy and radiation therapy). DNA breaks are the most lethal damage for cells, as one single double-strand break if unrepaired is sufficient to lead to cell death. *Dbait* enhances the efficacy of the DNA damaging agents as demonstrated with radiation therapy and/or chemotherapy in multiple *in vivo* experimental models such as melanoma, glioblastoma and colorectal cancer. Preclinical proof of concept of the synergic effect of the clinical candidate, *DT01*, with radiation therapy lead to a first-in-human Phase I, to evaluate the tolerance and efficacy of local *DT01* administration in association with RT in patients suffering from in-transit metastases of melanoma. Encouraging results were published in May 2016.

Satellite DNA

Satellite DNA is mainly present in heterochromatin or the tightly packed regions of chromosomes in centromeres, telomeres, and sometimes even in the euchromatin region (active region of the genome). Although conventionally satellite DNA has been known to be 'non-coding' (i.e. it does not encode protein), recent evidence suggests that some of the satellite DNA does undergo transcription.

Satellite DNA Structure

Satellite DNA consists of arrays of tandem repeats or repeats arranged side-by-side. These repeats can be as small as 1-2 bp long or as long as 10-60 bps long. The short tandem repeats (1-2 bp long) are called microsatellite or simple sequence repeats (SSRs), while the longer tandem repeats (10-60bp long) are called minisatellites or variable number tandem repeats (VNTRs).

The regions in between two simple sequence repeats or microsatellite are termed as 'inter simple sequence repeats' or ISSRs. Due to the presence of large number of repeats, the mutation rate is high in satellite DNA. It is speculated that as most of these sequences do not code for proteins, the detrimental consequences of high mutation rate is low. Thus, there is no selection pressure against it.

Reasons for Calling it Satellite DNA

The density of DNA can be calculated using density gradient centrifugation. When a DNA containing solution is spun at very high rotational speed, DNA sediments in the tube in a density-dependent manner. When the density of DNA was determined in this manner, it was found that satellite DNA formed a second 'satellite' band separate from the rest of the DNA.

The density of DNA is a function of its base and sequence, and satellite DNA with its highly repetitive DNA has a reduced or a characteristic density compared to the rest of the genome. Thus, the name 'satellite DNA' was coined.

Functions of Satellite DNA

Although popularly satellite DNA is thought to be part of 'junk DNA' or 'selfish DNA' which occupy the genome but have no effect on the fitness of an organism, recent studies show several distinct biological functions.

- They reside in the centromeric and pericentrometric regions and regulate centromere function.
- They are involved in the formation of heterochromatin. Periodic A-T distribution leads to the curvature of DNA, and satellite DNA with its AT-rich regions are considered to be important for packing of DNA in heterochromatin region.
- Transcripts of satellite RNA have been found in invertebrates, vertebrates, and plants where they are transcribed at a particular developmental stage in certain cells and tissues.

As the sequences of satellite DNA are highly diverse and variable, sequence-specific regulatory signals are speculated to be present in satellite DNA which fine tune gene expression

Applications

The short sequences of DNA stretches in satellite DNA vary from individual to individual. These variations in the length and sequence of satellite DNA are unique to each individual.

This uniqueness can be exploited to identify each individual based on their DNA map. This technique is called DNA fingerprinting, and it is used to identify criminals, perform paternity tests, and diagnose genetic disorders.

Bases of DNA

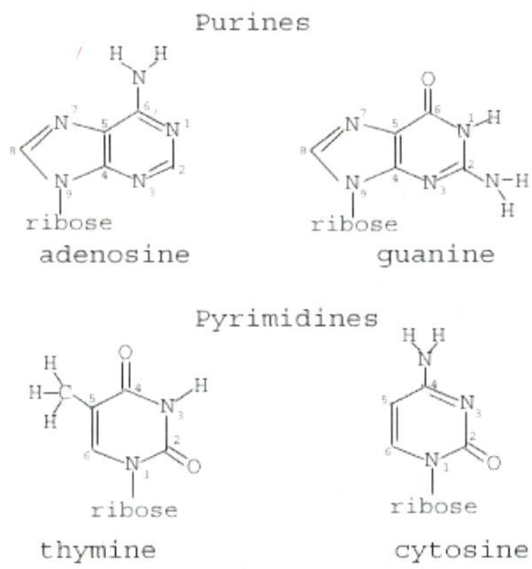

Figure: DNA Bases

The four nitrogen bases found in DNA are adenine, cytosine, guanine, and thymine. Each of these bases are often abbreviated a single letter: A (adenine), C (cytosine), G (guanine), T (thymine). The bases come in two categories: thymine and cytosine are pyrimidines, while adenine and guanine are purines.

The pyrimidine structure is produced by a six-membered, two-nitrogen molecule; purine refers to a nine-membered, four-nitrogen molecule. As you can see, each constituent of the ring making up the base is numbered to help with specificity of identification.

Base Pairing in DNA

The nitrogen bases form the double-strand of DNA through weak hydrogen bonds. The nitrogen bases, however, have specific shapes and hydrogen bond properties so that guanine and cytosine only bond with each other, while adenine and thymine also bond exclusively. This pairing off of the nitrogen bases is called complementarity. In order for hydrogen bonding to occur at all, a hydrogen bond donor must have a complementary hydrogen bond acceptor in the base across from it. Common hydrogen bond donors include primary and secondary amine groups or hydroxyl groups. Common acceptor groups are carbonyls and tertiary amines.

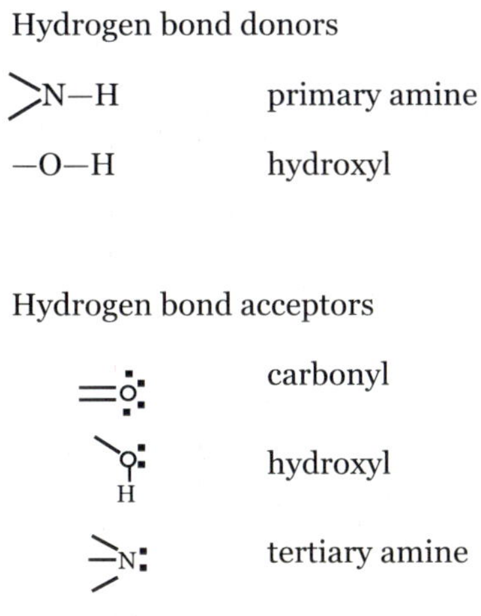

Figure: Common hydrogen bond donors and acceptors

There are three hydrogen bonds in a G: C base pair. One hydrogen bond forms between the 6' hydrogen bond accepting carbonyl of the guanine and the 4' hydrogen bond accepting primary amine of the cytosine. The second between the 1' secondary amine on guanine and the 3' tertiary amine on cytosine. And the third between the 2' primary amine on guanine and the 2' carbonyl on cytosine.

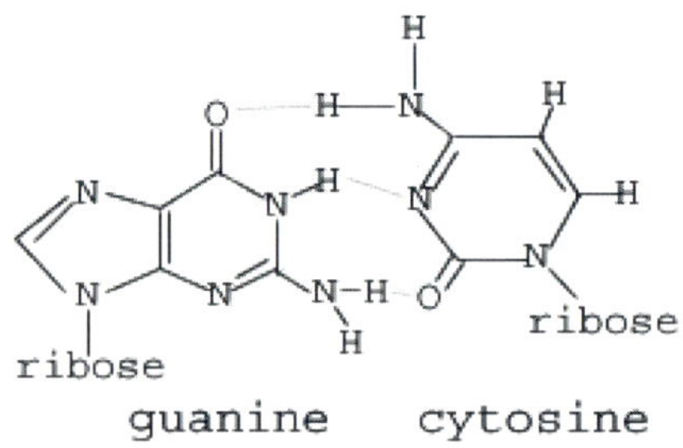

Figure: Guanine : Cytosine Base Pair

Between an A:T base pair, there are only two hydrogen bonds. One is found between the 6' primary amine of adenine and the 4' carbonyl of thymine. The other between the 1' tertiary amine of adenine and the 2' secondary amine of thymine.

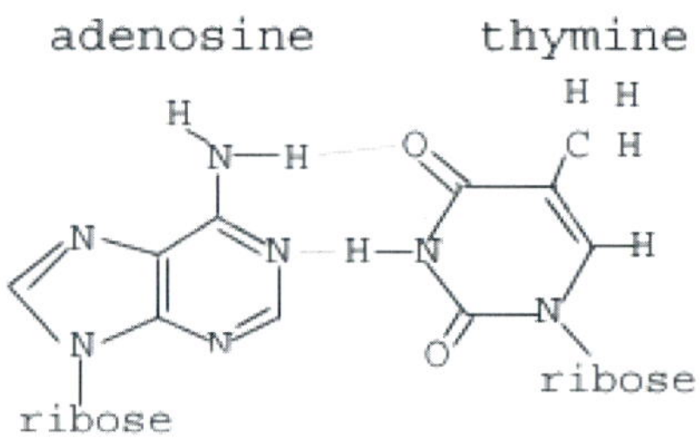

Figure: Adenine: Thymine Base Pair

Thymine

Thymine is an interesting base because it is the only one of the four bases found exclusively inside of DNA. The other bases are also found in RNA, which is often thought of DNA's cousin because of the close relationship and joint assistance the two often share genetic information transfer process.

Reactions of Thymine

- Thymine combined with deoxyribose (a monosaccharide, the most basic units of biologically important carbohydrates) creates the nucleoside (are glycosylamines, a biochemical compound which consists of a nucleobase bound to a ribose, an organic compound) deoxythymidine (DNA nucleoside), which is identical with the term 'Thymidine' (a chemical compound).
- Phosphorylation: Thymidine can be phosphorylated with one, two, or three phosphoric acid groups, creating, respectively, TMP, TDP, or TTP (thymidine mono-, di-, or triphosphate).
- Oxidation: Thymine bases are frequently oxidized to 'hydantoins' over time after the death of an organism. Hydantoin, which is also known as 'glycolylurea', is a heterocyclic (has atoms of at least two different elements as members of its rings) organic compound.

Role of Thymine in DNA mutation

A common mutation of DNA involves two adjacent thymines or cytosine, which, in presence of ultraviolet light, may form thymine dimers (damage to the structure of a biological molecule such as DNA formed from thymine via photochemical reactions), causing "kinks" in the DNA molecule that inhibits normal function. This mutation is responsible for melanoma formation. Melanoma is a form of skin cancer that often arises in a mole.

Stabilization of Thymine in the Nucleic Acid Structure

- In DNA, thymine (T) binds to adenine (A) via two hydrogen bonds, thus stabilizing the nucleic acid structures. A DNA molecule is made up of two strands of nucleotides that spiral around each other to form a double helix. The nucleotide backbone is created by the sugar of one nucleotide bonding with the phosphate group of the next. The two strands are held together by hydrogen bonds between the bases of the opposite nucleotides. This hydrogen bonding is very specific and only occurs between complementary base pairs. There are two main restrictions on how the cross steps between the DNA strands can be formed in order for the hydrogen bonds to form and the regular coiling of the double helix to occur.

- Initially, purine bases only bond with pyrimidine bases. By only having purine bases bond with pyrimidine bases, the length of the cross step (rung) between the DNA strands will remain steady. If purine bases could bond with purine bases or pyrimidine bases with pyrimidine bases, the length of the cross rung would change causing the DNA molecule to bow in and out.

- As the next specific condition, adenine only bonds with thymine and cytosine only bonds with guanine. When adenine bonds with thymine, two hydrogen bonds are formed. Three hydrogen bonds are formed between cytosine and guanine. Only these two pairs are capable of forming the necessary hydrogen bonds to maintain the stability of the DNA molecule.

Thymine is exclusive amongst the four bases as it only occurs in DNA molecules. Adenine, cytosine and guanine are also found in nucleotides that make up ribonucleic acid, or RNA. Inside an RNA molecule, thymine is replaced by uracil.

Aftermath of Cell Division

The order that the bases appear is insignificant in the DNA molecule. This means that there can be four different cross rungs – adenine with thymine, thymine with adenine, cytosine with guanine, and guanine with cytosine. This is biologically noteworthy since it means that the base sequence of one strand of a DNA molecule specifies the base sequence of the other strand. In other words, the two strands can be separated and exact copies are made each time a cell divides.

Guanine

Guanine is one of the four main nucleobases found in the nucleic acids DNA and RNA, the others being adenine, cytosine, and thymine (uracil in RNA). In DNA, guanine is paired with cytosine. The guanine nucleoside is called guanosine.

With the formula $C_5H_5N_5O$, guanine is a derivative of purine, consisting of a fused pyrimidine-imidazole ring system with conjugated double bonds. Being unsaturated, the bicyclic molecule is planar.

Properties

Guanine, along with adenine and cytosine, is present in both DNA and RNA, whereas thymine is usually seen only in DNA, and uracil only in RNA. Guanine has two tautomeric forms, the major keto form and rare enol form.

It binds to cytosine through three hydrogen bonds. In cytosine, the amino group acts as the hydrogen bond donor and the C-2 carbonyl and the N-3 amine as the hydrogen-bond acceptors. Guanine has the C-6 carbonyl group that acts as the hydrogen bond acceptor, while a group at N-1 and the amino group at C-2 act as the hydrogen bond donors.

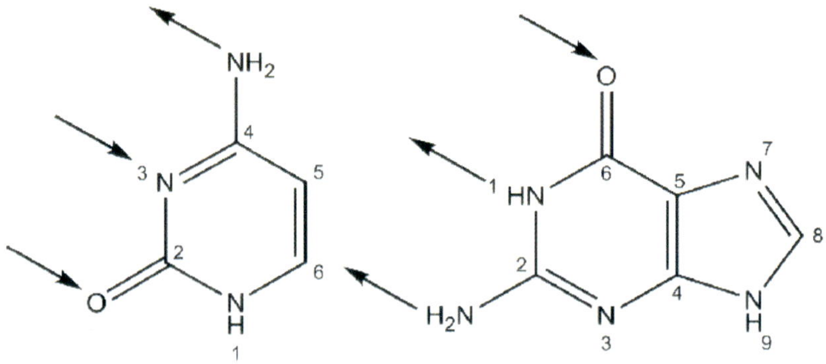

Cytosine Guanine

Guanine can be hydrolyzed with strong acid to glycine, ammonia, carbon dioxide, and carbon monoxide. First, guanine gets deaminated to become xanthine. Guanine oxidizes more readily than adenine, the other purine-derivative base in DNA. Its high melting point of 350°C reflects the intermolecular hydrogen bonding between the oxo and amino groups in the molecules in the crystal. Because of this intermolecular bonding, guanine is relatively insoluble in water, but it is soluble in dilute acids and bases.

Syntheses

Trace amounts of guanine form by the polymerization of ammonium cyanide (NH_4CN). Two experiments conducted by Levy et al. showed that heating 10 mol·L^{-1} NH_4CN at 80° C for 24 hours gave a yield of 0.0007%, while using 0.1 mol·L^{-1} NH_4CN frozen at -20° C for 25 years gave a 0.0035% yield. These results indicate guanine could arise in frozen regions of the primitive earth. In 1984, Yuasa reported a 0.00017% yield of guanine after the electrical discharge of NH_3, CH_4, C_2H_6, and 50 mL of water, followed by a subsequent acid hydrolysis. However, it is unknown whether the presence of guanine was not simply a resultant contaminant of the reaction.

$$10NH_3 + 2CH_4 + 4C_2H_6 + 2H_2O \rightarrow 2C_5H_8N_5O \text{ (guanine)} + 25H_2$$

A Fischer-Tropsch synthesis can also be used to form guanine, along with adenine, uracil, and thymine. Heating an equimolar gas mixture of CO, H_2, and NH_3 to 700 °C for 15 to 24 minutes, followed by quick cooling and then sustained reheating to 100 to 200 °C for 16 to 44 hours with an alumina catalyst, yielded guanine and uracil:

$$10CO + H_2 + 10NH_3 \rightarrow 2C_5H_8N_5O \text{ (guanine)} + 8H_2O$$

Another possible abiotic route was explored by quenching a 90% N_2–10%CO–H_2O gas mixture high-temperature plasma.

Traube's synthesis involves heating 2,4,5-triamino-1,6-dihydro-6-oxypyrimidine (as the sulfate) with formic acid for several hours.

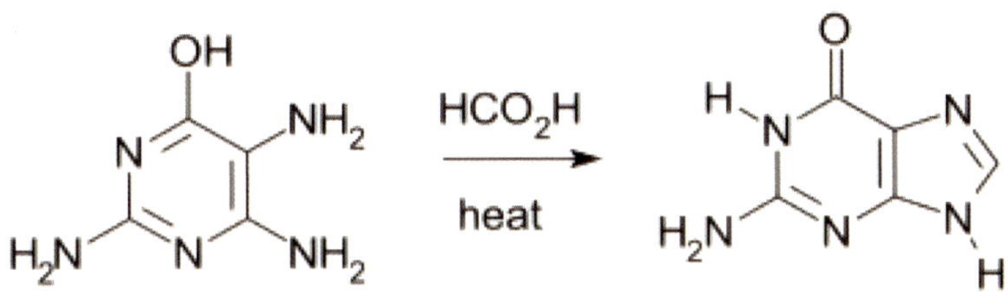

Other Occurrences/Biological uses

The word guanine derives from the Spanish loanword *guano* ("bird/bat droppings"), which itself is from the Quechua word *wanu*, meaning "dung". Guanine is "A white amorphous substance obtained abundantly from guano, forming a constituent of the excrement of birds".

In 1656 in Paris, a Mr. Jaquin extracted from the scales of the fish *Alburnus alburnus* so-called "pearl essence", which is crystalline guanine. In the cosmetics industry, crystalline guanine is used as an additive to various products (e.g., shampoos), where it provides a pearly iridescent effect. It is also used in metallic paints and simulated pearls and plastics. It provides shimmering luster to eye shadow and nail polish. Facial treatments using the droppings, or guano, from Japanese nightingales have been used in Japan and elsewhere, reportedly because the guanine in the droppings produces a clear, "bright" skin tone that users desire. Guanine crystals are rhombic platelets composed of multiple transparent layers, but they have a high index of refraction that partially reflects and transmits light from layer to layer, thus producing a pearly luster. It can be applied by spray, painting, or dipping. It may irritate the eyes. Its alternatives are mica, faux pearl (from ground shells), and aluminium and bronze particles.

Guanine has a very wide variety of biological uses that include a range of functions ranging in both complexity and versatility. These include camouflage, display, and vision among other purposes.

Spiders, scorpions, and some amphibians convert ammonia, as a product of protein metabolism in the cells, to guanine, as it can be excreted with minimal water loss.

Guanine is also found in specialized skin cells of fish called iridocytes (e.g., the sturgeon), as well as being present in the reflective deposits of the eyes of deep-sea fish and some reptiles, such as crocodiles.

On 8 August 2011, a report, based on NASA studies with meteorites found on Earth, was published suggesting building blocks of DNA and RNA (guanine, adenine and related organic molecules) may have been formed extra-terrestrially in outer space.

Cytosine

Cytosine is one of the four main bases found in DNA and RNA, along with adenine, guanine, and thymine (uracil in RNA). It is a pyrimidine derivative, with a heterocyclic aromatic ring and two substituents attached (an amine group at position 4 and a keto group at position 2). The nucleoside of cytosine is cytidine. In Watson-Crick base pairing, it forms three (3) hydrogen bonds with guanine.

Chemical Reactions

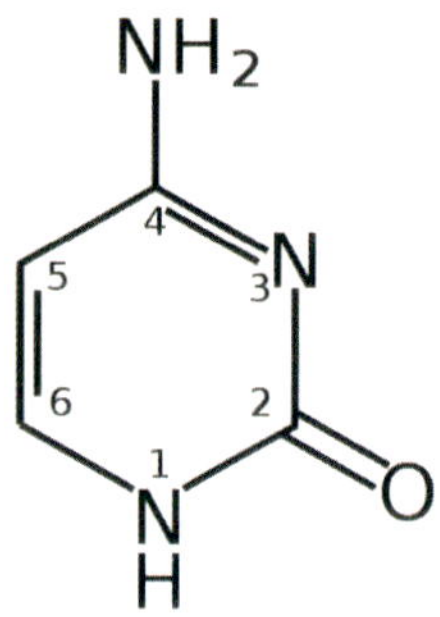

Figure: Cytosine with numbered components. Methylation occurs on carbon number 5

Cytosine can be found as part of DNA, as part of RNA, or as a part of a nucleotide. As cytidine triphosphate (CTP), it can act as a co-factor to enzymes, and can transfer a phosphate to convert adenosine diphosphate (ADP) to adenosine triphosphate (ATP).

In DNA and RNA, cytosine is paired with guanine. However, it is inherently unstable, and can change into uracil (spontaneous deamination). This can lead to a point mutation if not repaired by the DNA repair enzymes such as uracil glycosylase, which cleaves a uracil in DNA.

When found third in a codon of RNA, cytosine is synonymous with uracil, as they are interchangeable as the third base. When found as the second base in a codon, the third is always interchangeable. For example, UCU, UCC, UCA and UCG are all serine, regardless of the third base.

Cytosine can also be methylated into 5-methylcytosine by an enzyme called DNA methyltransferase or be methylated and hydroxylated to make 5-hydroxymethylcytosine. Active enzymatic deamination of cytosine or 5-methylcytosine by the APOBEC family of cytosine deaminases could have both beneficial and detrimental implications on various cellular processes as well as on organismal evolution. The implications of deamination on 5-hydroxymethylcytosine, on the other hand, remains less understood.

Cytosine has not been found in meteorites, which suggests the first strands of RNA and DNA had to look elsewhere to obtain this building block. Cytosine likely formed within some meteorite parent bodies, however did not persist within these bodies due to an effective deamination reaction into uracil.

Adenine

Adenine is one of the nucleobases present in deoxyribonucleic acid (DNA) and ribonucleic acid

(RNA), the genetic information stored within organisms. It is a substance often studied in biochemistry because of its many important roles in the bodies of organisms. It has the chemical formula $C_5H_5N_5$. It is a purine, meaning that it is a kind of organic compound that is composed of carbon and nitrogen atoms arranged in the form of two rings.

DNA and RNA are extraordinarily important nucleic acids because they contain the genetic information used for the growth, repair, development, and reproduction of all organisms. They are each made up of four nucleobases: DNA is composed of adenine, thymine, guanine, and cytosine; RNA is composed of the same, but with uracil instead of thymine. The arrangements of these nucleobases determine the exact nature of the genetic code contained in the DNA or RNA. Adenine is one of these nucleobases, so it is of the utmost importance to the genetic structure of all living organisms.

In DNA, adenine bonds only to thymine. It does so with two strong hydrogen bonds, so the bond is difficult to break and the code is difficult to damage. In RNA, adenine bonds with uracil; the particular kinds of reactions that RNA is involved in favor uracil to thymine. In both cases, the particular arrangement of nucleobases determines the genetic properties of the nucleic acid.

It was initially thought that Adenine was actually vitamin B_4. It is not considered to be a direct part of the B vitamin family anymore, though some B vitamins do bind with it with varying effects. This is most notably true of niacin and riboflavin, which bind with it to form cofactors, which are required for some proteins to function properly.

Adenine is not exclusively found in nucleic acids; many different substances, such as some blends of tea, actually contain the nucleobase. It can also form a variety of compounds that are very common in nature and in some foods and drinks. Cobalamim, more commonly referred to as vitamin B_{12}, is actually a compound of adenine known for its energizing effects and is a natural antidepressant. Adenosine triphosphate (ATP) is another compound that contains adenine; it is known for its role as a major energy source that is derived from cellular respiration. Glucose is broken down into ATP, which is a very significant energy-containing molecule used by a vast variety of organisms.

DNA Sequencing

DNA sequencing is a method used to determine the precise order of the four nucleotide bases – adenine, guanine, cytosine and thymine - that make up a strand of DNA. These bases provide the underlying genetic basis (the genotype) for telling a cell what to do, where to go and what kind of cell to become (the phenotype). Nucleotides are not the only determinants of phenotypes, but are essential to their formation. Each individual and organism has a specific nucleotide base sequence.

Importance

DNA sequencing played a pivotal role in mapping out the human genome, completed in 2003, and is an essential tool for many basic and applied research applications today. It has for example provided an important tool for determining the thousands of nucleotide variations associated with specific genetic diseases, like Huntington's, which may help to better understand these diseases and advance treatment.

DNA sequencing also underpins pharmacogenomics. This is a relatively new field which is leading the way to more personalized medicine. Pharmacogenomics looks at how a person's individual genome variations affect their response to a drug. Such data is being used to determine which drug gives the best outcome in particular patients. Over 140 drugs approved by the FDA now include pharmacogenomics information in their labeling. Such labeling is not only important in terms of matching patients to their most appropriate drug, but also for working out what their drug dose should be and their level of risk in terms of adverse events. Individual genetic profiling is already being used routinely to prescribe therapies for patients with HIV, breast cancer, lymphoblastic leukemia and colon cancer and in the future will be used to tailor treatments for cardiovascular disease, cancer, asthma, Alzheimer's disease and depression. Drug developers are also using pharmacogenomics data to design drugs which can be targeted at subgroups of patients with specific genetic profiles.

First-generation Sequencing Technology

So-called first-generation sequencing technologies, which emerged in the 1970s, included the Maxam-Gilbert method, discovered by and named for American molecular biologists Allan M. Maxam and Walter Gilbert, and the Sanger method (or dideoxy method), discovered by English biochemist Frederick Sanger. In the Sanger method, which became the more commonly employed of the two approaches, DNA chains were synthesized on a template strand, but chain growth was stopped when one of four possible dideoxy nucleotides, which lack a 3' hydroxyl group, became incorporated, thereby preventing the addition of another nucleotide. A population of nested, truncated DNA molecules was produced that represented each of the sites of that particular nucleotide in the template DNA. The molecules were separated according to size in a procedure called electrophoresis, and the inferred nucleotide sequence was deduced by a computer. Later, the method was performed by using automated sequencing machines, in which the truncated DNA molecules, labeled with fluorescent tags, were separated by size within thin glass capillaries and detected by laser excitation.

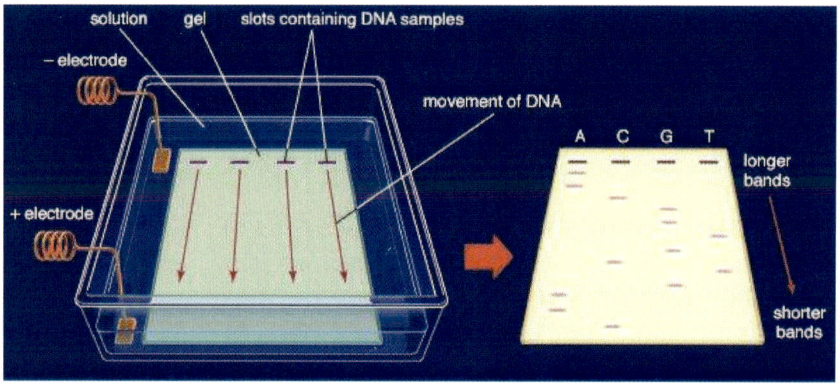

Figure: Gel Electrophoresis

In gel electrophoresis, an electric field is applied to a buffer solution covering an agarose gel, which has slots at one end containing DNA samples. The negatively charged DNA molecules travel through the gel toward a positive electrode and are separated based on size as they advance.

Next-generation Sequencing Technology

Next-generation (massively parallel, or second-generation) sequencing technologies have largely supplanted first-generation technologies. These newer approaches enable many DNA fragments (sometimes on the order of millions of fragments) to be sequenced at one time and are more cost-efficient and much faster than first-generation technologies. The utility of next-generation technologies was improved significantly by advances in bioinformatics that allowed for increased data storage and facilitated the analysis and manipulation of very large data sets, often in the giga-base range (1 gigabase = 1,000,000,000 base pairs of DNA).

Application

DNA sequencing provides the means to know how nucleotide bases are arranged in a piece of DNA. Several methods have been developed for this process. These have four key steps. In the first instance DNA is removed from the cell. This can be done either mechanically or chemically. The second phase involves breaking up the DNA and inserting its pieces into vectors, cells that indefinitely self-replicate, for cloning. In the third phase the DNA clones are placed with a dye-labeled primer (a short stretch of DNA that promotes replication) into a thermal cycler, a machine which automatically raises and lowers the temperature to catalyze replication. The final phase consists of electrophoresis, whereby the DNA segments are placed in a gel and subjected to an electrical current which moves them. Originally the gel was placed on a slab, but today it is inserted into a very thin glass tube known as a capillary. When subjected to an electrical current the smaller nucleotides in the DNA move faster than the larger ones. Electrophoresis thus helps sort out the DNA fragments by their size. The different nucleotide bases in the DNA fragments are identified by their dyes which are activated when they pass through a laser beam. All the information is fed into a computer and the DNA sequence displayed on a screen for analysis.

The method developed by Sanger was pivotal to the international Human Genome Project. Costing over US$3 billion and taking 13 years to complete, this project provided the first complete Human DNA sequence in 2003. Data from the project provided the first means to map out the genetic mutations that underlie specific genetic diseases. It also opened up a path to more personalized medicine, enabling scientists to examine the extent to which a patient's response to a drug is determined by their genetic profile. The genetic profile of a patient's tumor, for example, can now be used to work out what is the most effective treatment for an individual. It is also hoped that in the future that knowing the sequence of a person's genome will help work out a person's predisposition to certain diseases, such as heart disease, cancer and type II diabetes, which could pave the way to better preventative care.

Data from the Human Genome Project has also helped fuel the development of gene therapy, a type of treatment designed to replace defective genes in certain genetic disorders. In addition, it has provided a means to design drugs that can target specific genes that cause disease.

Beyond medicine, DNA sequencing is now used for genetic testing for paternity and other family relationships. It also helps identify crime suspects and victims involved in catastrophes. The technique is also vital to detecting bacteria and other organisms that may pollute air, water, soil and food. In addition the method is important to the study of the evolution of different population groups and their migratory patterns as well as determining pedigree for seed or livestock.

Importance

DNA sequencing played a pivotal role in mapping out the human genome, completed in 2003, and is an essential tool for many basic and applied research applications today. It has for example provided an important tool for determining the thousands of nucleotide variations associated with specific genetic diseases, like Huntington's, which may help to better understand these diseases and advance treatment.

DNA sequencing also underpins pharmacogenomics. This is a relatively new field which is leading the way to more personalized medicine. Pharmacogenomics looks at how a person's individual genome variations affect their response to a drug. Such data is being used to determine which drug gives the best outcome in particular patients. Over 140 drugs approved by the FDA now include pharmacogenomics information in their labeling. Such labeling is not only important in terms of matching patients to their most appropriate drug, but also for working out what their drug dose should be and their level of risk in terms of adverse events. Individual genetic profiling is already being used routinely to prescribe therapies for patients with HIV, breast cancer, lymphoblastic leukemia and colon cancer and in the future will be used to tailor treatments for cardiovascular disease, cancer, asthma, Alzheimer's disease and depression. Drug developers are also using pharmacogenomics data to design drugs, which can be targeted at subgroups of patients with specific genetic profiles.

DNA Damage

Damage to cellular DNA is involved in mutagenesis and the development of cancer. The DNA in a human cell undergoes several thousand to a million damaging events per day, generated by both external (exogenous) and internal metabolic (endogenous) processes. Changes to the cellular genome can generate errors in the transcription of DNA and ensuing translation into proteins necessary for signaling and cellular function. Genomic mutations can also be carried over into daughter generations of cells if the mutation is not repaired prior to mitosis. Once cells lose their ability to effectively repair damaged DNA, there are three possible responses:

1. The cell may become senescent, i.e., irreversibly dormant. In 2005, multiple laboratories reported that senescence could occur in cancer cells in vivo as well as in vitro, stopping mitosis and preventing the cell from evolving further.
2. The cell may become apoptotic. Sufficient DNA damage may trigger an apoptotic signaling cascade, forcing the cell into programmed cell death.
3. The cell may become malignant, i.e., develop immortal characteristics and begin uncontrolled division.

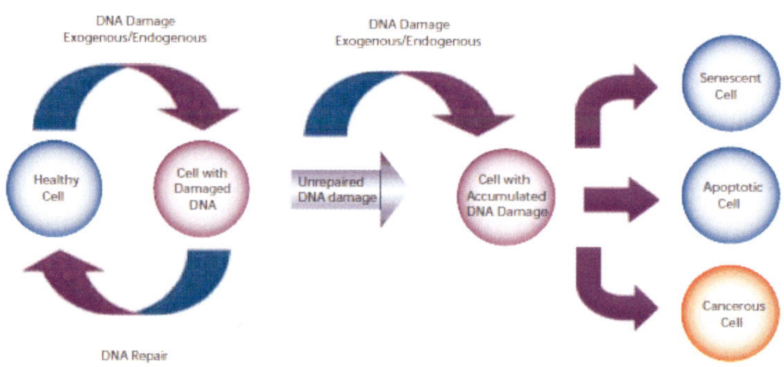

Figure: The pathway of cellular DNA damage and repair that leads to senescence, apoptosis, or cancer

To compensate for the degree and types of DNA damage that occur, cells have developed multiple repair processes including mismatch, base excision, and nucleotide excision repair mechanisms, with little process redundancy. Cells may have evolved to proceed into apoptosis or senescence if overwhelming damage occurs rather than expend energy to effectively repair the damage. The rate at which a cell is able to make repairs is contingent on factors including cell type and cell age.

Sources of DNA Damage

For many years, exogenous sources of damage have been thought to be the primary cause of DNA mutations leading to cancer. However, Jackson and Loeb proposed that endogenous sources of DNA damage also contribute significantly to mutations that lead to malignancy. Both environmental and cellular sources can result in similar types of DNA damage.

DNA can be attacked by physical and chemical mutagens. Physical mutagens are primarily radiation sources, including UV (200-300 nm wavelength) radiation from the sun. UV radiation produces covalent bonds that crosslink adjacent pyrimidine (cytosine and thymine) bases in the DNA strand. Ionizing radiation (X-rays) initiates DNA mutations by generating free radicals within the cell that create reactive oxygen species (ROS) and result in single-strand and double-strand breaks in the double helix. Chemical mutagens can attach alkyl groups covalently to DNA bases; nitrogen mustard compounds that can methylate or ethylate the DNA base are examples of DNA alkylating agents. Procarcinogens are chemically inert precursors that are metabolically converted into highly reactive carcinogens. These carcinogens can react with DNA by forming DNA adducts, i.e., chemical entities attached to DNA. Benzo[a]pyrene, a polyaromatic heterocycle, is not itself carcinogenic. It undergoes two sequential oxidation reactions mediated by cytochrome P450 enzymes, which results in benzo[a]pyrenediol epoxide (BPDE), the carcinogenic metabolite that is able to form a covalent DNA adduct.

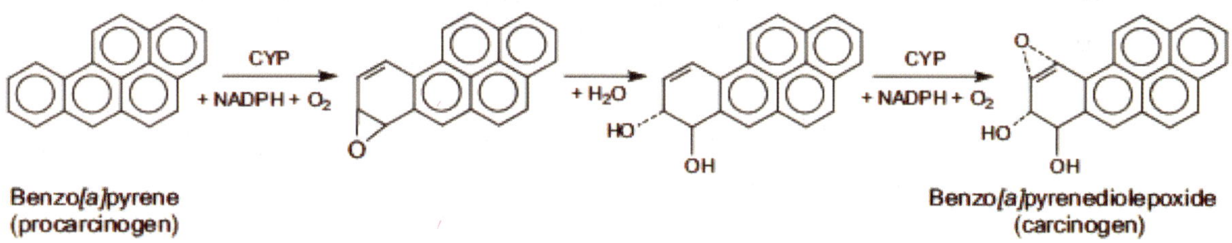

Figure: Benzo[a]pyrene is oxidized by P450 enzymes to create the highly carcinogenic benzo[a]pyrenediolepoxide

DNA damage can also result from endogenous metabolic and biochemical reactions, some of which are not well understood. Hydrolysis reactions can partially or completely cleave the nucleotide base from the DNA strand. The chemical bond connecting a purine base (adenine or guanine) to the deoxyribosyl phosphate chain can spontaneously break in the process known as depurination. An estimated 10,000 depurination events occur per day in a mammalian cell. Depyrimidination (loss of pyrimidine base from thymine or cytosine) also occurs, but at a rate 20 to 100-fold lower than depurination.

Deamination occurs within the cell with the loss of amine groups from adenine, guanine, and cytosine rings, resulting in hypoxanthine, xanthine, and uracil, respectively. DNA repair enzymes are able to recognize and correct these unnatural bases. However, an uncorrected uracil base may be misread as a thymine during subsequent DNA replication and generate a C→T point mutation.

DNA methylation, a specific form of alkylation, occurs within the cell due to a reaction with S-adenosyl methionine (SAM). SAM is an intracellular metabolic intermediate that contains a highly reactive methyl group. In mammalian cells, methylation occurs at the 5-position of the cytosine ring of a cytidine base (C) that is 5' to a guanosine base (G), i.e., sequence CpG. A significant source of mutation error is the spontaneous deamination of the 5-methylcytosine product of methyl-ation. Loss of the amine group results in a thymine base, which is not detected by DNA repair enzymes as an unnatural base. The resulting substitution is retained in DNA replication, creating a C→T point mutation.

Normal metabolic processes generate reactive oxygen species (ROS), which modify bases by oxidation. Both purine and pyrimidine bases are subject to oxidation. The most common mutation is guanine oxidized to 8-oxo-7,8-dihydroguanine, resulting in the nucleotide 8-oxo-deoxy guano sine (8-oxo-dG). The 8-oxo-dG is capable of base pairing with deoxyadenosine, instead of pairing with deoxycytotidine as expected. If this error is not detected and corrected by mismatch repair enzymes, the DNA subsequently replicated will contain a C→A point mutation. ROS may also cause depurination, depyrimidination, and single-strand or double strand breaks in the DNA.

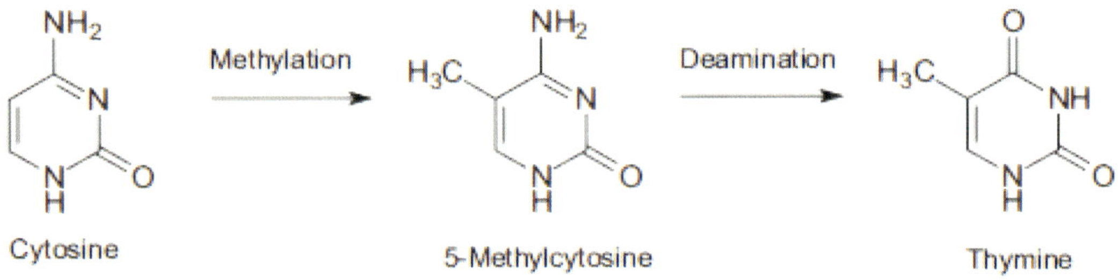

Figure: The 2-phase mutation of cytosine results in thymine, creating a C→T point mutation

Other genomic mutations may be introduced during DNA replication in the S phase of the cell cycle. Polymerases that duplicate template DNA have a small but significant error rate, and may incorporate an incorrect nucleotide based on Watson-Crick pairing versus the template DNA. Chemically altered nucleotide precursors may be incorporated into the generated DNA by the polymerase, instead of normal bases. In addition, polymerases are prone to "stuttering" when copying sections of DNA that contain a large number of repeating nucleotides or repeating sequences (microsatellite regions). This enzymatic "stuttering" is due to a strand slippage, when the template and replicated strands of DNA slip out of proper alignment. As a result, the polymerase fails to insert the correct number of nucleotides indicated by the template DNA, resulting in too few or too many nucleotides in the daughter strand.

Single strand and double strand cleavage of the DNA may occur. Single strand breaks may result from damage to the deoxyribose moiety of the DNA deoxyribosylphosphate chain. Breaks also result as an intermediate step of the base excision repair pathway after the removal of deoxyribose phosphate by AP-endonuclease 1. When a single strand break occurs, both the nucleotide base and the deoxyribose backbone are lost from the DNA structure. Double strand cleavage most often occurs when the cell is passing through S-phase, as the DNA may be more susceptible to breakage while it is unraveling for use as a template for replication.

DNA Repair

DNA, like any other molecule, can undergo a variety of chemical reactions. Because DNA uniquely serves as a permanent copy of the cell genome, however, changes in its structure are of much greater consequence than are alterations in other cell components, such as RNAs or proteins. Mutations can result from the incorporation of incorrect bases during DNA replication. In addition, various chemical changes occur in DNA either spontaneously or as a result of exposure to chemicals or radiation. Such damage to DNA can block replication or transcription, and can result in a high frequency of mutations—consequences that are unacceptable from the standpoint of cell reproduction. To maintain the integrity of their genomes, cells have therefore had to evolve mechanisms to repair damaged DNA. These mechanisms of DNA repair can be divided into two general classes:

1. Direct reversal of the chemical reaction responsible for DNA damage, and
2. Removal of the damaged bases followed by their replacement with newly synthesized DNA. Where DNA repair fails, additional mechanisms have evolved to enable cells to cope with the damage.

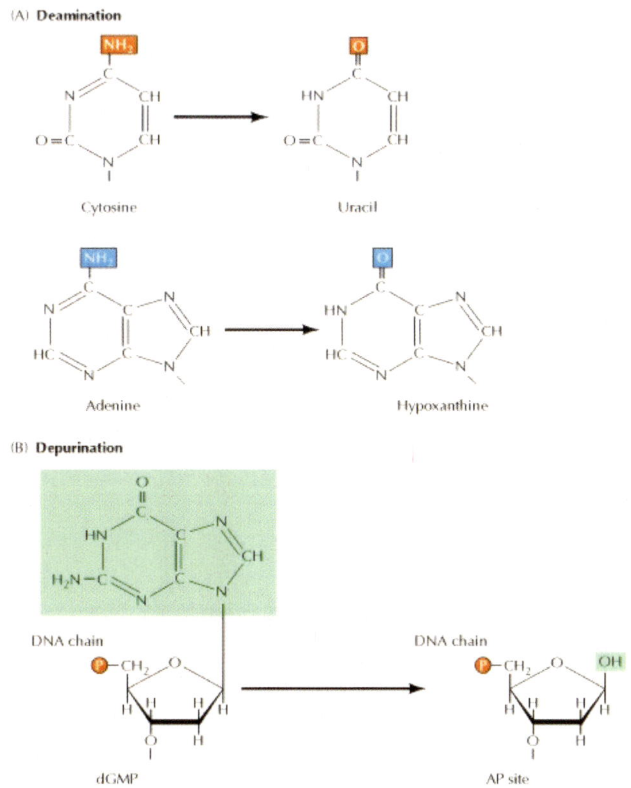

Figure: Spontaneous damage to DNA

There are two major forms of spontaneous DNA damage: (A) deamination of adenine, cytosine, and guanine, and (B) depurination (loss of purine bases) resulting from cleavage of the bond between the purine bases and deoxyribose, leaving an apurinic (AP) site in DNA. dGMP = deoxyguanosine monophosphate.

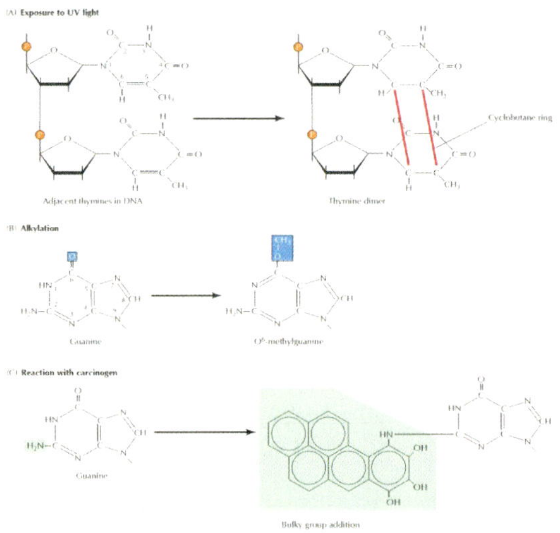

Figure: Examples of DNA damage induced by radiation and chemicals

(A) UV light induces the formation of pyrimidine dimers, in which two adjacent pyrimidines (e.g., thymines) are joined by a cyclobutane ring structure. (B) Alkylation is the addition of methyl or ethyl groups to various positions on the DNA bases. In this example, alkylation of the O^6 position of guanine results in formation of O^6-methylguanine. (C) Many carcinogens (e.g., benzo-(*a*)pyrene) react with DNA bases, resulting in the addition of large bulky chemical groups to the DNA molecule.

Direct Reversal of DNA Damage

Most damage to DNA is repaired by removal of the damaged bases followed by resynthesis of the excised region. Some lesions in DNA, however, can be repaired by direct reversal of the damage, which may be a more efficient way of dealing with specific types of DNA damage that occur frequently. Only a few types of DNA damage are repaired in this way, particularly pyrimidine dimers resulting from exposure to ultraviolet (UV) light and alkylated guanine residues that have been modified by the addition of methyl or ethyl groups at the O^6 position of the purine ring.

UV light is one of the major sources of damage to DNA and is also the most thoroughly studied form of DNA damage in terms of repair mechanisms. Its importance is illustrated by the fact that exposure to solar UV irradiation is the cause of almost all skin cancer in humans. The major type of damage induced by UV light is the formation of pyrimidine dimers, in which adjacent pyrimidines on the same strand of DNA are joined by the formation of a cyclobutane ring resulting from saturation of the double bonds between carbons 5 and 6. The formation of such dimers distorts the structure of the DNA chain and blocks transcription or replication past the site of damage, so their repair is closely correlated with the ability of cells to survive UV irradiation. One mechanism of repairing UV-induced pyrimidine dimers is direct reversal of the dimerization reaction. The process is called photoreactivation because energy derived from visible light is utilized to break the cyclobutane ring structure. The original pyrimidine bases remain in DNA, now restored to their normal state. As might be expected from the fact that solar UV irradiation is a major source of DNA damage for diverse cell types, the repair of pyrimidine dimers by photoreactivation is common to a variety of prokaryotic and eukaryotic cells, including *E. coli*, yeasts, and some species of plants and animals. Curiously, however, photoreactivation is not universal; many species (including humans) lack this mechanism of DNA repair.

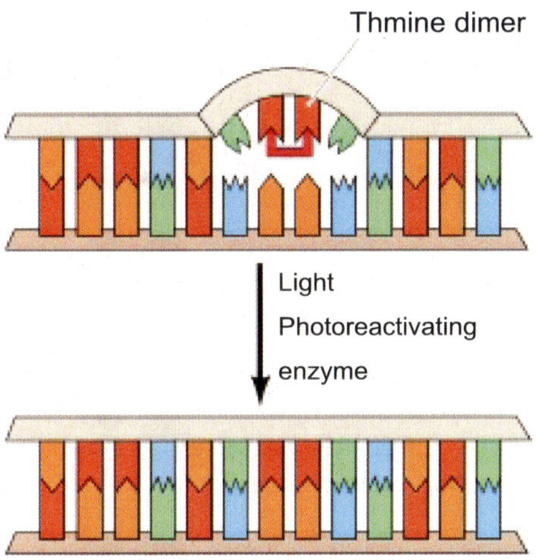

Figure: Direct repair of thymine dimers

UV-induced thymine dimers can be repaired by photoreactivation, in which energy from visible light is used to split the bonds forming the cyclobutane ring.

Another form of direct repair deals with damage resulting from the reaction between alkylating agents and DNA. Alkylating agents are reactive compounds that can transfer methyl or ethyl groups to a DNA base, thereby chemically modifying the base. A particularly important type of damage is methylation of the O^6 position of guanine, because the product, O^6-methylguanine, forms complementary base pairs with thymine instead of cytosine. This lesion can be repaired by an enzyme (called O^6-methylguanine methyltransferase) that transfers the methyl group from O^6-methylguanine to a cysteine residue in its active site. The potentially mutagenic chemical modification is thus removed, and the original guanine is restored. Enzymes that catalyze this direct repair reaction are widespread in both prokaryotes and eukaryotes, including humans.

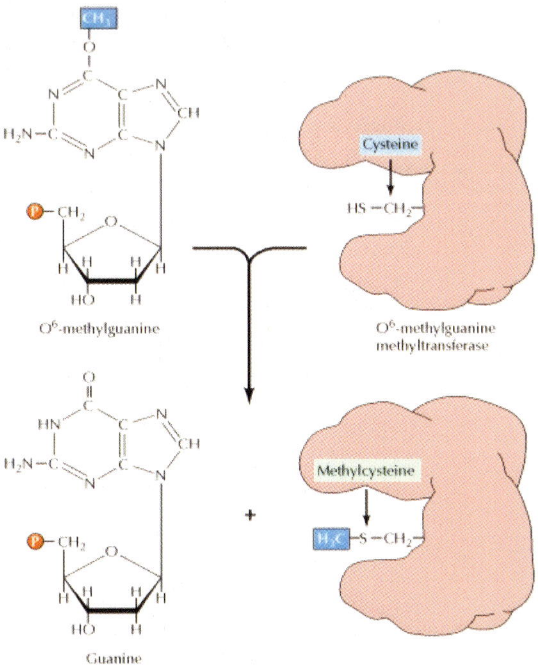

Figure: Repair of O^6-methylguanine

O^6-methylguanine methyltransferase transfers the methyl group from O^6-methylguanine to a cysteine residue in the enzyme's active site.

Excision Repair

Although direct repair is an efficient way of dealing with particular types of DNA damage, excision repair is a more general means of repairing a wide variety of chemical alterations to DNA. Consequently, the various types of excision repair are the most important DNA repair mechanisms in both prokaryotic and eukaryotic cells. In excision repair, the damaged DNA is recognized and removed, either as free bases or as nucleotides. The resulting gap is then filled in by synthesis of a new DNA strand, using the undamaged complementary strand as a template. Three types of excision repair—base-excision repair, nucleotide-excision repair, and mismatch repair—enable cells to cope with a variety of different kinds of DNA damage.

The repair of uracil-containing DNA is a good example of base-excision repair, in which single damaged bases are recognized and removed from the DNA molecule. Uracil can arise in DNA by two mechanisms:

1. Uracil (as dUTP [deoxyuridine triphosphate]) is occasionally incorporated in place of thymine during DNA synthesis, and
2. Uracil can be formed in DNA by the deamination of cytosine. The second mechanism is of much greater biological significance because it alters the normal pattern of complementary base pairing and thus represents a mutagenic event. The excision of uracil in DNA is catalyzed by DNA glycosylase, an enzyme that cleaves the bond linking the base (uracil) to the deoxyribose of the DNA backbone. This reaction yields free uracil and an apyrimidinic site—a sugar with no base attached. DNA glycosylases also recognize and remove other abnormal bases, including hypoxanthine formed by the deamination of adenine, pyrimidine dimers, alkylated purines other than O^6-alkylguanine, and bases damaged by oxidation or ionizing radiation.

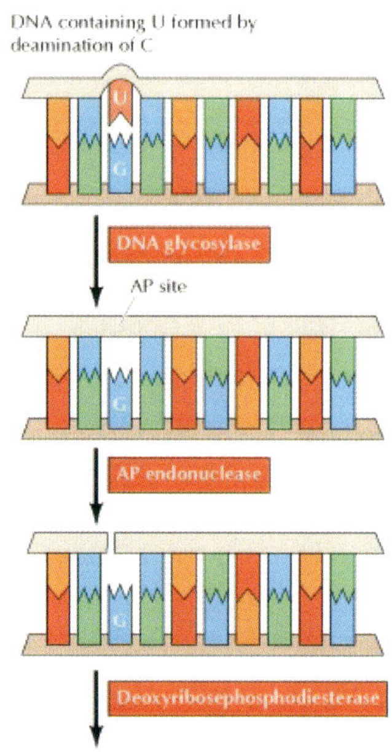

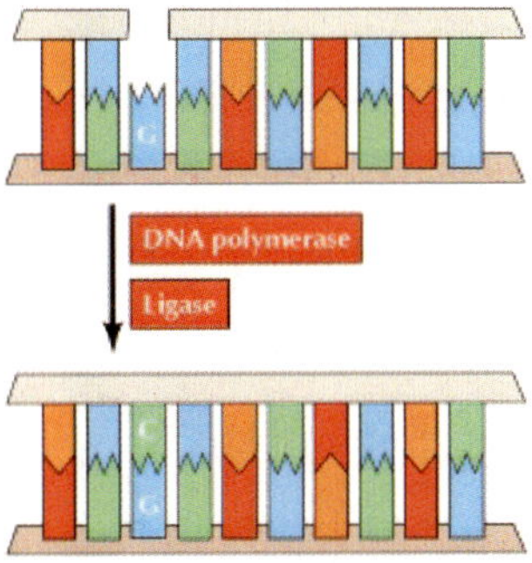

Figure: Base-excision repair

In this example, uracil (U) has been formed by deamination of cytosine (C) and is therefore opposite a guanine (G) in the complementary strand of DNA. The bond between uracil and the deoxyribose is cleaved by a DNA glycosylase, leaving a sugar with no base attached in the DNA (an AP site). This site is recognized by AP endonuclease, which cleaves the DNA chain. The remaining deoxyribose is removed by deoxyribosephosphodiesterase. The resulting gap is then filled by DNA polymerase and sealed by ligase, leading to incorporation of the correct base (C) opposite the G.

The result of DNA glycosylase action is the formation of an apyridiminic or apurinic site (generally called an AP site) in DNA. Similar AP sites are formed as the result of the spontaneous loss of purine bases, which occurs at a significant rate under normal cellular conditions. For example, each cell in the human body is estimated to lose several thousand purine bases daily. These sites are repaired by AP endonuclease, which cleaves adjacent to the AP site. The remaining deoxyribose moiety is then removed, and the resulting single-base gap is filled by DNA polymerase and ligase.

Whereas DNA glycosylases recognize only specific forms of damaged bases, other excision repair systems recognize a wide variety of damaged bases that distort the DNA molecule, including UV-induced pyrimidine dimers and bulky groups added to DNA bases as a result of the reaction of many carcinogens with DNA. This widespread form of DNA repair is known as nucleotide-excision repair, because the damaged bases (e.g., a thymine dimer) are removed as part of an oligonucleotide containing the lesion.

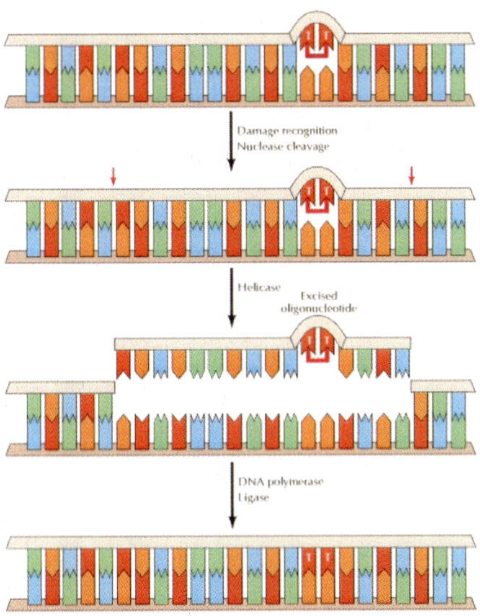

Figure: Nucleotide-excision repair of thymine dimers

Damaged DNA is recognized and then cleaved on both sides of a thymine dimer by 3′ and 5′ nucleases. Unwinding by a helicase results in excision of an oligonucleotide containing the damaged bases. The resulting gap is then filled by DNA polymerase and sealed by ligase.

In *E. coli*, nucleotide-excision repair is catalyzed by the products of three genes (*uvrA*, *B*, and *C*) that were identified because mutations at these loci result in extreme sensitivity to UV light. The protein UvrA recognizes damaged DNA and recruits UvrB and UvrC to the site of the lesion. UvrB and UvrC then cleave on the 3′ and 5′ sides of the damaged site, respectively, thus excising an oligonucleotide consisting of 12 or 13 bases. The UvrABC complex is frequently called an excinuclease, a name that reflects its ability to directly *excis*e an oligonucleotide. The action of a helicase is then required to remove the damage-containing oligonucleotide from the double-stranded DNA molecule, and the resulting gap is filled by DNA polymerase I and sealed by ligase.

Nucleotide-excision repair systems have also been studied extensively in eukaryotes, particularly in yeasts and in humans. In yeasts, as in *E. coli*, several genes involved in DNA repair (called *RAD* genes for *rad*iation sensitivity) have been identified by the isolation of mutants with increased sensitivity to UV light. In humans, DNA repair genes have been identified largely by studies of individuals suffering from inherited diseases resulting from deficiencies in the ability to repair DNA damage. The most extensively studied of these diseases is xeroderma pigmentosum (XP), a rare genetic disorder that affects approximately one in 250,000 people. Individuals with this disease are extremely sensitive to UV light and develop multiple skin cancers on the regions of their bodies that are exposed to sunlight. In 1968, James Cleaver made the key discovery that cultured cells from XP patients were deficient in the ability to carry out nucleotide-excision repair. This observation not only provided the first link between DNA repair and cancer, but also suggested the use of XP cells as an experimental system to identify human DNA repair genes. The identification of human DNA repair genes has been accomplished by studies not only of XP cells, but also of two other human diseases resulting from DNA repair defects (Cockayne's syndrome and trichothiodystrophy) and of UV-sensitive mutants of rodent cell lines. The availability of mammalian cells with defects in DNA repair has allowed the cloning of repair genes based on the ability of wild-type alleles to restore normal UV sensitivity to mutant cells in gene transfer assays, thereby opening the door to experimental analysis of nucleotide-excision repair in mammalian cells.

Molecular cloning has now identified seven different repair genes (designated *XPA* through *XPG*) that are mutated in cases of xeroderma pigmentosum, as well as in some cases of Cockayne's syndrome, trichothiodystrophy, and UV-sensitive mutants of rodent cells.Table given below lists the enzymes encoded by these genes. Some UV-sensitive rodent cells have mutations in yet another repair gene, called *ERCC1* (for *e*xcision *r*epair *c*ross *c*omplementing), which has not been found to be mutated in known human diseases. It is notable that the proteins encoded by these human DNA repair genes are closely related to proteins encoded by yeast *RAD* genes, indicating that nucleotide-excision repair is highly conserved throughout eukaryotes.

Human	Yeast	Function
XPA	RAD14	Damage recognition
XPB	RAD25	Helicase
XPC	RAD4	DNA binding
XPD	RAD3	Helicase

Human	Yeast	Function
XPF	RAD1	5′ nuclease
XPG	RAD2	3′ nuclease
ERCC1	RAD10	Dimer with XPF

Table: Enzymes involved in nucleotide-excision repair

With cloned yeast and human repair genes available, it has been possible to purify their encoded proteins and develop *in vitro* systems to study the repair process. Although some steps remain to be fully elucidated, these studies have led to the development of a basic model for nucleotide-excision repair in eukaryotic cells. In mammalian cells, the XPA protein (and possibly also XPC) initiates repair by recognizing damaged DNA and forming complexes with other proteins involved in the repair process. These include the XPB and XPD proteins, which act as helicases that unwind the damaged DNA. In addition, the binding of XPA to damaged DNA leads to the recruitment of XPF (as a heterodimer with ERCC1) and XPG to the repair complex. XPF/ERCC1 and XPG are endonucleases, which cleave DNA on the 5′ and 3′ sides of the damaged site, respectively. This cleavage excises an oligonucleotide consisting of approximately 30 bases. The resulting gap then appears to be filled in by DNA polymerase δ or ε (in association with RFC and PCNA) and sealed by ligase.

An intriguing feature of nucleotide-excision repair is its relationship to transcription. A connection between transcription and repair was first suggested by experiments showing that transcribed strands of DNA are repaired more rapidly than nontranscribed strands in both *E. coli* and mammalian cells. Since DNA damage blocks transcription, this transcription-repair coupling is thought to be advantageous by allowing the cell to preferentially repair damage to actively expressed genes. In *E. coli*, the mechanism of transcription-repair coupling involves recognition of RNA polymerase stalled at a lesion in the DNA strand being transcribed. The stalled RNA polymerase is recognized by a protein called transcription-repair coupling factor, which displaces RNA polymerase and recruits the UvrABC excinuclease to the site of damage.

Although the molecular mechanism of transcription-repair coupling in mammalian cells is not yet known, it is noteworthy that the XPB and XPD helicases are components of a multisubunit transcription factor (called TFIIH) that is required to initiate the transcription of eukaryotic genes. Thus, these helicases appear to be required for the unwinding of DNA during both transcription and nucleotide-excision repair, providing a direct biochemical link between these two processes. Patients suffering from Cockayne's syndrome are also characterized from a failure to preferentially repair transcribed DNA strands, suggesting that the proteins encoded by the two genes known to be responsible for this disease (*CSA* and *CSB*) function in transcription-coupled repair. In addition, one of the genes responsible for inherited breast cancer in humans (*BRCA1*) appears to encode a protein specifically involved in transcription-coupled repair of oxidative DNA damage, suggesting that defects in this type of DNA repair can lead to the development of one of the most common cancers in women.

A third excision repair system recognizes mismatched bases that are incorporated during DNA replication. Many such mismatched bases are removed by the proofreading activity of DNA polymerase. The ones that are missed are subject to later correction by the mismatch repair system, which scans newly replicated DNA. If a mismatch is found, the enzymes of this repair system are able to identify and excise the mismatched base specifically from the newly replicated DNA strand, allowing the error to be corrected and the original sequence restored.

In *E. coli*, the ability of the mismatch repair system to distinguish between parental DNA and newly synthesized DNA is based on the fact that DNA of this bacterium is modified by the methylation of adenine residues within the sequence GATC to form 6-methyladenine. Since methylation occurs after replication, newly synthesized DNA strands are not methylated and thus can be specifically recognized by the mismatch repair enzymes. Mismatch repair is initiated by the protein MutS, which recognizes the mismatch and forms a complex with two other proteins called MutL and MutH. The MutH endonuclease then cleaves the unmethylated DNA strand at a GATC sequence. MutL and MutS then act together with an exonuclease and a helicase to excise the DNA between the strand break and the mismatch, with the resulting gap being filled by DNA polymerase and ligase.

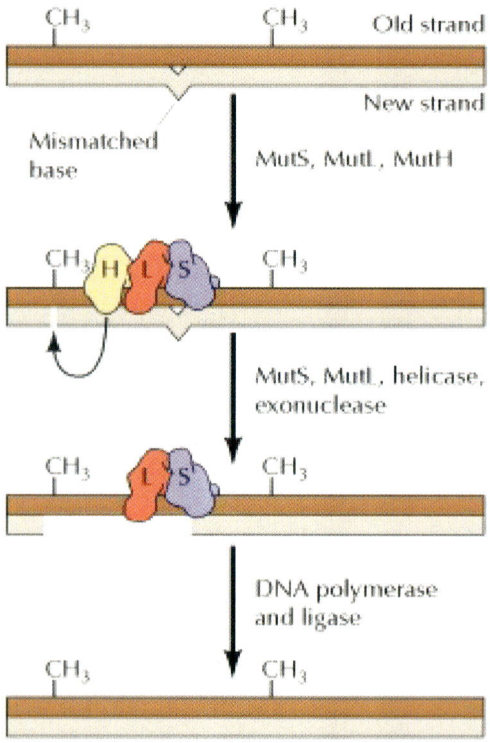

Figure: Mismatch repair in *E. coli*

The mismatch repair system detects and excises mismatched bases in newly replicated DNA, which is distinguished from the parental strand because it has not yet been methylated. MutS binds to the mismatched base, followed by MutL. The binding of MutL activates MutH, which cleaves the unmodified strand opposite a site of methylation. MutS and MutL, together with a helicaseand an exonuclease, then excise the portion of the unmodified strand that contains the mismatch. The gap is then filled by DNA polymerase and sealed by ligase.

Eukaryotes have a similar mismatch repair system, although the mechanism by which eukaryotic cells identify newly replicated DNA differs from that used by *E. coli*. In mammalian cells, it appears that the strand-specificity of mismatch repair is determined by the presence of single-strand breaks (which would be present in newly replicated DNA) in the strand to be repaired. The eukaryotic homologs of MutS and MutL then bind to the mismatched base and direct excision of the DNA between the strand break and the mismatch, as in *E. coli*. The importance of this repair system is dramatically illustrated by the fact that mutations in the human homologs of *MutS*and *MutL* are responsible for a common type of inherited colon cancer (hereditary nonpolyposis colorectal cancer, or HNPCC). HNPCC is one of the most common inherited diseases; it affects as many as one in 200 people and is responsible for about 15% of all colorectal cancers in this country. The relationship between HNPCC and defects in mismatch repair was discovered in 1993, when two groups of researchers cloned the human homolog of *MutS* and found that mutations in this gene were responsible for about half of all HNPCC cases. Subsequent studies have shown that most of the remaining cases of HNPCC are caused by mutations in one of three human genes that are homologs of *MutL*.

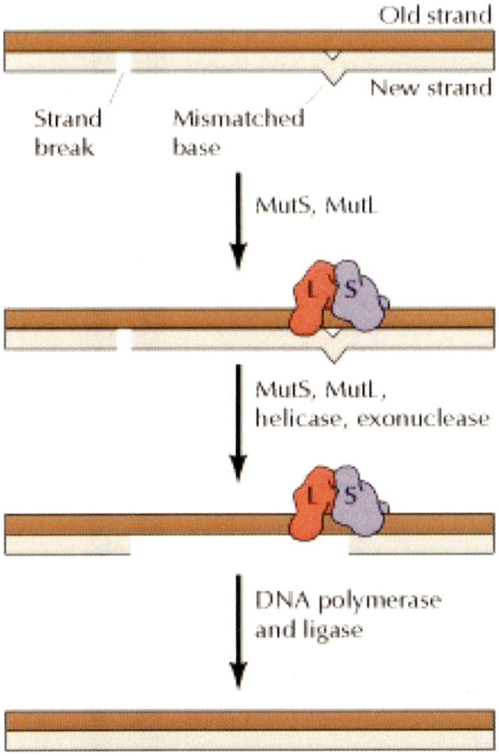

Figure: Mismatch repair in mammalian cells

Mismatch repair in mammalian cells is similar to E. coli, except that the newly replicated strand is distinguished from the parental strand because it contains strand breaks. MutS and MutL bind to the mismatched base and direct excision of the DNA between the strand break and the mismatch.

Postreplication Repair

The direct reversal and excision repair systems act to correct DNA damage before replication, so that replicative DNA synthesis can proceed using an undamaged DNA strand as a template. Should these systems fail, however, the cell has alternative mechanisms for dealing with damaged DNA at the replication fork. Pyrimidine dimers and many other types of lesions cannot be copied by the normal action of DNA polymerases, so replication is blocked at the sites of such damage. Downstream of the damaged site, however, replication can be initiated again by the synthesis of an Okazaki fragment and can proceed along the damaged template strand. The result is a daughter strand that has a gap opposite the site of damage to the parental strand. One of two types of mechanisms may be used to repair such gaps in newly synthesized DNA: recombinational repair or error-prone repair.

The presence of a thymine dimer blocks replication, but DNA polymerase can bypass the lesion and reinitiate replication at a new site downstream of the dimer. The result is a gap opposite the dimer in the newly synthesized DNA strand. In recombinational repair, this gap is filled by recombination with the undamaged parental strand. Although this leaves a gap in the previously intact parental strand, the gap can be filled by the actions of polymerase and ligase, using the intact daughter strand as a template. Two intact DNA molecules are thus formed, and the remaining thymine dimer eventually can be removed by excision repair.

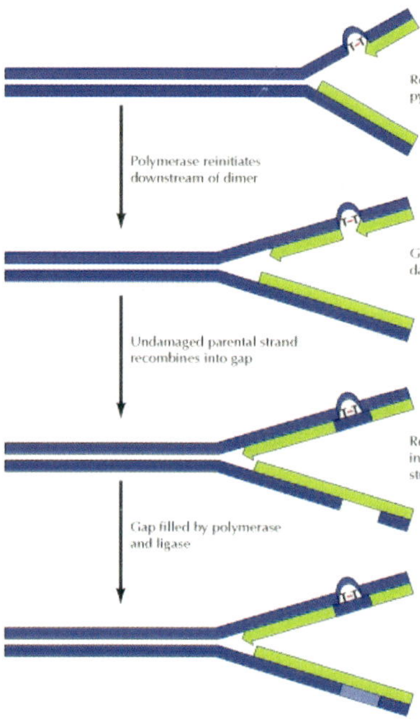

Figure: Postreplication repair

Recombinational repair depends on the fact that one strand of the parental DNA was undamaged and therefore was copied during replication to yield a normal daughter molecule. The undamaged parental strand can be used to fill the gap opposite the site of damage in the other daughter molecule by recombination between homologous DNA sequences. Because the resulting gap in the previously intact parental strand is opposite an undamaged strand, it can be filled in by DNA polymerase. Although the other parent molecule still retains the original damage (e.g., a pyrimidine dimer), the damage now lies opposite a normal strand and can be dealt with later by excision repair. By a similar mechanism, recombination with an intact DNA molecule can be used to repair double strand breaks, which are frequently introduced into DNA by radiation and other damaging agents.

In error-prone repair, a gap opposite a site of DNA damage is filled by newly synthesized DNA. Since the new DNA is synthesized from a damaged template strand, this form of DNA synthesis is very inaccurate and leads to frequent mutations. It is used only in bacteria that have been subjected to potentially lethal conditions, such as extensive UV irradiation. Such treatments induce the SOS response, which may be viewed as a mechanism for dealing with extreme environmental stress. The SOS response includes inhibition of cell division and induction of repair systems to cope with a high level of DNA damage. Under these conditions, error-prone repair mechanisms are used, presumably as a way of dealing with damage so extensive that cell death is the only alternative.

DNA Profiling

The technique of DNA profiling was developed by Alec Jefferys in the mid-1980s and is based on the analysis of markers in DNA known as microsatellites or short tandem repeats (STRs). These markers are found at specific points (also called loci) in everyone's DNA and they're motifs of two-six bases (the units that make up our genes) that are repeated numerous times. The exact number of times these markers are repeated differs between individuals, but members of a family will share the same or a similar number of repeated markers, depending on how closely related they are.

Therefore, when the markers in two samples are analyzed, the number of times that they're repeated can be compared and the statistical likelihood that they came from the same person or from two closely related individuals can be calculated. This is why DNA profiling can be used to establish biological relationships, as well as to connect DNA evidence with a criminal suspect.

Something to be aware of if you're considering taking a DNA profiling test is that this isn't the same as whole genome sequencing (aka DNA sequencing). A DNA profile will only report on the number of STRs that you possess at certain loci in your DNA, so that it can be compared to a second DNA profile to prove or disprove a match.

Whole genome sequencing is the most accurate representation of your DNA that you can buy, as it provides you with the details of every single base in your DNA (more than 5 billion). This is unsurprisingly much more expensive than DNA profiling, and isn't necessary if you are looking for a profile for identification purposes.

Uses of DNA Profiling

A DNA profile or fingerprint represents a small proportion of a person's overall DNA, but it's enough for two profiles to be compared to prove or disprove that they came from the same person (or from related persons). Therefore, DNA profiles are commonly used for DNA identification.

A DNA profile can also be used in posthumous disputes, inheritance issues for example. One of the reasons for this is that DNA is much more difficult to forge than other forms of identification, and the coded information it contains is highly resilient.

In addition, because a DNA profile provides a 'genetic fingerprint', this can be used to identify perpetrators of crimes. This is because profiles can be produced from DNA samples found at crime scenes, and compared to the DNA profiles of suspects to prove or disprove a match.

Genetic Profiling for Identification

For individuals working in high risk professions, a DNA profile can ensure that in the event of a fatal accident, their body is identified. This is especially important if the person has a job where any other forms of identification may be destroyed when the accident occurs.

Using a DNA profile for identification means that the distress of the person's family and friends is minimized, unnecessary search efforts aren't undertaken, and life insurance claims can be expedited so that loved ones receive the security they may need.

As discussed, DNA is much more resilient than the items traditionally used to determine someone's identity, such as passports, licenses or dog tags. In addition, a tiny DNA sample is often enough to produce a complete DNA profile, whereas paper or digital records can become difficult to interpret with even small amounts of damage. DNA profiling for DNA identification therefore offers a quicker and more conclusive method of identification than other approaches.

References

- Tropp BE (2012). Molecular Biology (4th ed.). Sudbury, Mass.: Jones and Barlett Learning. ISBN 978-0-7637-8663-2
- Mandelkern M, Elias JG, Eden D, Crothers DM (October 1981). "The dimensions of DNA in solution". Journal of Molecular Biology. 152 (1): 153–61. doi:10.1016/0022-2836(81)90099-1. PMID 7338906
- Mitochondrial-dna-mtdna: embryo.asu.edu, Retrieved 18 July 2018
- Carr S (1953). "Watson-Crick Structure of DNA". Memorial University of Newfoundland. Archived from the original on 19 July 2016. Retrieved 13 July 2016
- Johnson TB, Coghill RD (1925). "Pyrimidines. CIII. The discovery of 5-methylcytosine in tuberculinic acid, the nucleic acid of the tubercle bacillus". Journal of the American Chemical Society. 47: 2838–44
- What-is-Satellite-DNA, life-sciences: news-medical.net, Retrieved 15 March 2018
- Alberts B, Johnson A, Lewis J, Raff M, Roberts K, Walter P (2014). Molecular Biology of the Cell (6th ed.). Garland. p. Chapter 4: DNA, Chromosomes and Genomes. ISBN 9780815344322. Archived from the original on 14 July 2014
- What-is-dna-profiling, news: dnatestingchoice.com, Retrieved 18 April 2018
- Gregory SG, Barlow KF, McLay KE, Kaul R, Swarbreck D, Dunham A, et al. (May 2006). "The DNA sequence and biological annotation of human chromosome 1". Nature. 441(7091): 315–21. Bibcode:2006Natur.441..315G. doi:10.1038/nature04727. PMID 16710414
- What-thymine, health: innovateus.net, Retrieved 31 March 2018
- Designation of the two strands of DNA Archived 24 April 2008 at the Wayback Machine. JCBN/NC-IUB Newsletter 1989. Retrieved 7 May 2008
- What-is-adenine: wisegeek.com, Retrieved 28 June 2018
- Kuo TT, Huang TC, Teng MH (1968). "5-Methylcytosine replacing cytosine in the deoxyribonucleic acid of a bacteriophage for Xanthomonas oryzae". Journal of Molecular Biology. 34 (2): 373–5. PMID 5760463

Chapter 4

Human Chromosomes

A chromosome is a DNA molecule, which contains the genetic information of an organism. Humans have 23 pairs of chromosomes. They are of two types, autosomes and allosomes. This chapter has been carefully written to provide a comprehensive study of human chromosomes through topics such as centromere, telomere, autosome, chromatin, etc.

Chromosomes are thread-like structures located inside the nucleus of animal and plant cells. Each chromosome is made of protein and a single molecule of deoxyribonucleic acid (DNA). Passed from parents to offspring, DNA contains the specific instructions that make each type of living creature unique.

The term chromosome comes from the Greek words for color (Chroma) and body (soma). Scientists gave this name to chromosomes because they are cell structures, or bodies, that are strongly stained by some colorful dyes used in research.

The human chromosome is the basic building block of life and is one of the most important components of the cell to be transmitted from generation to generation. It is essentially an organized structure of DNA that exists within the nucleus of all human cells and comprises a single chain of DNA that is coiled and super coiled to form dense thread like pieces.

Like all other eukaryotes, humans contain a fixed number of chromosomes within each of the nuclei in all their cells. There are essentially two types of chromosomes as characterized by karyotyping at the metaphase of cell division. These include:

Autosomes - There are 22 pairs of autosomes in humans. These code for most of the genetic traits in the body.

Gonosomes or sex chromosomes - Humans contain two types of sex chromosomes including X and Y. While males have an X and a Y chromosome, females possess two X chromosomes.

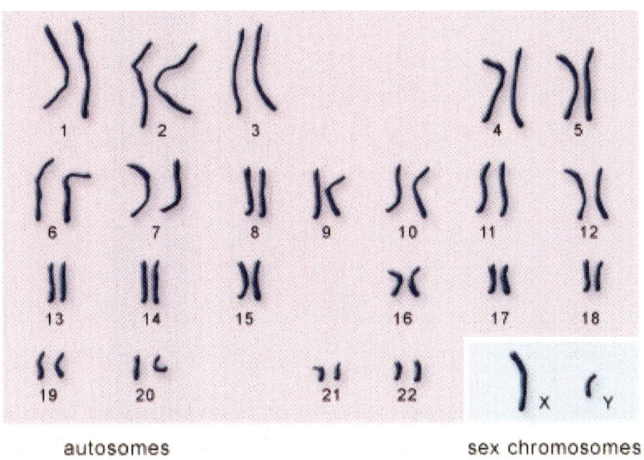

The 22 autosomes are numbered by size. The other two chromosomes, X and Y, are the sex chromosomes. This picture of the human chromosomes lined up in pairs is called a karyotype.

Each human cell thus contains 46 chromosomes in 23 pairs. The gametes or ovum produced by the female ovaries and the sperm produced by the male testicles, however, contain only 23 chromosomes. This ensures that when the egg and the sperm get fertilized to form a baby, it contains 23 pairs and restores the total chromosomal count to 46.

Apart from chromosomes present in the nuclei, humans also possess hundreds of copies of the mitochondrial genome, present in the mitochondria of cells. Mitochondria are normally responsible for generating the energy required by the cell to perform functional processes.

Each of the chromosomes contains highly condensed and coiled DNA consisting of millions of gene sequences.

Chromosome	Genes	Total bases	Sequenced bases
1	4,220	247,199,719	224,999,719
2	1,491	242,751,149	237,712,649
3	1,550	199,446,827	194,704,827
4	446	191,263,063	187,297,063
5	609	180,837,866	177,702,766
6	2,281	170,896,993	167,273,993
7	2,135	158,821,424	154,952,424
8	1,106	146,274,826	142,612,826
9	1,920	140,442,298	120,312,298
10	1,793	135,374,737	131,624,737
11	379	134,452,384	131,130,853
12	1,430	132,289,534	130,303,534
13	924	114,127,980	95,559,980
14	1,347	106,360,585	88,290,585
15	921	100,338,915	81,341,915
16	909	88,822,254	78,884,754
17	1,672	78,654,742	77,800,220
18	519	76,117,153	74,656,155
19	1,555	63,806,651	55,785,651
20	1,008	62,435,965	59,505,254
21	578	46,944,323	34,171,998
22	1,092	49,528,953	34,893,953
X (sex chromosome)	1,846	154,913,754	151,058,754
Y (sex chromosome)	454	57,741,652	25,121,652
Total	32,185	3,079,843,747	2,857,698,560

Function of Human Chromosome

The chromosome holds not only the genetic code, but many of the proteins responsible for helping express it. Its complex form and structure dictate how often genes can be translated into proteins, and which genes are translated. This process is known as gene expression and is responsible for creating organisms. Depending on how densely packed the chromosome is at certain point determines how often a gene gets expressed. Less active genes will be more tightly packed than genes undergoing active transcription. Cellular molecules that regulate genes and transcription often work by activating or deactivating these proteins, which can contract or expand the chromosome. During cell division, all the proteins are activated and the chromatin becomes densely packed into distinct chromosomes. These dense molecules have a better chance of withstanding the pulling forces that occur when chromosomes are separated into new cells.

Structure and Regions Recognized in Chromosomes

Structurally, each chromosome is differentiated into three parts—

(a) Pellicle,

(b) Matrix

(c) Chromonemata.

(a) Pellicle

It is the outer envelope around the substance of chromosome. It is very thin and is formed of achromatic substances. Certain scientists Darlington and Ris have denied its presence.

(b) Matrix

It is the ground substance of chromosome which contains the chromonemata. It is also formed of nongenic materials.

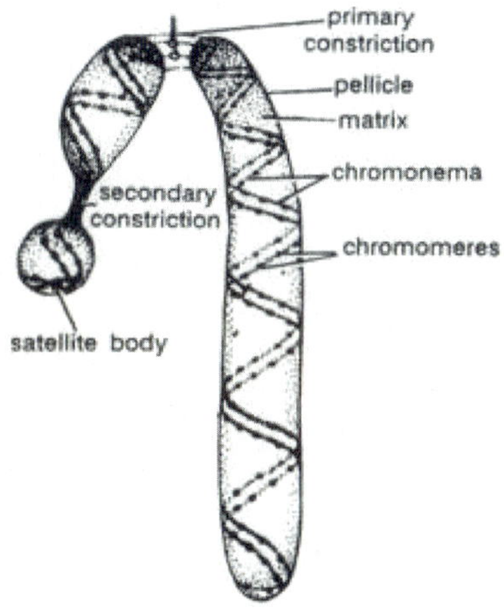

Figure: Structure of Chromosome at anaphase stage of mitosis

(c) Chromonemata

Embedded in the matrix of each chromosome are two identical, spirally coiled threads, the chromonemata. The two chromonemata are also tightly coiled together that they appear as single thread of about 800A thickness. Each chromonemata consists of about 8 microfibrils, each of which is formed of a double helix of DNA.

Chromomeres

In favorable preparations, chromomeres in the form of small dense masses are observed at regular intervals on the chromonemata. These are more distinct in the prophase stage when chromonemata are less coiled and most clearly visible during leptotene and zygotene stages of meiotic prophase.

The thin and lightly stained parts between the adjacent chromosomes are termed as inter-chromomeres. The position of chromomeres on chromonemata is constant for a given chromosome.

While pairing during zygotene of meiotic prophase the homologous chromosomes pair chromomere to chromomere. Chromomeres are regions of tightly folded DNA and are believed to correspond to the units of genetic function in the chromosomes.

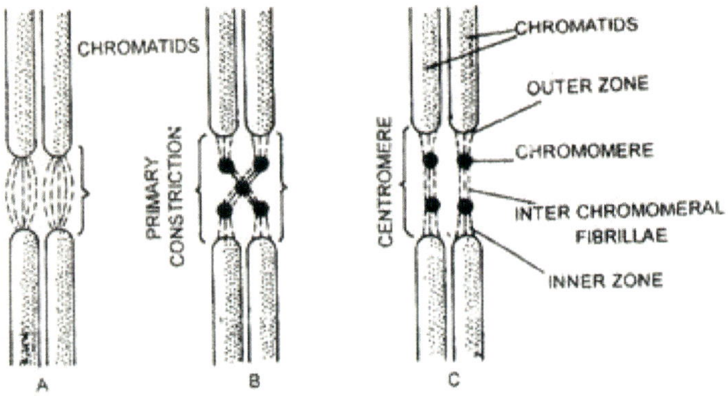

Figure: Structure of Centomere

Chromatid

At mitotic metaphase each chromosome consists of two symmetrical structures called chromatids. Each chromatid contains a single DNA molecule. Both chromatids are attached to each other only by the centromere and become separated at the beginning of anaphase, when the sister chromatids of a chromosome migrate to the opposite poles.

Centromere

A part of the chromosome is recognized as permanent. It is a small structure in the chromonema and is marked by a constriction. At this point the two chromonemata are joined together. This is known as centromere or kinetochore or primary constriction. Its position is constant for a given type of chromosome and forms a feature of identification.

In thin electron microscopic sections, the kinetochore shows a trilaminar structure, i.e., a 10 nm thick dense outer protein aceous layer, a middle layer of low density and a dense inner layer tightly bound to the centromere.

The chromosomes are attached to spindle fibers at this region during cell division. The part of the chromosome which lies on either side of the centromere represents arms which may be equal or unequal depending upon the position of centromere.

Depending upon the number of centromeres, the chromosomes may be:

1. Monocentric with one centromere.
2. Dicentric with two centromeres.

3. Polycentric with more than two centromeres as in Luzula.
4. Acentric without centromere. Such chromosomes represent freshly broken segments of chromosomes which do not survive for long.
5. Diffused or non-located with indistinct centromere diffused throughout the length of chromosome.

Depending upon the location of centromere the chromosomes are categorized into:

1. Telocentric are rod-shaped chromosomes with centromere occupying the terminal position, so that the chromosome has just one arm.
2. Acrocentric are also rod-shaped chromosomes with centromere occupying a sub-terminal position. One arm is very long and the other is very short.
3. Sub metacentric chromosomes are with centromere slightly away from the mid-point so that the two arms are unequal.
4. Metacentric are V-shaped chromosomes in which centromere lies in the middle of chromosome so that the two arms are almost equal.

Centromere controls the orientation and movement of the chromosomes on the spindle. It is the point where force is exerted when the chromosomes move apart during anaphase.

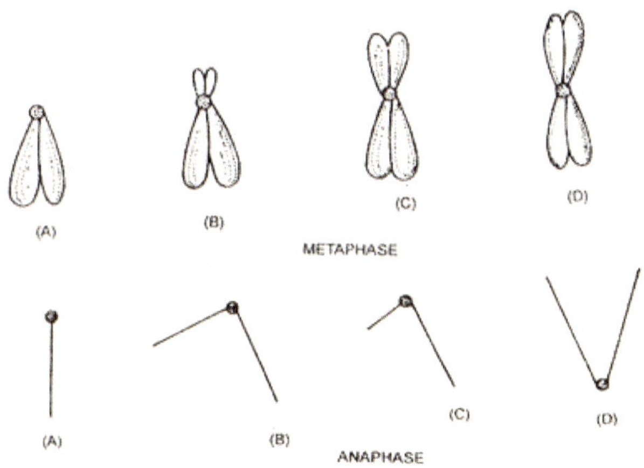

Figure: Metaphase and anaphase configurations of the four classes of chromosomes: (A) Telocentric, (B) Acrocentric, (C) Submetacentric, (D) Metacentric.

Secondary Constriction or Nucleolar Organizer

The chromosome besides having the primary constriction or the centromere possesses secondary constriction at any point of the chromosome. Constant in their position and extent, these constrictions are useful in identifying particular chromosomes in a set.

Secondary constrictions can be distinguished from primary constriction or centromere, because chromosome bends only at the position of centromere during anaphase. The chromosome region distal to the secondary constriction i.e., the region between the secondary constriction and the nearest telomere is known as satellite.

Therefore, chromosomes having secondary constrictions are called satellite chromosomes or sat-chromosomes. The number of sat-chromosomes in the genome varies from one species to the other.

Nucleolus is always associated with the secondary constriction of sat-chromosomes. Therefore, secondary constrictions are also called nucleolus organizer region (NOR) and sat-chromosomes are often referred to as nucleolus organizer chromosomes. NOR of each sat-chromosome contains several hundred copies of the gene coding for ribosomal RNA (rRNA).

Telomeres

These are specialized ends of a chromosome which exhibits physiological differentiation and polarity. Each extremity of the chromosome due to its polarity prevents other chromosomal segments to be fused with it. The chromosomal ends are known as the telomeres. If a chromosome breaks, the broken ends can fuse with each other due to lack of telomeres.

Karyotype and Idiogram

A group of plants and animals comprising a species is characterized by a set of chromosomes, which have certain constant features such as chromosome number, size and shape of individual chromosomes. The term karyotype has been given to the group of characteristics that identifies a particular set of chromosomes. A diagrammatic representation of a karyotype of a species is called idiogram. Generally, in an idiogram, the chromosomes of a haploid set of an organism are ordered in a series of decreasing size.

Uses of Karyotypes:

1. The karyotypes of different groups are sometimes compared and similarities in karyotypes are presumed to present evolutionary relationship.
2. Karyotype also suggests primitive or advanced feature of an organism. A karyotype showing large differences between smallest and largest chromosome of the set and having fewer metacentric chromosomes, is called asymmetric karyotype, which is considered to be a relatively advanced feature when compared with symmetric karyotype which has all metacentric chromosomes of the same size. Levitzky suggested that in flowering plants there is a prominent trend towards asymmetric karyotypes.

Material of the Chromosomes

The material of the chromosomes is the chromatin. Depending on their staining properties with basic dyes (particularly the Feulgen reagent), the following two types of chromatin may be distinguished in the interphase nucleus.

1. Euchromatin: Portions of chromosomes that stain lightly are only partially condensed; this chromatin is termed euchromatin. It represents most of the chromatin that disperse after mitosis has been completed. Euchromatin contains structural genes which replicate and transcribe during G_1 and S phase of interphase. It is considered genetically active chromatin, since it has a role in the phenotype expression of the genes. In euchromatin, DNA is found packed in 3 to 8 nm fiber.

2. Heterochromatin: In the dark-staining regions, the chromatin remains in the condensed state and is called heterochromatin. In 1928, Heitz defined it as those regions of the chromosome that remain condensed during interphase and early prophase and form the so-called chromocentre.

Heterochromatin is characterized by its especially high content of repetitive DNA sequences and contains very few, if any, structural genes. It is late replicating (i.e., it is replicated when the bulk of DNA has already been replicated) and is not transcribed. It is thought that in heterochromatin the DNA is tightly packed in the 30 nm fiber. It is established now that genes in heterochromatic region are inactive.

During early and mid-prophase stages, the heterochromatic regions are constituted into three structures namely chromomeres, centromeres and knobs. Chromomeres may not represent true heterochromatin since they are transcribed.

Centromeric regions invariably contain heterochromatin; in salivary glands, these regions of all the chromosomes fuse to form a large heterochromatic mass called chromocentre. Knobs are spherical heterochromatin bodies, usually several times the diameter of the concerned chromosomes, present in certain chromosomes of some species, e.g.,

Maize; knobs are more clearly observable during pachytene stage in maize. Where present, knobs serve as valuable chromosome markers.

Heterochromatin is classified into two groups: (i) Constitutive and (ii) Facultative.

(i) Constitutive heterochromatin remains permanently in the heterochromatic state, i.e., it does not revert to euchromatic state, e.g., centromeric regions. It contains short repeated sequences of DNA, called satellite DNA.

(ii) Facultative heterochromatin is essentially euchromatin that has undergone heterochromatinization which may involve a segment of chromosome, a whole chromosome (e.g. one X chromosome of human females and females of other mammals), or one whole haploid set of chromosomes (e.g., in some insects, such as mealy bugs).

Chemical Composition

Chromatin is composed of DNA, RNA and protein. The protein of chromatin is of two types: the histones and the non-histones. Purified chromatin isolated from interphase nuclei consists of about 30-40% DNA, 50-65% protein and 0.5-10% RNA: but there is a considerable variation due to species and tissues of the same species.

DNA

The amount of DNA present in normal somatic cells of a species is constant for that species; any variation in DNA from this value is strictly correlated with a corresponding variation at the chromosome level. Gametes of a species contain only half of the amount of DNA present in its somatic cells. The amount of DNA present in somatic cells also depends on the phase of cell cycle.

Protein

Proteins associated with chromosomes may be classified into two broad groups: (i) basic proteins or histones and (ii) non-histone proteins.

Histones constitute about 80% of the total chromosomal protein; they are present in an almost 1:1 ratio with DNA (weight/weight). Their molecular weight ranges from 10,000-30,000 and they are completely devoid of tryptophan. Histones are a highly heterogenous class of proteins separable in 5 different fractions designated as H_1 H_2a, H_2b, H_3 and H_4.

Fraction H_1 is lysin-rich, H_2a and H_2b are slightly lysine rich, while H_3 and H_4 are arginine-rich. These five fractions are present in all cell types of eukaryotes, except in the sperm of some animal species where they are replaced by another class of smaller molecule basic proteins called protamines.

Histones play a primary function in chromosome organization where H_2a, H_2b, H_3 and $H_4$are involved in the structural organization of chromatin fibers, while fraction H_1 holds together the folded chromatin fibers of chromosomes.

Non-histone proteins make up about 20% of the total chromosome mass, but their amount is variable and there is no definite ratio between the amounts of DNA and non-histones present in chromosomes.

There may be 12 to more than 20 different types of non-histone proteins which show variation from one species to the other and even in different tissues of the same organism. This class of proteins includes many important enzymes, such as DNA and RNA polymerases etc.

Ultrastructure of Chromosomes

Electron microscopic studies have demonstrated that chromosomes have very fine fibrils having a thickness of 2 nm-4nm. Since DNA is 2 nm wide, there is possibility that a single fibril corresponds to a single DNA molecule. Several models of chromosome structure have been proposed from time to time based on various types of data on chromosomes.

Folded Fiber Model of Chromosomes

This model was proposed by Du Praw in 1965 and is widely accepted. According to this model, chromosomes are made up of chromatin fibers of about 230Å diameter. Each chromatin fiber contains only one DNA double helix which is in a coiled state; this DNA coil is coated with histone and non-histone proteins.

Thus the 230Å chromatin fiber is produced by coiling of a single DNA double helix, the coils of which are stabilized by proteins and divalent cations (Ca^{++} and Mg^{++}). Each chromatid contains a single long chromatin fibers; the DNA of this fiber replicates during interphase producing two sister chromatin fibers, it remains unreplicated in the centromeric region so that the two sister fibers remain joined in the region.

Subsequently, the chromatin fiber undergoes replication in the centromeric region as well so that the sister chromatin fiber are separated in this region also. During cell division the two sister chromatin fibers undergo extensive folding separately in an irregular manner to give rise to two sister chromatids.

Folding of the chromatin fibers drastically reduces their length and increases their stainability and thickness. This folded structure normally undergoes supercoiling which further increases the thickness of chromosomes and reduces the length. Most of the available evidence supports this model.

Overwhelming evidence from a variety of studies supports the theory that each chromatid contains a single giant DNA molecule. The strongest evidence in the support of the unineme model (single stranded chromatid) is provided by studies on lamp-brush chromosomes.

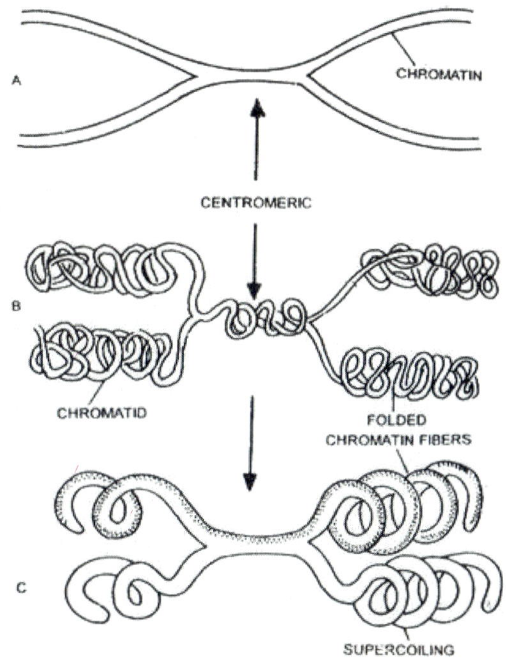

Figure: The folded-fiber model of chromosome organization.
Each chromatin fiber consists of one DNA molecule and has the average diameter of 230 Å.

Organization of Chromatin Fibers

Any model of chromatin fiber structure has to account for (i) packaging of a very long DNA molecule into a unit length of fiber; (ii) production of very thick (230-300Å) fibers from very thin (20Å) DNA molecules and (iii) the beads-on-a-string ultrastructure of chromatin fibers observed particularly during replication. Two clearly different models of chromatin fiber structure have been proposed:

I. Coiled DNA Model:

This is the simplest model of chromatin fiber organization and was given by Du Praw. According to this model, the single DNA molecule of a chromatin fiber is coiled in a manner similar to the wire in a spring; the coils being held together by histone bridges produced by binding histone molecules in the large groove of DNA molecules. Such a coiled structure that would be stabilized as a single histone molecule would bind to several coils of DNA.

This coiled structure is coated with chromosomal proteins to yield the basic structure of chromatin fibers (type A fiber), which may undergo supercoiling to produce the type B fiber of DuPraw, which is akin to the beads seen in electron micrographs of chromatin fibers.

II. Nucleosome-Solenoid Model:

This model was proposed by Romberg and Thomas and is the most widely accepted. According to this model, chromatin is composed of a repeating unit called nucleosome. Nucleosomes are the fundamental packing unit particles of the chromatin and give chromatin a "beads-on-a string" appearance in electron micrographs that unfold higher-order packing. One complete nucleosome consists of a nucleosome core, linker DNA, an average of one molecule of H1 histone and other associated chromosomal proteins.

Nucleosome Core

It consists of a histone octamer composed of two molecules, each of histones H_2a, H_2b, H_3 and H_4. In addition, a 146 bp long DNA molecule is wound round this histone octamer in 13/4 turns; this segment of DNA is nuclease resistant.

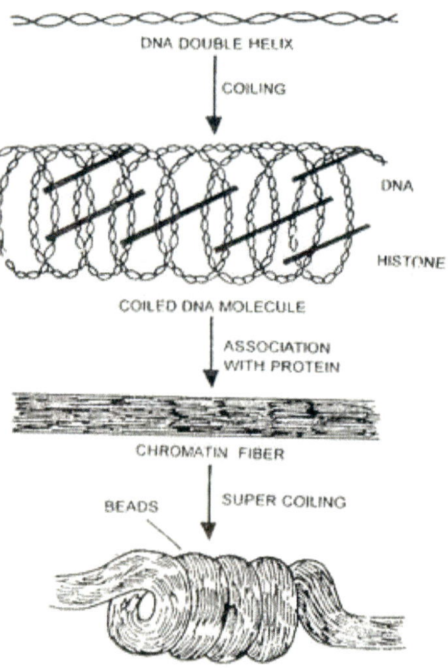

Figure: The coiled DNA model of chromatin fiber organization

Linker DNA

Its size varies from 8bp to 114 bp depending on the species. This DNA forms the string part of the beads-on-a string chromatin fiber, and is nuclease susceptible; and the beads are due to nucleosome cores. Thus, linker DNA joins two neighboring nucleosomes.

H_1 Histone

Each nucleosome contains, on an average, one molecule of HI histone, although its uniform distribution throughout the length of chromatin fibers is not clearly know. Some studies suggest that the molecules of H_1 histone are involved in stabilizing the supercoils of nucleosome chromatin fibers. Other studies suggest that HI is associated on the outside of each nucleosome core, and that one H_1 molecule stabilizes about 166 bp long DNA molecule.

Other Chromosomal Proteins

Both linker DNA and nucleosome are associated with other chromosomal proteins. In native chromatin, the beads are about 110Å in diameter, 60Å high and ellipsoidal in shape. Each bead corresponds to a single nucleosome core. Under some conditions, nucleosomes pack together without any linker DNA, which produces the 100Å thick chromatin fiber called nucleosome fiber which may then supercoil to give rise to the 300Å chromatin fiber called solenoid.

The nucleosome model of chromatin fiber structure is consistent with almost all of the evidence accumulated so far.

Special Chromosomes

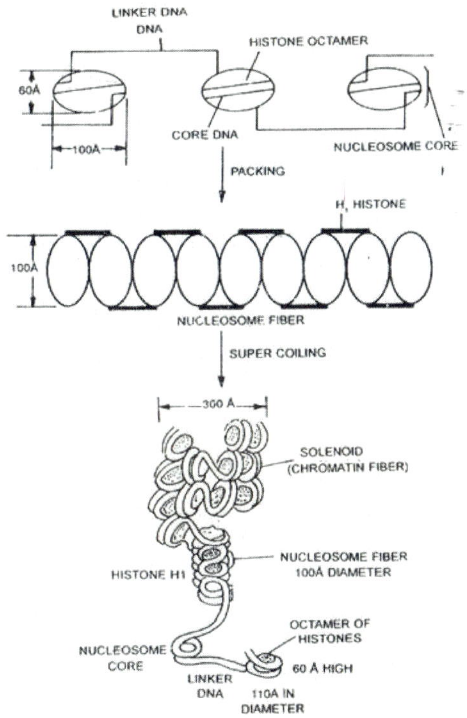

Figure: The nucleosome-solenoid model of chromatin fiber organization

Some tissues of certain organisms contain chromosomes which differ significantly from normal ones in terms of either morphology or function; such chromosomes are referred to as special chromosomes. The following types of chromosomes may be included under this category:

1. Lampbrush chromosomes,
2. Giant chromosomes or salivary gland chromosomes and
3. Accessory or B chromosomes.

Lampbrush Chromosomes

Lampbrush chromosomes are found in oocytes of many invertebrates and all vertebrates, except mammals; they have also been reported in human and rodent oocytes. But they have been the most extensively studied in amphibian oocytes.

These chromosomes are most distinctly observed during the prolonged diplotene stage of oocytes. During diplotene, the homologous chromosomes begin to separate from each other, remaining in contact only at several points along their length.

Each chromosome of a pair has several chromomeres distributed over its length; from each of a majority of the chromomeres generally a pair of lateral loops extends in the opposite directions perpendicular to the main axis of the chromosome.

In some cases, more than one pair, even upto 9 pairs of loops may emerge from a single chromomere. These lateral loops give the chromosomes the appearance of a lampbrush which is the reason for their name 'lamp-brush chromosomes.'

These chromosomes are extremely long, in some cases being 800-1000^ in length. The size of loops may range from an average of 9.5ja in frog to 200|i in newt. The pairs of loops are produced due to uncoiling of the two chromatin fibers (hence the two sister chromatids) present in a highly coiled state in the chromosomes; this makes their DNA available for transcription (RNA synthesis).

Thus each loop represents one chromatid of a chromosome and is composed of one DNA double helix. One end of each loop is thinner (thin end) than the other end (thick end). There is extensive RNA synthesis at the thin ends of loops, while there is little or no RNA synthesis at the thick end.

The chromatin fiber of the chromomere is progressively uncoiled towards the thin end of a loop; the DNA in this region supports active RNA synthesis but later becomes associated with RNA and protein to become markedly thicker.

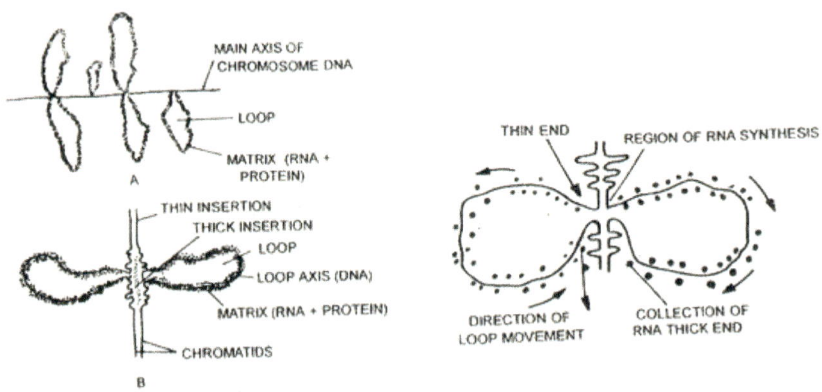

Figure: Lampbrush chromosome.

The DNA at the thick end of a loop is progressively withdrawn and reassembled into the chromomere. The number of pairs of loops gradually increases in meiosis till it reaches maximum in diplotene. As meiosis proceeds further, number of loops gradually decreases and the loops ultimately disappear due to disintegration rather than reabsorption back into the chromomere.

Loops represent the sites of gene action (transcription), and the function of lampbrush chromosomes is to produce the large numbers and quantities of proteins and RNA's stored in eggs.

Giant Chromosomes

Giant chromosomes are found in certain tissues, e.g., salivary glands of larvae, gut epithelium, Malphigian tubules and some Diptera, e.g., Drosophila, Chironomous, Sciara, Rhyncosciara etc. These chromosomes are very long (upto 200 times their size during mitotic metaphase in the case of Drosophila) and very thick, hence they are known as giant chromosomes.

They were first discovered by Balbiani in dipteran salivary glands, giving them the commonly used name salivary gland chromosomes. The giant chromosomes are somatically paired. Consequently, the number of these giant chromosomes in the salivary gland cells always appear to be half that in the normal somatic cells.

The giant chromosomes have a distinct pattern of transverse banding which consists of alternate chromatic and achromatic regions. The bands occasionally form reversible puffs, known as chromosome puffs or Balabiani rings, which are associated with active RNA synthesis.

The giant chromosomes represent a bundle of fibrils which arise by repeated cycles of endo- reduplication (replication of chromatin without cell-division) of single chromatids. This is why these chromosomes are also popularly known as polytene chromosomes and the condition is described as polyteny. The number of chromonemata (fibrils) per chromosomes may reach upto 2000 in extreme cases some workers placed this figure as high as 16,000.

In D melanogaster, the giant chromosomes radiate as five long and one short arms from a single more or less amorphous mass known as chromocentre. The chromocentre is formed by fusion of the centromeric regions of all the chromosomes and, in males, of the entire Y chromosomes.

The short arm radiating from the chromocentre represents chromosome IV, one of the long arms is due to the X-chromosome, while the remaining four long arms represent the arms of chromosome II and III. The total length of D. melanogaster giant chromosomes is about 2000μ.

Accessory Chromosomes

In many species, one too many extra chromosomes in addition to the normal somatic complement are found; these extra chromosomes are called accessory chromosomes, B-chromosomes or supernumerary chromosomes.

About 600 plant species and more than 100 animal species are reported to possess B- chromosomes. B-chromosomes are generally smaller in size than the chromosomes of the normal somatic complement but in some species they may be larger (e.g., in Sciara).

One of- the most important features of these chromosomes is that their numbers may vary considerably among individuals of the same species; in maize as many as 25-30 B-chromosomes may become accumulated in some individuals without any marked effect on their phenotype. These chromosomes are generally gained by and lost from the individuals of a species without any apparent adverse or beneficial effect.

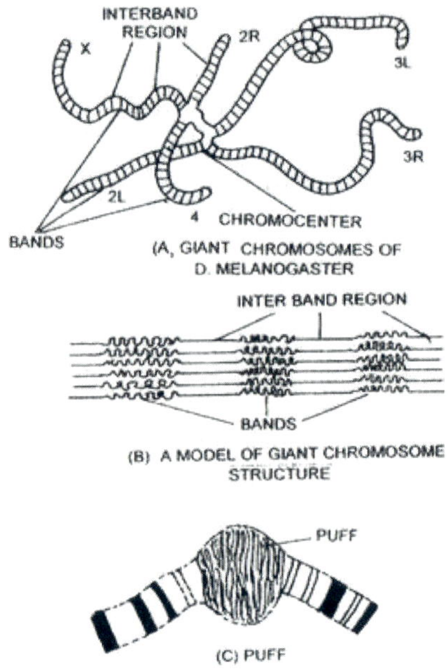

Figure: Giant chromosomes of D. melangster salivary glands.

However, the presence of several B-chromosomes often leads to some reduction in vigour and fertility in maize. In most cases, they are largely heterochromatic, while in some species (e.g. maize) they are partly heterochromatic, and in some other (e.g., Tradescantia) they are entirely euchromatic. They are believed to be generally inactive genetically, but they may not be completely devoid of genes.

The origin of B-chromosomes in most species is unknown. In some animals they may arise due to fragmentation of the heterochromatic Y chromosome. In maize, morphological features and pairing behavior of B-chromosome clearly shows that they do not have any segment which is homologous to a segment of any chromosome of the normal somatic complement.

B-chromosomes are relatively unstable; in many species they tend to be eliminated from somatic tissues due to lagging and non-disjunction and they frequently change in morphology through fragmentation. Further, they may also show irregular distribution during meiosis, but they are invariably maintained in the reproductive tissue.

Centromere

The centromere is the point on a chromosome where mitotic spindle fibers attach to pull sister chromatids apart during cell division.

When a cell seeks to reproduce itself, it must first make a complete copy of each of its chromosomes, to ensure that their daughter cell receives a full complement of the parent cell's DNA.

The two copies of each chromosome often remain stuck together until they are separated, with one copy going to each daughter cell. While stuck together, these two copies are called "sister chromatids."

As a cell prepares to divide, the sister chromatids begin to become unstuck from each other until they are almost completely separated. They remain joined, however, at the centromere – a special region that plays a vital role in cell division.

At the centromere, elements of the cell's cytoskeleton assemble and attach. First, a complex of proteins called the kinetochore assembles around the centromere region of DNA; then, mitotic spindle fibers attach to the kinetochore. The other end of these fibers are anchored to opposite ends of the parent cell, which will shortly split to become new daughter cells.

When the spindle fibers begin to contract, the chromatids are pulled to opposite ends of the parent cell. In this way, when the parent cells splits in two during cytokinesis, each sister chromatid becomes a chromosome of the new daughter cell.

To understand this process, it is important to remember that each sister chromatid is actually a full copy of the parent cell's chromosome.

The two sister chromatids combined are often referred to as a single chromosome because they are packaged tightly together – but each contains all the information of the original chromosome, so when they split, each becomes a complete chromosome containing all of the information contained in the parent cell's original chromosome.

The image below provides a visual illustration of the cell's preparations to undergo cell division. Note that in phase 2 the nuclear envelope dissolves, leaving the chromosomes free in the cytoplasm.

In stages 3 and 4, the DNA condenses into tightly-packed chromosomes, in which sister chromatids are paired up and joined at their centromere. In stage 5 pictured below, the sister chromatids are pulled apart to opposite sides of the cell.

In stage 6, at last the cell splits in two, separating the sisters into daughter cells.

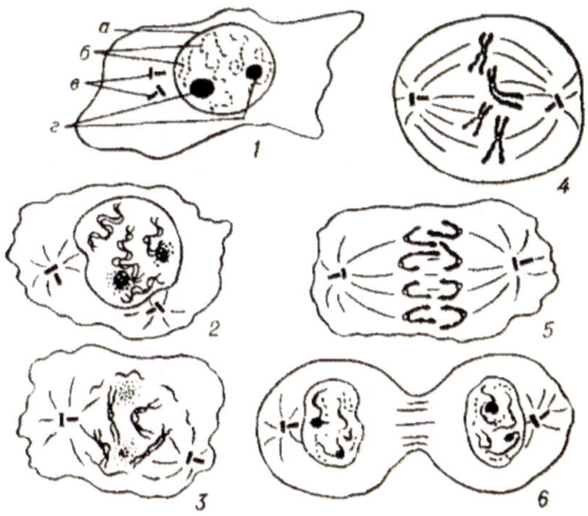

Function of Centromere

All living things are made up of cells. In order for cells to grow or reproduce, cell division must occur. In cell division, one "parent" cell splits in two, with each of the resulting cells being "daughter" cells.

For each daughter cell to survive, it is essential that they get a copy of each of their parent cells' chromosomes.

When this does not happen, and daughter cells receive incomplete information, or too many copies of one chromosome, serious disease or cell death can result.

To ensure that a full copy of its DNA is given to each daughter cell, a cell first makes a complete copy of its DNA. The two copies stick together, ultimately condensing to form sister chromatids, until they are pulled apart during cell division.

The centromere of the chromosome provides a binding site for the mitotic spindle fiber that will attach to each sister chromatid and pull them to opposite ends of the parent cell, which will ultimately become the cytoplasm of the two daughter cells.

In cases where centromeres do not function properly, cells cannot successfully divide. Any attempt to do so results in daughter cells which do not have the genetic instructions they need to survive.

Centromere dysfunction leading to problems with chromosome sorting is believed to play a role in many instances of miscarriage, in which inherited centromere disorders may result in early embryonic death. Centromere dysfunction is also suspected to play a role in cancer cells, which display massive chromosome imbalance of the type that would be expected if the sorting of chromosomes during cell division failed.

Types of Centromeres

Point Centromeres

Point centromeres are centromeres where mitotic spindle fibers are attracted to specific sequences of DNA. In these cases, the cell has proteins that bind to these specific DNA sequences, and these proteins form the basis for the binding of the mitotic spindle fibers.

In these cases, mitotic spindle fibers will typically appear anywhere that the DNA sequence of the point centromere appears. The protein that begins the creation of the mitotic spindle fiber complex will bind to that DNA sequence without regard for its location or other factors.

Regional Centromeres

Humans and most eukaryotic cells use regional centromeres. These are centromeres where mitotic spindle binding is determined, not by a precise sequence of DNA, but by a combination of characteristics working together to signal the location of a centromere.

In regional centromeres, it is thought that epigenetic marks tell the proteins that begin to build the mitotic spindle complex where to bind.

"Epigenetic marks" are chemical changes made to DNA by enzymes, which can change the DNA's chemical properties and other properties. Epigenetic marks can be added or removed without changing information contained in the DNA.

Telomere

Telomeres are distinctive structures found at the ends of our chromosomes. They consist of the same short DNA sequence repeated over and over again. Telomeres are sections of DNA Found at the ends of each of our chromosomes. They consist of the same sequence of bases repeated over and over. In humans the telomere sequence is TTAGGG. This sequence is usually repeated about 3,000 times and can reach up to 15,000 base pairs in length.

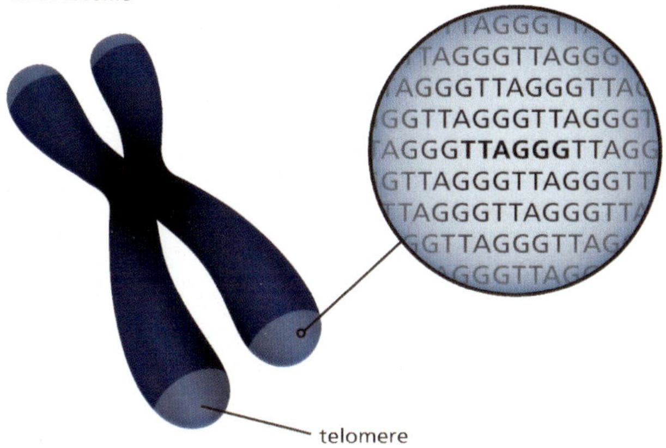

Figure: Illustration showing the position of telomeres at the end of our chromosomes.

Functions of Telomeres

Telomeres serve three major purposes:

1. They help to organize each of our 46 chromosomes in the nucleus (control center) of our cells.
2. They protect the ends of our chromosomes by forming a cap, much like the plastic tip on shoelaces. If the telomeres were not there, our chromosomes may end up sticking to other chromosomes.
3. They allow the chromosome to be replicated properly during cell division:

Every time a cell carries out DNA replication the chromosomes are shortened by about 25-200 bases (A, C, G, or T) per replication.

However, because the ends are protected by telomeres, the only part of the chromosome that is lost is the telomere, and the DNA is left undamaged.

Without telomeres, important DNA would be lost every time a cell divides (usually about 50 to 70 times).

Changes that Occurs in Telomeres as we Age

Each time a cell divides, 25-200 bases are lost from the ends of the telomeres on each chromosome. Two main factors contribute to telomere shortening during cell division:

- The “end replication problem” during DNA replication: Accounts for the loss of about 20 base pairs per cell division.
- Oxidative stress: Accounts for the loss of between 50-100 base pairs per cell division. The amount of oxidative stress in the body is thought to be affected by lifestyle factors such as diet, smoking and stress.

When the telomere becomes too short, the chromosome reaches a ‘critical length’ and can no longer be replicated. This ’critical length’ triggers the cell to die by a process called apoptosis, also known as programmed cell death.

Length of Telomere

Telomerase is an enzyme that adds the TTAGGG telomere sequence to the ends of chromosomes. Telomerase is only found in very low concentrations in our somatic cells. Because these cells do not regularly use telomerase they age leading to a reduction in normal function. The result of ageing cells, is an ageing body. Telomerase is found in high levels in germline cells (egg and sperm) and stem cells. In these cells telomere length is maintained after DNA replication and the cells do not show signs of ageing. Telomerase is also found in high levels in cancer cells. This enables cancer cells to be immortal and continue replicating themselves. If telomerase activity was switched off in cancer cells, their telomeres would shorten until they reached a ‘critical length’. This would, prevent the cancer cells from dividing uncontrollably to form tumors. The action of telomerase allows cells to keep multiplying and avoid ageing.

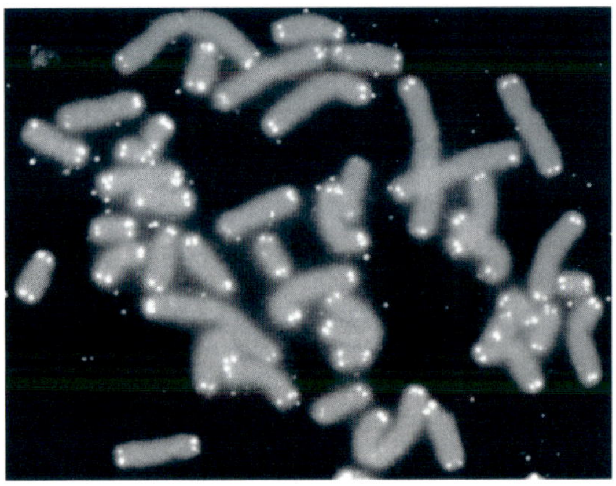

Telomere caps

Use of Telomeres in Medicine

Research on telomeres and the role of telomerase could uncover valuable information to combat ageing and fight cancer. The medical relevance of telomeres is uncertain. Human cells cultured in the lab have been observed to stop dividing when telomerase is inactivated, because the length of telomeres is not maintained after cell division. The cells then enter a state of inactivity called senescence. However, once telomerase is reactivated, the cells are able to continue dividing. If telomerase can be used to help human cells live forever, it may also be possible to mass produce cells for transplantation. These cells could help to treat a range of conditions, from severe burns to diabetes.

Telomeres and Ageing

Mice models lacking the enzyme telomerase were found to show signs of premature ageing. However, it is not certain whether telomere shortening is responsible for ageing in humans or whether it is just a sign of ageing, like grey hair. There are several indications that telomere length is a good predictor of lifespan. Newborn babies tend to have telomeres ranging in length from around 8,000 to 13,000 base pairs.

It has been observed that this number tends to decline by around 20-40 base pairs each year. So, by the time someone is 40 years old they could have lost up to 1,600 base pairs from their telomeres. However, looking at the bigger picture, the overall shortening of our telomeres is not significant, even in very old people. Cells that divide rapidly, such as germ cells and stem cells, are among the few cell types in our bodies containing active telomerase. This means that in these cells telomere length is maintained or even lengthened over time. However, there are a number of other factors that have an effect on the length of our telomeres that all need to be considered, such as smoking and obesity.

Telomeres and Cancer

Telomeres and telomerase present a number of potential targets for the design of new cancer therapies. Cancer cells contain active telomerase to enable them to become 'immortal' and continue dividing uncontrolled. Cancer is a disease characterized by the rapid and uncontrolled division of cells.

Without telomerase activity, these cells would become inactive, stop dividing and eventually die. Drugs that inhibit telomerase activity, or kill telomerase-producing cells, may potentially stop and kill cancer cells in their tracks. However, blocking telomerase activity could affect cells where telomerase activity is important, such as sperm, eggs, platelets and immune cells. Disrupting telomerase in these cell types could affect fertility, wound healing and the ability to fight infections.

However, telomerase activity in somatic cells is very low. These cells would therefore be largely unaffected by anti-telomerase therapy. Scientists hope this would result in fewer side effects for the patient, compared to current cancer therapies. Telomere biology is incredibly important in human cancer and scientists are working hard to understand the best way to exploit their knowledge of it to advance the treatment of cancer.

Autosome

An autosome is a chromosome in a eukaryotic cell that is not a sex chromosome.

Unlike prokaryotic cells, eukaryotic cells have many chromosomes in which they package their DNA. This allows eukaryotes to store much more genetic information.

Most eukaryotic organisms reproduce through sexual reproduction – meaning that each individual has two copies of each chromosome. One copy is inherited from one parent, while the other is inherited from the other parent.

This system enhances genetic diversity and protects against some diseases, since it enables individuals to inherit immune system genes from two different parents, and having two copies of a gene often enables a healthy copy inherited from one parent to "cover for" a copy of a gene that has been corrupted through harmful mutation.

It's normal for diploid eukaryotic organisms (those which have a full set of chromosomes inherited through sexual reproduction) to have two copies of each autosome.

Sex chromosomes are considered separately from autosomes, since their inheritance pattern works differently. In humans, the sex chromosomes are referred to as the X chromosome and the Y chromosome. Other animals, like birds, use a different system of sex chromosomes.

During the process of meiosis which creates eukaryotic sex cells, the sex cells "remix" DNA between their two copies of each autosome in the process of crossing over. The result is a unique set of chromosomes which has a mix of material from both of the individual's parents. This process is illustrated below.

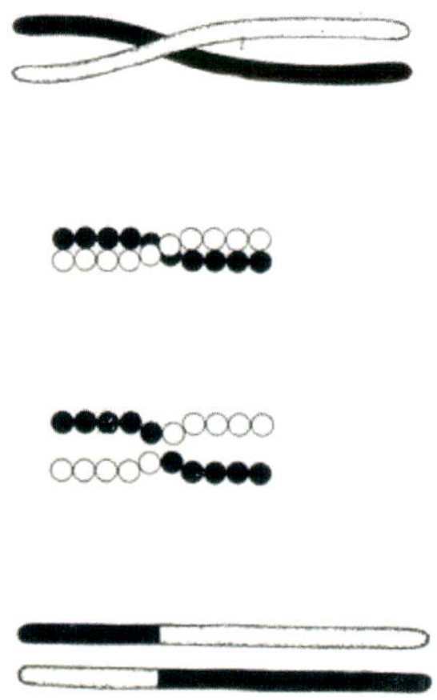

Crossing over of the chromosomes

The sex cell then discards one of each of the resulting remixed autosomes, resulting in a gamete cell that has only one copy of each autosome.

When two gametes combine, they produce a cell which will grow into a new individual which will possess a copy of each chromosome from each parent. The individual's unique genetic profile will include DNA from each of its four grandparents.

During the growth of a multicellular organism, it's normal for a cell to make a full copy of each of its chromosomes, and give one copy to each daughter cell.

When errors are made in distributing chromosomes during meiosis or early in embryonic development, serious diseases can result due to many cells in an individual's body having the wrong number of chromosomes.

Because each chromosome contains thousands of genes, having too many or too few chromosomes can result in serious imbalances in gene expression. In humans, many pregnancies that do not survive the first trimester are cases where the embryo inherited the wrong number of chromosomes and was not able to survive.

Other errors in chromosome replication can cause more mild syndromes such as Down syndrome, which is caused by inheriting an extra copy of chromosome 21 from one parent.

Function of Autosomes

Each autosome stores many thousands genes, each of which performs a unique function in the organism's cells.

Under normal circumstances, each chromosome follows a "map" that is shared across individuals in the species. This allows cells to "know" where to start gene expression when they want to express a certain gene. It is thought that factors which effect gene expression use this "map" to accurately respond to a cell's needs.

When autosomes are healthy, this enables cells to perform an awesome array of functions. Each of hundreds of subtly differing cell types in a eukaryotic organism express a different combination of genes in the right place at the right time, enabling the huge array of cellular functions we see in eukaryotic organisms like ourselves.

Each of our cells contain the necessary compliment of genes to reproduce our whole bodies. Differences between brain cells, skin cells, and muscle cells are made by cells transcribing the right genes in the right places at the right times.

Our bodies get it right almost all the time! But biologists often learn how something works by watching cases where it breaks, and seeing what happens when the mechanism is not working properly.

In the case of autosomes and their carefully arranged "map" that allows for the complexity of our bodies, problems can arise when chromosomes break and their pieces end up in the wrong place.

This event, called "translocation," can cause genes to the expression of the wrong genes at the wrong times. Some types of cancer may be caused by translocations leading to errors in cell development and reproduction.

Examples of Autosomal Disorders

Trisomy 21 (Down Syndrome)

Down syndrome occurs when a person inherits all or part of an extra copy of chromosome 21 from one parent. This usually occurs due to a one-time error in meiosis and is not passed down through the generations.

People with Down syndrome have a variety of unusual traits and symptoms related to skeletal tissue (unusual skeletal shape, weak ligaments), nerve tissue (cognitive disabilities, poor muscle tone), and have a higher risk of some diseases due to extra expression of material from chromosome 21.

Due to the range of symptoms seen in Down syndrome, some people with Down syndrome can complete regular schooling and have independent careers, while others may need special education classes and may not be able to function independently in the workplace.

The only known risk factor for Down syndrome is having older parents, which can increase the chance that parents' bodies will incorrectly sort chromosomes during meiosis.

Cri du Chat

Cri du chat, also known as “chromosome 5p deletion syndrome” or “Lejeune’s syndrome,” happens when a person inherits just one copy of part of chromosome 5. Some people with cri du chat also have extra copies of other parts of chromosome 5.

As with Down syndrome, cri du chat usually occurs due to an error in the sorting of the parents’ chromosomes during meiosis.

The syndrome’s name comes from French for “cry of the cat,” in reference to the unusual catlike cry that babies with cri du chat have due to their unusual skeletal and neurological traits.

Like people with Down syndrome, people with cri du chat can have unusual skeletons, weak muscles, and cognitive impairment due to the under-expression of the 5p chromosomal section.

People with cri du chat may also have hearing loss, heart problems, and microcephaly (a small head).

Philadelphia Chromosome

The Philadelphia chromosome is a chromosome found in many leukemia cancer cells, which may give a clue as to how the cancer gets started.

In the Philadelphia chromosome, chromosome 9 and chromosome 22 have swapped some genetic material. The specific place where the two are joined creates a fusion protein – that is, a protein coded for by a fusion of two different genes, one from chromosome 9 and one from chromosome 22.

This gene turns cell replication to “always on,” and as a result leads to uncontrolled replication of cells which never mature and become properly functional. Leukemia occurs when these non-functioning cells multiply out of control and destroy healthy, functioning tissue.

Some scientists believe that chromosomal translocations are a common cause of cancer. At least fifteen different kinds of cancer have been found to frequently involve chromosomal translocation, often resulting in the creation of fusion proteins.

Allosome

An allosome is a sex chromosome that differs from an ordinary autosome in form, size, or behavior. The human sex chromosomes are a typical pair of allosomes. The X chromosome is present in the ovum, while either X or Y chromosomes can be present in sperm.

The chromosomes which determine the sex (maleness or femaleness) of an individual in sexually producing organisms are called sex chromosomes or allosomes or idiosomes. In humans an individual whose cells contain XX chromosomes (homo or isogametic) becomes a female, while one whose cells contains XY chromosomes (heterogametic) becomes a male.

X Chromosome

The X chromosome is one of the two sex chromosomes in humans (the other is the Y chromosome). The sex chromosomes form one of the 23 pairs of human chromosomes in each cell. The X chromosome spans about 155 million DNA building blocks (base pairs) and represents approximately 5 percent of the total DNA in cells.

Each person normally has one pair of sex chromosomes in each cell. Females have two X chromosomes, while males have one X and one Y chromosome. Early in embryonic development in females, one of the two X chromosomes is randomly and permanently inactivated in cells other than egg cells. This phenomenon is called X-inactivation or lyonization. X-inactivation ensures that females, like males, have one functional copy of the X chromosome in each body cell. Because X-inactivation is random, in normal females the X chromosome inherited from the mother is active in some cells, and the X chromosome inherited from the father is active in other cells.

Some genes on the X chromosome escape X-inactivation. Many of these genes are located at the ends of each arm of the X chromosome in areas known as the pseudo autosomal regions. Although many genes are unique to the X chromosome, genes in the pseudo autosomal regions are present on both sex chromosomes. As a result, men and women each have two functional copies of these genes. Many genes in the pseudo autosomal regions are essential for normal development.

Functions of the X Chromosome

In women, the X chromosome represents almost 5% of the total DNA and in men, who have only one X chromosome, it represents about 2.5% of the total DNA.

Men inherit the X chromosome they have from their mother and the Y chromosome from their father, while women inherit one X chromosome from the mother and the other from the father.

There are around 2000 genes located on the X chromosome and genetic research is focused on identifying these genes. This compares with 78 genes on the Y chromosome out of approximately 20,000 to 25,000 present in the human genome.

When X chromosomal genes are mutated, they may give rise to genetic conditions and these are termed X-linked disorders.

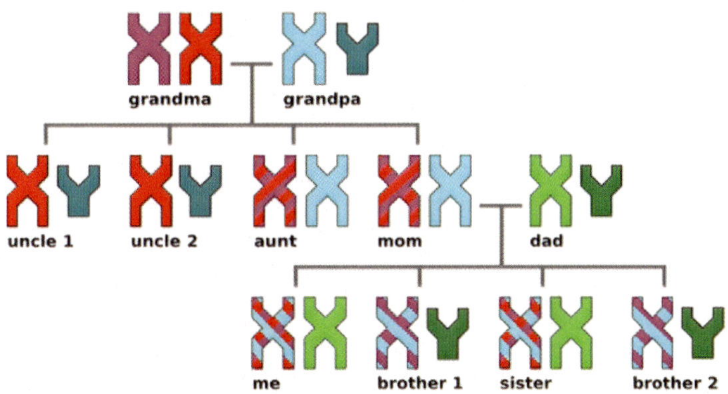

Genetic disorders that arise from missing, additional or malformed copies of the X chromosome are termed numerical disorders. Examples include Klinefelter's syndrome where a male has one or more extra copies; Triple X syndrome, where a female has one extra copy and Turner syndrome where a female has one normal X chromosome and one missing or abnormal one.

Y Chromosome

The Y chromosome is one of the two sex chromosomes in humans (the other is the X chromosome). The sex chromosomes form one of the 23 pairs of human chromosomes in each cell. The Y chromosome spans more than 59 million building blocks of DNA (base pairs) and represents almost 2 percent of the total DNA in cells.

Each person normally has one pair of sex chromosomes in each cell. The Y chromosome is present in males, who have one X and one Y chromosome, while females have two X chromosomes.

Structure of the Y Chromosome

The Y is one of the smallest chromosomes in the human genome and represent around 2%–3% of a haploid genome. Cytogenetic observations based on chromosome-banding studies allowed different Y regions to be identified: the pseudo autosomal portion (divided into two regions: PAR1 and PAR2) and the euchromatic and heterochromatic regions.

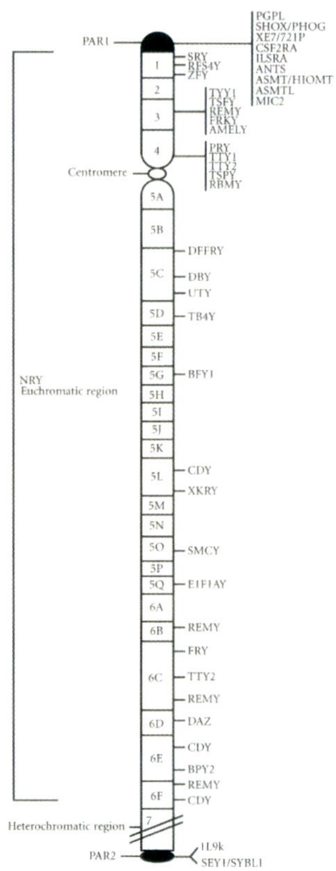

Figure: Schematic representation of the Y chromosome

Genes in the two pseudo autosomal regions (PAR1 and PAR2) as well as those in the nonrecombining Y region (NRY) are illustrated.

The Pseudo autosomal regions (PAR): PAR1 is located at the terminal region of the short arm (Yp), and the PAR2 at the tip of the long arm. PAR1 and PAR2 cover approximately 2600 and 320 kb of DNA, respectively. The pseudo autosomal regions, and in particular PAR1, are where the Y chromosome pairs and exchanges genetic material with the pseudo autosomal region of the X chromosome during male meiosis. Consequently, genes located within the PAR are inherited in the same manner as autosomal genes. The euchromatic region is distal to the PAR1 and consists of the short arm paracentromeric region, the centromere and the long arm paracentromeric region. Finally, the heterochromatic region comprises distal Yq corresponding to Yq12. This region is assumed to be genetically inert and polymorphic in length in different male populations, since it is composed mainly of two highly repetitive sequences families, DYZ1 and DYZ2, containing about 5000 and 2000 copies of each respectively.

Whereas PAR1 and PAR2 represent the 5% of the entire chromosome, the majority of the length of the Y (95%) is made by the so-called "Non-Recombining Y" (NRY). This includes the euchromatic and heterochromatic regions of the chromosome. Whereas the heterochromatic region is considered genetically inert, the euchromatic region has numerous highly repeated sequences but also contains some genes responsible for important biological functions that we will review here.

Physical and Molecular Mapping

The physical mapping of the Y chromosome has mainly depended on naturally occurring deletions on this chromosome. The creation of a deletion map, and the resultant ordering of DNA loci along the chromosome, is very useful not only in locating genes but also in studying the structural diversity of the Y within and among human populations and primates. This allows information on the evolution of human species through paternal lineages to be obtained.

First attempts at mapping the Y were based on cytogenetically detectable deletions on this chromosome and suffer, then, from the limited accuracy and resolution of chromosome banding patterns. However, these preliminary studies led, for the first time, to the hypothesis that a gene or genes located on Yq were related to spermatogenic failure. Similar studies defined also a region associated with sex determination.

Vergnaud performed the first molecular map of the Y in 1986. By using different Y-specific probes on patients with microscopically detectable Y anomalies, they subdivided the Y chromosome into 7 intervals, corresponding with naturally occurring deletions of this chromosome. Later in 1992, Vollrath constructed a more precise deletion map of the Y chromosome based on the detection of about 200 sequence-tagged sites (STS's). The presence or absence of these STS's on a large set of patients with a wide range of Y anomalies subdivided the euchromatic into 43 ordered intervals, all defined by naturally occurring chromosomal breakpoints. These 43 deletion intervals further refined the seven-interval map of Vergnaud. This collection of ordered STS's along the Y chromosome have been extensively used in order to define shortest deleted regions associated with particular phenotypes and then, in identifying Y chromosomal genes and exploring the origin of Y chromosome disorders. Moreover, the same group in Boston led by David Page prepared a library of yeast artificial clones (YAC) from a human XYYYY male. The clones were screened with the Y-specific STS's in order to identify those containing the corresponding sequences. Finally, an essentially complete physical map of the Y chromosome was generated with 196 overlapping DNA clones, which covered 98 percent of the euchromatic region. These Y physical maps have certainly accelerated the search for new genes and made it much easier to explore the biology of this chromosome.

Genes on the Y Chromosome

Compared to the other human chromosomes, the Y chromosome has a limited number of genes. The Y gene poverty may have been the result of the known the tendency of Y chromosome's genes to degenerate during evolution, being nowadays the relic of an ancient common ancestry with the X chromosome. Both mammalian X and Y chromosomes evolved from ancestral autosomes. The most ancestral gene functions were retained on the nascent X chromosome but deteriorated on NRY portion of the emerging Y giving females with two copies but males with only one copy of many genes. The gene dosage problem has been solved through inactivation of one X chromosome in females.

In spite of the limited make-up of genes, different transcription units or families of closely related transcription units have been identified in the NRY region during the past decade. Recently, Lahn and Page identified 12 novel genes or gene families and assessed their expression in diverse human tissues. The different genes identified so far throughout both the NRY region and the two pseudoautosomal regions are summarised in, together with some information on their location and possible pathological implications. According to the same authors, all NRY genes can be divided into two different categories. The first comprises those genes which are ubiquitously expressed, have X homologues, appear in a single copy on the NRY and exhibit housekeeping cell functions. The second category includes genes expressed specifically in the testes, exist in multiple copies (with the exception of SRY) on the NRY and encode proteins, which more specialised functions. It is worth mentioning the finding of X-homologous NRY genes, which suggest an alternative solution for the gene dosage compensation. It has been proposed that these genes should escape X-inactivation and encoded proteins functionally interchangeable.

Gene symbol	Location	Gene name	Associate pathology/function	X-homologs
CSFR2Rα	PAR1	GM-CSF receptor α subunit	unknown	+
SHOX	"	Short stature homebox-containing	short stature, Lerf-Weill syndrome	+
IL3RA	"	Interleukin-3 receptor α subunit	unknown	+
ANT3	"	Adenine nucleotide translocase	unknown	+
ASMTL	"	Acetylserotonine methyltransferase-like	unknown	+
ASMT	"	Acetylserotonine methyltransferase	unknown	+
XE7	"	X-escapee	unknown	+
PGPL	"	Pseudoautosomal GTP-binding protein-like	unknown	+
MIC2	"		unknown	+
SRY★	Yp: 1A1A	Sex Reversal Y	Sex reversal	–
RPS4Y	Yp: 1A1B	Ribosomal protein S4, Y	Turner syndrome?	+

ZFY	Yp: 1A2	Zinc-finger Y	Turner syndrome?	+
PRKY	Yp: 3C-4A	protein kinase, Y	unknown	+
TTY1★	Yp: 4A	testis transcript, Y1	unknown	−
TSPY★	Yp: 3C + 5	testis-specific protein, Y	gonadoblastoma?	−
AMELY	Yp: 4A	Amelogenin, Y	unknown	+
PRY★	Y: 4A, 6E	putative tyrosine phosphatase protein-related Y	infertility?	+
TTY2★	Y: 4A, 6C	testis transcript, Y2	unknown	−
USP9Y (or DFFRY)	Yq: 5C	ubiquitin-specific protease (or Drosophila fat-facets related, Y)	azoospermia?	+
DBY	Yq: 5C	DEAD box, Y	infertility?	+
UTY	Yq: 5C	Ubitiquitous TRY motif, Y	infertility	+
TB4Y	Yq: 5D	Thymosin ?4, Y isoform	infertility	+
BPY1★	Yq: 5G	basic protein, Y1	Turner?	+
CDY	Yq: 5L, 6F	chromodomain, Y	infertility?	−
XKRY★	Yq: 5L	XK-related, Y	infertility?	−
RBM★	Yp + q	RNA-binding motif, Y	infertility?	−
SMCY	Yq: 5P	Selected Mouse cDNA, Y	unknown	+
EIF1AY	Yq: 5Q	Translation initiation factor 1A, Y	infertility?	+
DAZ★	Yq: 6F	Deleted in azoospermia	infertility?	−
VCY2	Yq: 6A	variably charged protein, Y2	infertility	−
IL9R	PAR2	Interleukin 9 receptor	unknown	+
SYBL1	"	Synaptobrevin-like 1	unknown	+
HSPRY3	"	Human-sprouty 3	unknown	+
CXYorf1	"	CXYorf1	unknown	+

Table: Genes of the human Y chromosome PAR1, PAR2, and NRY.

Biological Functions of the Human Y Chromosome

Several phenotypes have been associated with the nonrecombining portion of the Y chromosome. For obvious reasons, most of these are male-specific and make the Y a specialized chromosome during human evolution. The most characterizing features of this chromosome remain its implication in human sex determination and in male germ cell development and maintenance.

SRY Gene and Sex Determination

The first indices that the Y chromosome was involved in male sex determination came from the observation that XY or XYY (Klinefelter syndrome) individuals develop testes whereas XX or XO (Turner's syndrome) individuals develop ovaries. Later, studies showing that mice XX presenting a male phenotype carried a small portion of the Y chromosome supported the proposition that a master gene involved in male sex determination was carried by the Y chromosome. In 1990, the gene responsible for testicular determination, named SRY (Sex-determining Region on the Y chromosome), was finally identified. *SRY* was cloned by isolation of small fragments of translocated Y on XX sex-reversed patients. This gene is located on the short arm of the Y chromosome close to the pseudo autosomal boundary. It comprises a single exon encoding a protein of 204 amino acids which presents conserved DNA-binding domain (the HMG-box: High Mobility Group), suggesting this protein regulates gene expression. This gene has been shown to be essential for initiating testis development and the differentiation of the indifferent, bipotential, gonad into the testicular pathway. Moreover, *SRY* has been proposed to be the master gene regulating the cascade of testis determination. Although many genes and loci have been proposed to interact with *SRY* protein, such as *WT-1* (Wilm's tumor gene), *SF-1* (Steroidogenic Factor 1) and *SOX-9*, the question of how these genes are regulated, if so, by SRY is still unanswered.

Anti-Turner Syndrome Effect

Turner syndrome is characterized by a female 45 X karyotype or monosomy X. The principal manifestations of this syndrome are growth failure, infertility, anatomic abnormalities, and selective cognitive deficits. This human genetic disorder is ascribed to haplo-insufficiency of genes of the X chromosome that are common to both X and Y. These genes must escape X-inactivation because otherwise no difference will be observed between 45, X and 46, XX females. Secondly, in 46, XY these genes must have a male counterpart on the Y responsible to simulate the effects of their X homologues. Although there is no formal identification of genes involved in Turner syndrome, there appear to be different loci on the X and Y chromosome associated with Turner characteristic features, such as *SHOX/PHOG* ZFX/*ZFY*, *GCY* and *TCY*.

Genes Controlling Spermatogenesis

Tiepolo and Zuffardi reported the occurrence of grossly cytogenetically detectable *de novo* deletions in six azoospermic individuals, describing for the first time the role of the Y chromosome in spermatogenesis. These observations led the authors to postulate the existence of a locus, called AZoospermia Factor (AZF), on Yq11 required for a complete spermatogenesis since the seminal fluid of these patients did not contain mature spermatozoa. The location of AZF in Yq11 was further confirmed by numerous studies at cytogenetic and molecular level. Once the molecular map by Vergnaud became available, AZF was localized to the deletion interval 6, a region in band q11.23. The publication of about 200 Y-specific STS, allowed by Vollrath, allowed a much simpler Y chromosome screening for microdeletions to be performed. Thus, the original AZF region was further subdivided into three different nonoverlapping subregions in Yq11 associated with male infertility, named AZF_a, AZF_b, and AZF_c . Each one of these regions contains several genes proposed as candidate genes involved in male infertility.

The AZFa region is located in proximal Yq within the deletion interval 5 and its molecular extension has been roughly estimated between 1 and 3Mb. Several genes have been identified in this region; Dead Box Y (*DBY*), Ubiquitous TPR motif Y (*UTY*), Tymosin B4Y isoform (*TB4Y*), and the homologue of the Drosophila Developmental gene Fats Facets (*DFFRY*). The first three genes have no apparent specialised functions and they seem to be involved in cellular "housekeeping." By contrast, the *DFFRY* gene has been proposed to play a role in gametogenesis. It encodes a protein involved in desubiquitination (the process by which proteins are tagged for degradation) and mutations in the Drosophila homologue of the gene causes a sterile phenotype.

The AZFb region is located between deletion interval 5 and proximal deletion interval 6, and its molecular extension has been estimated to be similar to that of the AZFa region (1–3 Mb). Five genes have been so far described within this interval; RNA-binding motif (RBM), Chromodomain Y (*CDY*), XK Related Y (*XKRY*), eukaryotic translation initiation factor 1A (*eIF-1A*), and Selected Mouse cDNA on the Y (*SMCY*). The *RBM* gene encodes germ cell specific nuclear proteins containing RNA-binding motif and it is present in multiple copies along the Y. However, not all of these copies are functional and most may be pseudogenes. It has been strongly proposed as a candidate infertility gene since its expression is testis-specific, it is recurrently deleted in azoospermic men and is seems to be specifically expressed in spermatogonia and primary spermatocytes. Other two genes are expressed specifically in adult testis and are recurrently deleted in infertile males, the *CDY* and the *XKRY*.

The AZFc region is located in the proximity of the heterochromatin region distal to Yq11 and its molecular extension is about 500 kb. This region contains the DAZ (Deleted in AZoospermia) gene cluster, two copies of the PTP-BL Related Y (*PRY*), Basic Protein Y2 (*BPY2*), as well as copies of CDY and RBM. DAZ encodes a testis-specific RNA binding protein and contains seven tandem repeats of 24 aa unit. DAZ is present in at least six to nine copies, all being located within AZFc. It is homologous to an autosomal gene on chromosome 3, with a single DAZ repeat, named DAZL1 (DAZ like-autosomal 1) which is also specifically expressed in the testis. It has been hypothesized that DAZ originated from a translocation and subsequent amplification of this ancestral autosomal gene. Cooke *et al.* described the homologue of the human Y-linked DAZ gene, named Dazla (DAZ like autosomal), in the mouse where is located on chromosome 17. Dazla presents an RNA-binding domain with 89% of homology with DAZ and is expressed specifically in the testis and ovaries. Knockout mice for this gene have been shown to be infertile in both the two sexes. These observations suggested Dazla as an important gene in mouse gametogenesis. Although DAZ has been proposed as the cause of the AZFc phenotype, other genes must be involved since deletions within AZFc region without including DAZ have been recently reported. The other genes identified within this region, PRY, BPY2, and TTY2, all also present a testis-specific expression and are present in multiple copies on the Y.

Many of the AZF genes have been proposed as candidate genes involved in human male fertility on the basis of their expression profiles and sterile phenotypes from targeted disruption of their homologues in mice. However, no direct relation between a Y chromosome gene and male infertility has been demonstrated. In a recent paper, Page and coworkers relate spematogenic failure to a single mutation in a Y-linked gene in AZFa: the USP9Y or, also called, DFFRY. They found a *de novo* 4bp deletion in a splice-donor site of this gene present in a patient with non obstructive azoospermia but absent in his fertile brother. This mutation causes protein truncation leading to spermatogenic arrest. These findings lead the authors to conclude that the USP9Y gene has a role in human spermatogenesis.

Oncogenic Role of the Y Chromosome

The implication of the Y chromosome in cancer remains still speculative. Y chromosome loss and rearrangements have been associated with different types of cancer, such as bladder cancer, male sex cord stroma tumors, lung cancer and esophageal carcinoma. Although loss and rearrangements of this chromosome are relatively frequent in different types of cancer, there is no direct evidence for a role of Y in tumor progression since no proto-oncogenes, tumor suppresser genes or mismatch repair genes have been localized to the Y chromosome.

However, it is well presumable that both oncogenes and tumor suppressor genes must lay on this chromosome, having a pathogenic significance mainly in male-specific organs such as testis. One cancer predisposition locus has been assigned to this chromosome, the gonadoblastoma locus on the Y chromosome (GBY). The gonadoblastoma is a rare form of cancer that consists of aggregates of germ cells and sex cord elements. It develops in more than 30% of dysgenetic gonads from sex-reversed females (Swyers syndrome) who harbor some Y-chromosomal material. This observation led to postulate the existence of a predisposing locus on the Y (GBY) that enhance dysgenetic gonads to develop gonadoblastoma. This locus could act as an oncogene in dysgenetic gonads, having a normal function in the testis, and it would have a pathogenic effect when is expressed out of its natural environment (normal testis). This locus would expand over a region of 1–2 Mb on the short arm of the Y chromosome, in the region 4A-4B Several genes have been proposed as candidates for GBY according to their location, function, and expression profile. Among them, the most likely candidate seems to be TSPY. This gene, present in several copies, is located in the critical region where GBY has been mapped and is expressed in gonadoblastoma, in spermatogonias at early stages of testicular tumorigenesis, in carcinoma *in situ* of the testis, in seminoma and prostate cancers. These observations strongly suggest that this Y-linked gene may predispose germ cells to other oncogenic events in the multistep process of tumorigenesis.

Chromosomal Sex Determination System

Primary and Secondary Sex Determination

Primary sex determination is the determination of the gonads. In mammals, primary sex determination is strictly chromosomal and is not usually influenced by the environment. In most cases, the female is XX and the male is XY. Every individual must have at least one X chromosome. Since the female is XX, each of her eggs has a single X chromosome. The male, being XY, can generate two types of sperm: half bear the X chromosome, half the Y. If the egg receives another X chromosome from the sperm, the resulting individual is XX, forms ovaries, and is female; if the egg receives a Y chromosome from the sperm, the individual is XY, forms testes, and is male. The Y chromosome carries a gene that encodes a testis-determining factor. This factor organizes the gonad into a testis rather than an ovary. Unlike the situation in Drosophila, the mammalian Y chromosome is a crucial factor for determining sex in mammals. A person with five X chromosomes and one Y chromosome (XXXXXY) would be male. Furthermore, an individual with only a single X chromosome and no second X or Y (i.e., XO) develops as a female and begins making ovaries, although the ovarian follicles cannot be maintained. For a complete ovary, a second X chromosome is needed.

In mammalian primary sex determination, there is no "default state." The formation of ovaries and testes are both active, gene-directed processes. Moreover, both diverge from a common precursor, the bipotential gonad.

Secondary sex determination affects the bodily phenotype outside the gonads. A male mammal has a penis, seminal vesicles, and prostate gland. A female mammal has a vagina, cervix, uterus, oviducts, and mammary glands. In many species, each sex has a sex-specific size, vocal cartilage, and musculature. These secondary sex characteristics are usually determined by hormones secreted from the gonads. However, in the absence of gonads, the female phenotype is generated. When Jost removed fetal rabbit gonads before they had differentiated, the resulting rabbits had a female phenotype, regardless of whether they were XX or XY. They each had oviducts, a uterus, and a vagina, and each lacked a penis and male accessory structures.

The general scheme of mammalian sex determination is shown in the figure. If the Y chromosome is absent, the gonadal primordia develop into ovaries. The ovaries produce estrogen, a hormone that enables the development of the Müllerian duct into the uterus, oviducts, and upper end of the vagina. If the Y chromosome is present, testes form and secrete two major hormones. The first hormone—anti-Müllerian duct hormone (AMH; also referred to as Müllerian-inhibiting substance, MIS)—destroys the Müllerian duct. The second hormone—testosterone—masculinizes the fetus, stimulating the formation of the penis, scrotum, and other portions of the male anatomy, as well as inhibiting the development of the breast primordia. Thus, the body has the female phenotype unless it is changed by the two hormones secreted by the fetal testes. We will now take a more detailed look at these events.

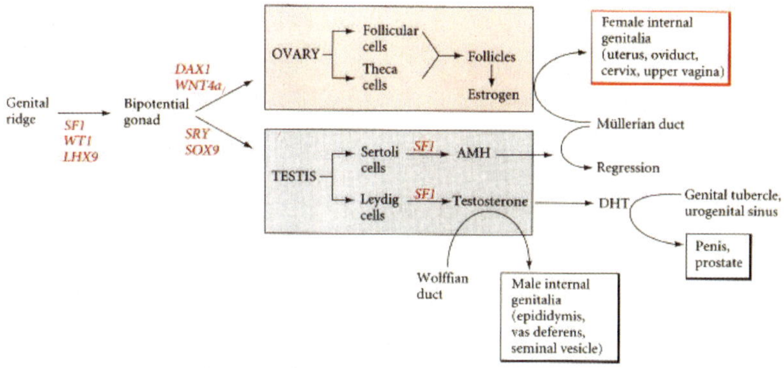

Figure: Postulated cascades leading to the formation of the sexual phenotypes in mammals.

The conversion of the genital ridge into the bipotential gonad requires the *LHX9*, *SF1* and *WT1* genes, since mice lacking either of these genes lack gonads. The bipotential gonad appears to be moved into the female pathway (ovary development) by the *WNT4* and *DAX1* genes and into the male pathway (testis development) by the *SRY* gene (on the Y chromosome) in conjunction with autosomal genes such as *SOX9*. The ovary makes thecal cells and granulosa cells, which together are capable of synthesizing estrogen. Under the influence of estrogen (first from the mother, then from the fetal gonads), the Müllerian duct differentiates into the female genitalia, and the offspring develops the secondary sex characteristics of a female. The testis makes two major hormones. The first, anti-Müllerian duct factor (AMH), causes the Müllerian duct to regress. The second, testosterone, causes the differentiation of the Wolffian duct into the male internal genitalia. In the urogenital region, testosterone is converted into dihydrotestosterone (DHT), and this hormone causes the morphogenesis of the penis and prostate gland.

Developing Gonads

The gonads embody a unique embryological situation. All other organ rudiments can normally differentiate into only one type of organ. A lung rudiment can become only a lung, and a liver rudiment can develop only into a liver. The gonadal rudiment, however, has two normal options. When it differentiates, it can develop into either an ovary or a testis. The path of differentiation taken by this rudiment determines the future sexual development of the organism. But, before this decision is made, the mammalian gonad first develops through a bipotential (indifferent) stage, during which time it has neither female nor male characteristics.

In humans, the gonadal rudiments appear in the intermediate mesoderm during week 4 and remains sexually indifferent until week 7. The gonadal rudiments are paired regions of the intermediate mesoderm; they form adjacent to the developing kidneys. The ventral portions of the gonadal rudiments are composed of the genital ridge epithelium. During the indifferent stage, the genital ridge epithelium proliferates into the loose connective mesenchymal tissue above it. These epithelial layers form the sex cords. The germ cells migrate into the gonad during week 6, and are surrounded by the sex cords. In both XY and XX gonads, the sex cords remain connected to the surface epithelium.

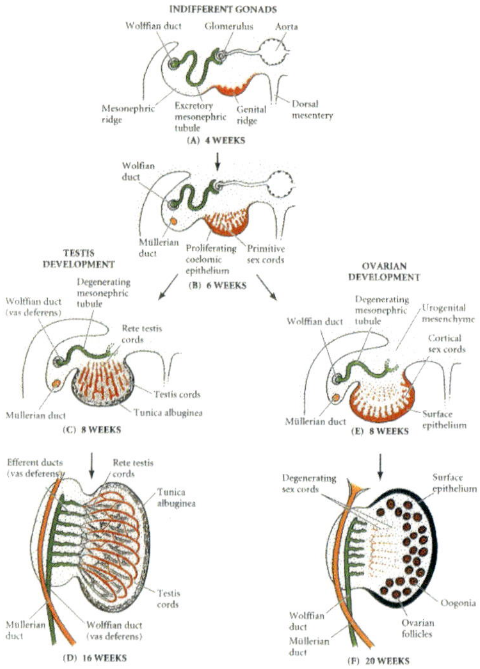

Figure: Differentiation of human gonads shown in transverse section

(A) Genital ridge of a 4-week embryo. (B) Genital ridge of a 6-week indifferent gonad showing primitive sex cords. (C) Testis development in the eighth week. The sex cords lose contact with the cortical epithelium and develop the rete testis. (D) By the sixteenth week of development, the testis cords are continuous with the rete testis and connect with the Wolffian duct. (E) Ovary development in an 8-week human embryo, as primitive sex cords degenerate. (F) The 20-week human ovary does not connect to the Wolffian duct, and new cortical sex cords surround the germ cells that have migrated into the genital ridge.

If the fetus is XY, the sex cords continue to proliferate through the eighth week, extending deeply into the connective tissue. These cords fuse, forming a network of internal (medullary) sex cords and, at its most distal end, the thinner rete testis . Eventually, the sex cords—now called testis cords—lose contact with the surface epithelium and become separated from it by a thick extracellular matrix, the tunica albuginea. Thus, the germ cells are found in the cords within the testes. During fetal life and childhood, the testis cords remain solid. At puberty, however, the cords will hollow out to form the seminiferous tubules, and the germ cells will begin to differentiate into sperm.

The cells of the seminiferous tubule are called Sertoli cells. The Sertoli cells of the testis cords nurture the sperm and secrete anti-Müllerian duct hormone. The sperm are transported from the inside of the testis through the rete testis, which joins the efferent ducts. These efferent tubules are the remnants of the mesonephric kidney, and they link the testis to the Wolffian duct, which used to be the collecting tube of the mesonephric kidney. In males, the Wolffian duct differentiates to become the epididymis (adjacent to the testis) and the vas deferens, the tube through which the sperm pass into the urethra and out of the body. Meanwhile, during fetal development, the interstitial mesenchyme cells of the testes differentiate into Leydig cells, which make testosterone.

Vade Mecum

Mammalian gonads. The histology of the mammalian ovary and testis can be seen in labeled photographs that show progressively smaller regions at higher magnifications.

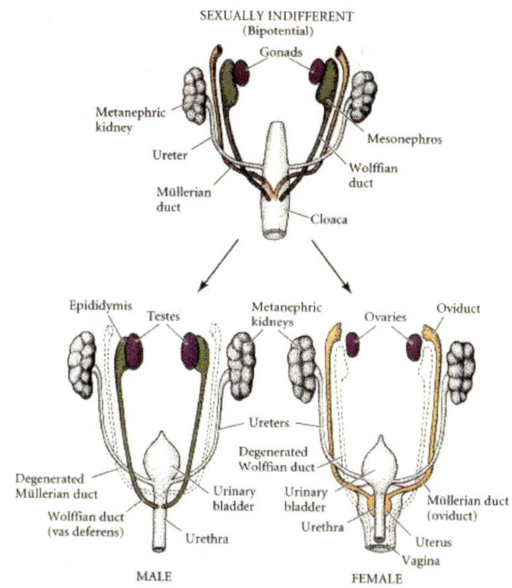

GONADS		
Gonadal type	Testis	Ovary
Sex cords	Medullary (internal)	Cortical (external)
DUCTS		
Remaining duct for germ cells	Wolffian	Müllerian
Duct differentiation	Vas deferens, epididymis, seminal vessicle	Oviduct, uterus, cervix, upper portion of vagina

Figure: Summary of the development of the gonads and their ducts in mammals

In females, the germ cells will reside near the outer surface of the gonad. Unlike the sex cords in males, which continue their proliferation, the initial sex cords of XX gonads degenerate. However, the epithelium soon produces a new set of sex cords, which do not penetrate deeply into the mesenchyme, but stay near the outer surface (cortex) of the organ. Thus, they are called cortical sex cords. These cords are split into clusters, with each cluster surrounding a germ cell . The germ cells will become the ova, and the surrounding cortical sex cords will differentiate into the granulosa cells. The mesenchyme cells of the ovary differentiate into the thecal cells. Together, the thecal and granulosa cells will form the follicles that envelop the germ cells and secrete steroid hormones. Each follicle will contain a single germ cell. In females, the Müllerian duct remains intact, and it differentiates into the oviducts, uterus, cervix, and upper vagina. The Wolffian duct, deprived of testosterone, degenerates. A summary of the development of mammalian reproductive systems is shown in figure.

The Mechanisms of Mammalian Primary Sex Determination

Several genes have been found whose function is necessary for normal sexual differentiation. Unlike those that act in other developing organs, the genes involved in sex determination differ extensively between phyla, so one cannot look at *Drosophila* sex-determining genes and expect to see their homologues directing mammalian sex determination. However, since the phenotype of mutations in sex-determining genes is often sterility, clinical studies have been used to identify those genes that are active in determining whether humans become male or female. Experimental manipulations to confirm the functions of these genes can be done in mice.

Sry: Y Chromosome Sex Determinant

In humans, the major gene for the testis-determining factor resides on the short arm of the Y chromosome. Individuals who are born with the short arm but not the long arm of the Y chromosome are male, while individuals born with the long arm of the Y chromosome but not the short arm are female. By analyzing the DNA of rare XX men and XY women, the position of the testis-determining gene has been narrowed down to a 35,000-base-pair region of the Y chromosome located near the tip of the short arm. In this region, Sinclair and colleague found a male-specific DNA sequence that could encode a peptide of 223 amino acids. This peptide is probably a transcription factor, since it contains a DNA-binding domain called the HMG (*h*igh-*m*obility *g*roup) box. This domain is found in several transcription factors and nonhistone chromatin proteins, and it induces bending in the region of DNA to which it binds. This gene is called *SRY* (*s*ex-determining *r*egion of the *Y* chromosome), and there is extensive evidence that it is indeed the gene that encodes the human testis-determining factor. *SRY* is found in normal XY males and in the rare XX males, and it is absent from normal XX females and from many XY females. Another group of XY females was found to have point or frameshift mutations in the *SRY* gene; these mutations prevent the SRY protein from binding to or bending DNA. It is thought that several testis-specific genes contain SRY-binding sites in their promoters or enhancers, and that the binding of SRY to these sites begins the developmental pathway to testis formation.

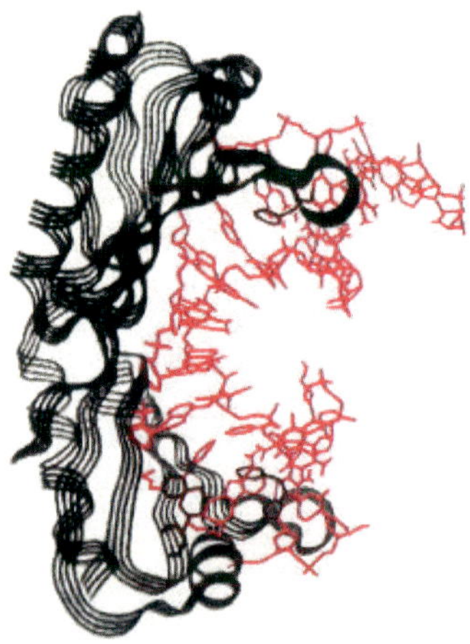

Figure: The black structure represents the HMG box of the SRY protein.
The red coil is the double helix of DNA specifically bound by SRY

If *SRY* actually does encode the major testis-determining factor, one would expect that it would act in the genital ridge immediately before or during testis differentiation. This prediction has been met in studies of the homologous gene found in mice. The mouse gene (*Sry*) also correlates with the presence of testes; it is present in XX males and absent in XY females. The *Sry* gene is expressed in the somatic cells of the bipotential mouse gonad immediately before or during its differentiating into a testis; its expression then disappears.

The most impressive evidence for *Sry* being the gene for testis-determining factor comes from transgenic mice. If *Sry* induces testis formation, then inserting *Sry* DNA into the genome of a normal XX mouse zygote should cause that XX mouse to form testes. Koopman and colleagues took the 14-kilobase region of DNA that includes the *Sry* gene (and presumably its regulatory elements) and microinjected this sequence into the pronuclei of newly fertilized mouse zygotes. In several instances, the XX embryos injected with this sequence developed testes, male accessory organs, and penises . (Functional sperm were not formed, but they were not expected, either, because the presence of two X chromosomes prevents sperm formation in XXY mice and men, and the transgenic mice lacked the rest of the Y chromosome, which contains genes needed for spermatogenesis.) Therefore, there are good reasons to think that *Sry/SRY* is the major gene on the Y chromosome for testis determination in mammals.

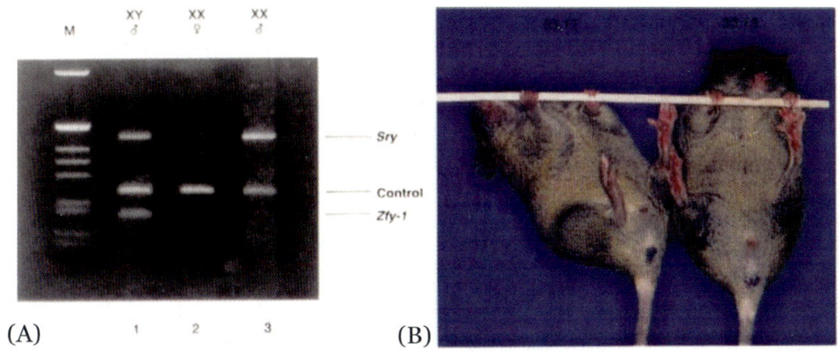

Figure: An XX mouse transgenic for Sry is male. (A) Polymerase chain reaction followed by electrophoresis shows the presence of the Sry gene in normal XY males and in a transgenic XX Sry mouse. The gene is absent in a female XX littermate. (B) The external genitalia of the transgenic mouse are male (right) and essentially the same as in an XY male (left).

Sry/SRY is necessary, but not sufficient, for the development of the mammalian testis. Studies on mice have shown that the *Sry* gene of some strains of mice failed to produce testes when placed into a different strain of mouse. When the Sry protein binds to its sites on DNA, it probably creates large conformational changes. It unwinds the double helix in its vicinity and bends the DNA as much as 80 degrees . This bending may bring distantly bound proteins of the transcription apparatus into close contact, enabling them to interact and influence transcription. The identities of these proteins are not yet known, but they, too, are needed for testis determination.

SRY may have more than one mode of action in converting the bipotential gonads into testes. It had been assumed for the past decade that SRY worked directly in the genital ridge to convert the epithelium into male-specific Sertoli cells. Recent studies, however, have suggested that SRY works via an indirect mechanism: SRY in the genital ridge cells induces the cells to secrete a chemotactic factor that permits the migration of mesonephric cells into the XY gonad. These mesonephric cells induce the gonadal epithelium to become Sertoli cells with male-specific gene expression patterns. The researchers found that when they cultured XX gonads with either XX or XY mesonephrons, the mesonephric cells did not enter the gonads. However, when they cultured XX or XY mesonephrons with XY gonads, or with gonads from XX mice containing the *Sry* transgene, the mesonephric cells did enter the gonads. There was a strict correlation between the presence of Sry in the gonadal cells, mesonephric cell migration, and the formation of testis cords. Tilmann and Capel showed that mesonephric cells are critical for testis cord formation and that the migrating mesonephric cells can induce XX gonadal cells to form testis cords. It appears, then, that Sry may function indirectly to create testes by inducing mesonephric cell migration into the gonad.

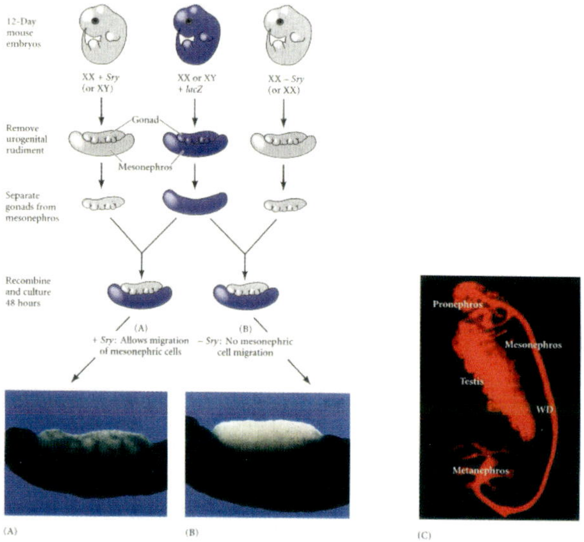

Figure: Migration of the mesonephric cells into Sry+ gonadal rudiments

In the experiment diagrammed, urogenital ridges (containing both the mesonephric kidneys and gonadal rudiments) were collected from 12-day embryonic mice. One of the mice was marked with a β-galactosidase transgene (lacZ) that is active in every cell. Thus every cell of this mouse turns blue when stained for β-galactosidase. The gonad and mesonephros were separated and recombined, using gonadal tissue from unlabeled mice and mesonephros from labeled mice. (A) Migration of mesonephric cells into the gonad was seen when the gonadal cells were XY or when they were XX with an Sry transgene. (B) No migration of mesonephric tissue into the bipotential gonad was seen when the gonad contained either XX cells or XXY cells in which the Y chromosome had a deletion in the Sry gene. The sex chromosomes of the mesonephros did not affect the migration. (C) Intimate relation between the mesonephric ducts and the developing gonad in the 16-day male mouse embryo. The duct tissue has been stained for cytokeratin-8. WD, Wolfian duct.

Sox9: Autosomal Sex Reversal

One of the autosomal genes involved in sex determination is *SOX9*, which encodes a putative transcription factor that also contains an HMG box. XX humans who have an extra copy of *SOX9* develop as males, even though they have no *SRY* gene. Individuals having only one functional copy of this gene have a syndrome called campomelic dysplasia, a disease involving numerous skeletal and organ systems. About 75% of XY patients with this syndrome develop as phenotypic females or hermaphrodites . It appears that *SOX9* is essential for testis formation. The mouse homologue of this gene, *Sox9,* is expressed only in male (XY) but not in female (XX) genital ridges. Moreover, *Sox9* expression is seen in the same genital ridge cells as *Sry,* and it is expressed just slightly after *Sry* expression. The Sox9 protein binds to a promoter site on the *Amh* gene, providing a critical link in the pathway toward a male phenotype .

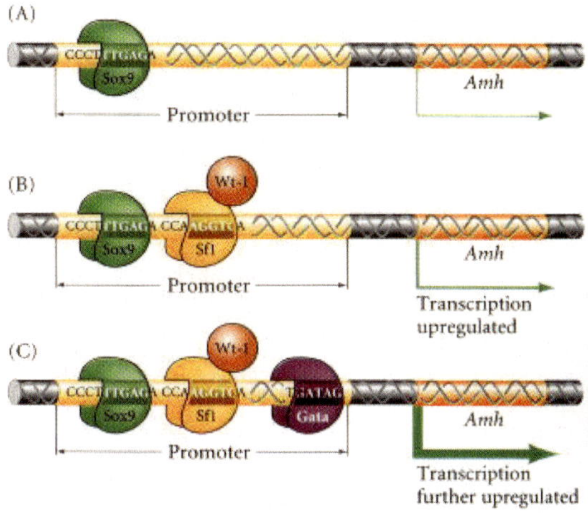

Figure: Synergism of Sox9 and Sf1 to activate the expression of the Amh gene

(A) The binding of Sox9 to the Amh promoter initiates transcription of the Amh gene in the Sertoli cells. (B) After Sox9 binds, the expression of AMH is upregulated by the binding of SF1 and Wt-1. AMH is believed to position SF1 on its DNA-binding site, and Wt-1 is joined to the Sf1 protein. (C) Gata (a transcription factor common to many cell types) upregulates Amh expression further. Neither Sf1 nor Gata can function if Sox9 is absent.

While Sry is found specifically in mammals, Sox9 is found throughout the vertebrates. Sox9 may be the older and more central sex determination gene, although in mammals it became activated by its relative, Sry.

Sf1: The Link between Sry and the Male Developmental Pathways

Another protein that may be directly or indirectly activated by SRY is the transcription factor SF1 (*s*teroidogenic *f*actor *1*). *Sf1* is necessary to make the bipotential gonad; but while Sf1 levels decline in the genital ridge of XX mouse embryos, the *Sf1* gene stays on in the developing testis. Sf1 appears to be active in masculinizing both the Leydig and the Sertoli cells. In the Sertoli cells, Sf1, working in collaboration with Sox9, is needed to elevate the levels of AMH transcription. In the Leydig cells, Sf1 activates the genes encoding the enzymes that make testosterone. The importance of SF1 for testis development and AMH regulation in humans is demonstrated by an XY patient who is heterozygous for *SF1*. Although the genes for SRY and SOX9 are normal, this individual has malformed fibrous gonads and retains fully developed Müllerian duct structures. It is thought that SRY (directly or indirectly) activates the *SF1* gene, and the SF1 protein then activates both components of the male sexual differentiation pathway (Sertoli AMH and Leydig testosterone).

Dax1: A Potential Ovary-determining Gene on the X Chromosome

In 1980, Bernstein and her colleagues reported two sisters who were genetically XY. Their Y chromosomes were normal, but they had a duplication of a small portion of the short arm of the X chromosome. Subsequent cases were found, and it was concluded that if there were two copies of this region on the active X chromosome, the SRY signal would be reversed proposed that this region contains a gene for a protein that competes with the SRY factor and that is important in directing the development of the ovary. In testicular development, this gene would be suppressed, but having two active copies of the gene would override this suppression. This gene, *DAX1*, has been cloned and shown to encode a member of the nuclear hormone receptor family. *Dax1* is expressed in the genital ridges of the mouse embryo, shortly after *Sry* expression. Indeed, in XY mice, *Sry* and *Dax1* are expressed in the same cells. DAX1 appears to antagonize the function of SRY, and it down-regulates SF1 expression. Thus, *DAX1* is probably a gene that is involved in ovary determination.

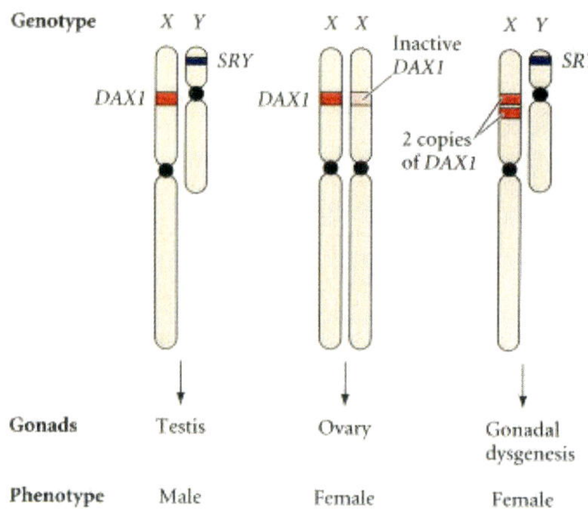

Figure : Phenotypic sex reversal in humans having two copies of the DAX1 locus

DAX1 (on the X chromosome) plus SRY (on the Y chromosome) produces testes. DAX1 without SRY (since the other DAX1 locus is on the inactive X chromosome) produces ovaries. Two active copies of DAX1 (on the active X chromosome) plus SRY (on the Y chromosome) lead to a poorly formed gonad. Since the gonad makes neither AMH nor testosterone, the phenotype is female.

Wnt4: A Potential Ovary-determining Gene on an Autosome

The *WNT4* gene is another gene that may be critical in ovary determination. This gene is expressed in the mouse genital ridge while it is still in its bipotential stage. *Wnt4* expression then becomes undetectable in XY gonads (which become testes), whereas it is maintained in XX gonads as they begin to form ovaries. In transgenic XX mice that lack the *Wnt4* genes, the ovary fails to form properly, and its cells express testis-specific markers, including AMH- and testosterone-producing enzymes. Sry may form testes by repressing *Wnt4* expression in the genital ridge, as well as by promoting *Sf1*. One possible model is shown in figure.

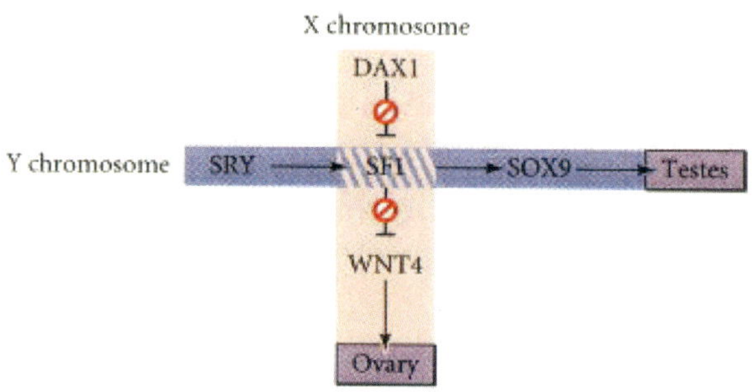

Figure: Possible mechanism for primary sex determination in mammals

While we do not know the specific interactions involved, this model attempts to organize the data into a coherent sequence. Other models are possible. In this model, SRY competes with the DAX1 protein to activate or repress the SF1 gene. If a single X and Y chromosome are present, the SRY will be favored, and the activation of SF1 will occur. If two copies of DAX1 are present on the X chromosome (or if there is no Y chromosome), the SF1 gene will not be activated. The SF1 protein is thought to activate the SOX9 gene, which instructs the sex cords to develop into the Sertoli cells of the testes, and may also repress WNT4. WNT4 would otherwise cause the differentiation of the gonad into an ovary. Most of the genes activated by WNT4 and SOX9 have not been identified, and the mechanisms by which SRY and DAX1 function are not yet known.

It should be realized that both testis and ovary development are active processes. In mammalian primary sex determination, neither is a "default state". Although remarkable progress has been made in recent years, we still do not know what the testis- or ovary-determining genes are doing, and the problem of primary sex determination remains (as it has since prehistory) one of the great unsolved problems of biology.

Secondary Sex Determination: Hormonal Regulation of the Sexual Phenotype

Primary sex determination involves the formation of either an ovary or a testis from the bipotential gonad. This, however, does not give the complete sexual phenotype. Secondary sex determination in mammals involves the development of the female and male phenotypes in response to hormones secreted by the ovaries and testes. Both female and male secondary sex determination have two major temporal phases. The first occurs within the embryo during organogenesis; the second occurs during adolescence.

As mentioned earlier, if the bipotential gonads are removed from an embryonic mammal, the female phenotype is realized: the Müllerian ducts develop while the Wolffian duct degenerates. This pattern also is seen in certain humans who are born without functional gonads. Individuals whose cells have only one X chromosome (and no Y chromosome) originally develop ovaries, but these ovaries atrophy before birth, and the germ cells die before puberty. However, under the influence of estrogen, derived first from the ovary but then from the mother and placenta, these infants are born with a female genital tract.

The formation of the male phenotype involves the secretion of two testicular hormones. The first of these hormones is AMH, the hormone made by the Sertoli cells that causes the degeneration of the Müllerian duct. The second is the steroid testosterone, which is secreted from the fetal Leydig cells. This hormone causes the Wolffian duct to differentiate into the epididymis, vas deferens, and seminal vesicles, and it causes the urogenital swellings to develop into the scrotum and penis.

The existence of these two independent systems of masculinization is demonstrated by people having androgen insensitivity syndrome. These XY individuals have the *SRY* gene, and thus have testes that make testosterone and AMH. However, they lack the testosterone receptor protein, and therefore cannot *respond* to the testosterone made by their testes . Because they are able to respond to estrogen made in their adrenal glands, they develop the female phenotype. However, despite their distinctly female appearance, these individuals do have testes, and even though they cannot respond to testosterone, they produce and respond to AMH. Thus, their Müllerian ducts degenerate. These people develop as normal but sterile women, lacking a uterus and oviducts and having testes in the abdomen.

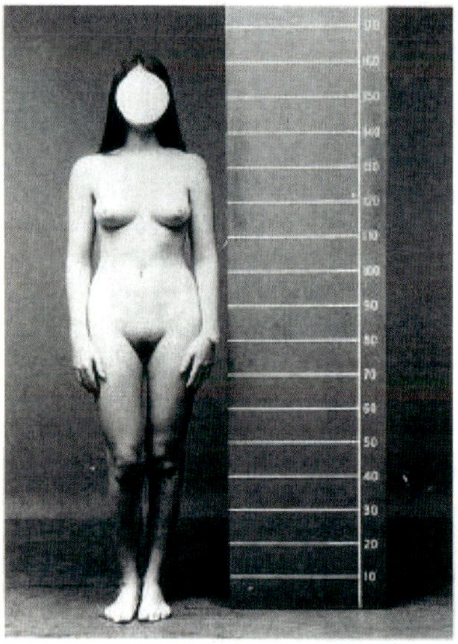

Figure: An XY individual with androgen insensitivity syndrome.

Despite the XY karyotype and the presence of testes, such individuals develop female secondary sex characteristics. Internally, however, these women lack the Müllerian duct derivatives and have undescended testes.

Testosterone and Dihydrotestosterone

Although testosterone is one of the two primary masculinizing hormones, there is evidence that it might not be the active masculinizing hormone in certain tissues. Testosterone appears to be responsible for promoting the formation of the male reproductive structures (the epididymis, seminal vesicles, and vas deferens) that develop from the Wolffian duct primordium. However, it does not directly masculinize the male urethra, prostate, penis, or scrotum. These latter functions are controlled by 5a-dihydrotestosterone showed that testosterone is converted to 5α-dihydrotestosterone in the urogenital sinus and swellings, but not in the Wolffian duct. 5α-dihydrotestosterone appears to be a more potent hormone than testosterone.

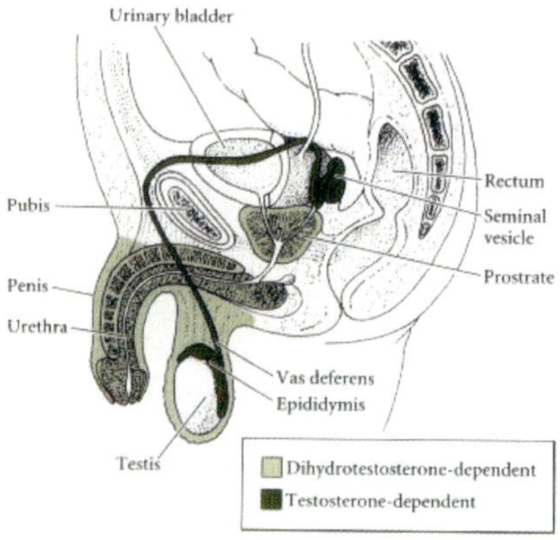

Figure: Testosterone- and dihydrotestosterone-dependent regions of the human male genital system

The importance of 5α-dihydrotestosterone was demonstrated by Imperato-McGinley and her colleagues. They found a small community in the Dominican Republic in which several inhabitants had a genetic deficiency of the enzyme 5α-ketosteroid reductase 2, the enzyme that converts testosterone to dihydrotestosterone. These individuals lack a functional gene for this enzyme. Although XY children with this syndrome have functioning testes, they have a blind vaginal pouch and an enlarged clitoris. They appear to be girls and are raised as such. Their internal anatomy, however, is male: they have testes, Wolffian duct development, and Müllerian duct degeneration. Thus, it appears that the formation of the external genitalia is under the control of dihydrotestosterone, whereas Wolffian duct differentiation is controlled by testosterone itself. Interestingly, when the testes of these children produce more testosterone at puberty, the external genitalia are able to respond to the higher levels of the hormone, and they differentiate. The penis enlarges, the scrotum descends, and the person originally thought to be a girl is shown to be a young man.

Anti-müllerian Duct Hormone

Anti-müllerian duct hormone (AMH), the hormone that causes the degeneration of the Müllerian duct, is a 560-amino acid glycoprotein secreted from the Sertoli cells. When fragments of fetal testes or isolated Sertoli cells are placed adjacent to cultured tissue segments containing portions of the Wolffian and Müllerian ducts, the Müllerian duct atrophies even though no change occurs in the Wolffian duct . AMH is thought to bind to the mesenchyme cells surrounding the Müllerian duct and to cause these cells to secrete a paracrine factor that induces apoptosis in the Müllerian duct epithelium.

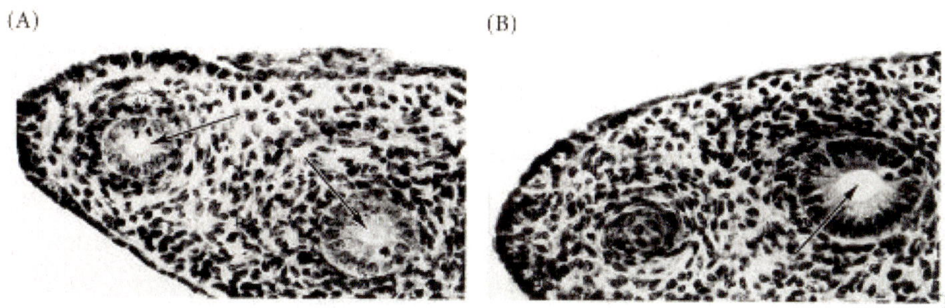

Figure: Assay for AMH activity in the anterior segment of a 14.5-day fetal rat reproductive tract

In figure (A) At the start of the experiment, both the Müllerian duct (arrow at left) and Wolffian duct (arrow at right) are open. (B) After 3 days in culture with AMH-secreting tissue, the Wolffian duct (arrow) is open, but the Müllerian duct has degenerated and closed.

Estrogen

Estrogen is needed for the complete development of both the Müllerian and the Wolffian ducts. In females, estrogen secreted from the fetal ovaries appears sufficient to induce the differentiation of the Müllerian duct into its various components: the uterus, oviducts, and cervix. The extreme sensitivity of the Müllerian duct to estrogenic compounds is demonstrated by the teratogenic effects of diethylstilbesterol (DES), a powerful synthetic estrogen that can cause infertility by changing the patterning of the Müllerian duct. In mice, DES can cause the oviduct epithelium to take on the appearance of the uterus, and the uterine epithelium to resemble that of the cervix.

In males, estrogen is actually needed for fertility. One of the functions of the efferent duct (vas eferens) cells is to absorb about 90% of the water from the lumen of the rete testis. This concentrates the sperm, giving them a longer lifespan and providing more sperm per ejaculate. This absorption of water is regulated by estrogen. If estrogen or its receptor is absent in mice, this water is not absorbed, and the mouse is sterile. While blood concentrations of estrogen are higher in females than in males, the concentration of estrogen in the rete testis is even higher than that in female blood.

Nucleosome

A nucleosome is a section of DNA that is wrapped around a core of proteins. Inside the nucleus, DNA forms a complex with proteins called chromatin, which allows the DNA to be condensed into a smaller volume. When the chromatin is extended and viewed under a microscope, the structure resembles beads on a string. Each of these tiny beads is a called a nucleosome and has a diameter of approximately 11 nm. The nucleosome is the fundamental subunit of chromatin. Each nucleosome is composed of a little less than two turns of DNA wrapped around a set of eight proteins called histones, which are known as a histone octamer. Each histone octamer is composed of two copies each of the histone proteins H2A, H2B, H3, and H4. The chain of nucleosomes is then compacted further and forms a highly organized complex of DNA and protein called a chromosome.

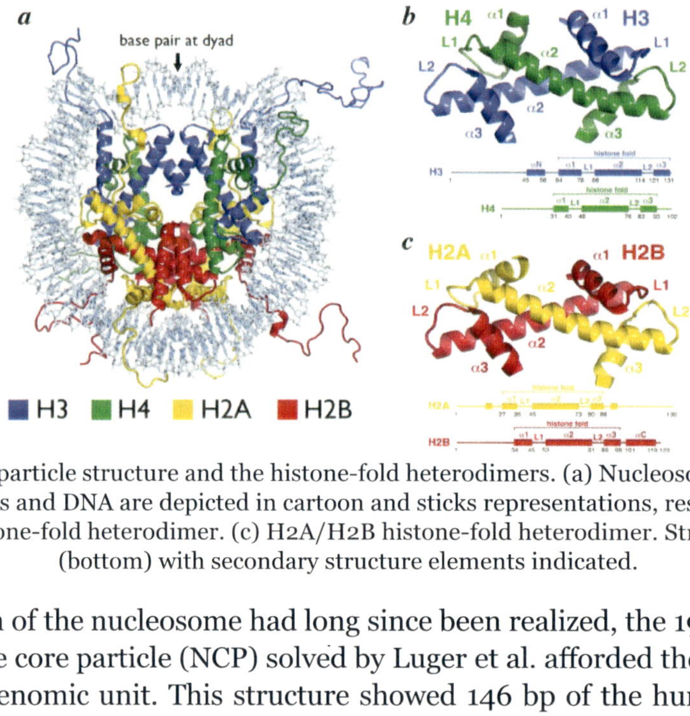

Figure: Nucleosome core particle structure and the histone-fold heterodimers. (a) Nucleosome core particle structure (PDB ID 1KX5). Histones and DNA are depicted in cartoon and sticks representations, respectively, and colored as indicated. (b) H3/H4 histone-fold heterodimer. (c) H2A/H2B histone-fold heterodimer. Structures (top) and schemes (bottom) with secondary structure elements indicated.

While the composition of the nucleosome had long since been realized, the 1997 2.8 Å crystal structure of the nucleosome core particle (NCP) solved by Luger et al. afforded the first atomic depiction of this fundamental genomic unit. This structure showed 146 bp of the human alpha-satellite sequence wrapped 1.65 times around an octameric scaffold of *Xenopus laevis* histone proteins in a left-handed superhelix. A single base pair is centered on the nucleosome dyad, which defines the pseudo-2-fold symmetry axis of the NCP. DNA locations are designated by superhelical locations (SHL) representing superhelical turns from the dyad (SHL 0) and ranging from SHL -7 to SHL 7. The central histone octamer contains two copies of each of the core histone proteins, H2A, H2B, H3, and H4 as established by Arents and Moudrianakis in the 1991 3.1 Å crystal structure of the histone octamer. The core histones are assembled into four histone-fold heterodimers (two each of H2A/H2B and H3/H4). Ten flexible tails protrude from the NCP at defined locations, one N-terminal tail from each of the eight core histone proteins and two additional C-terminal tails contributed by H2A.

Histone-fold Heterodimers

Each of the core histones contains a central α-helical region that forms a histone-fold motif, flanked by N- and C-terminal extensions. The histone-fold is constructed from three α helices connected by two intervening loops specified as α1-L1-α2-L2-α3. The two shorter α1 and α3 helices loop back to pack against the longer central α2 helix. Each histone-fold pairs with a complementary histone-fold, H3 pairs with H4 and H2A pairs with H2B, to form a histone-fold heterodimer handshake motif. The antiparallel arrangement of this heterodimer approximates the L1 loop from one histone-fold and the L2 loop of the complementary histone-fold, placing one L1L2 pair at each end of the heterodimer. The result is a crescent-shaped heterodimer with the convex surface including the L1L2 loops and the α1 helices and the concave surface including the α3 and central α2 helices. The convex surface of the H2A/H2B and H3/H4 heterodimers carries a strong positive charge and constitutes the primary DNA binding element of each histone-fold heterodimer.

Histone Octamer Architecture and DNA Binding

The histone octamer forms a spool for wrapping nucleosomal DNA. It is assembled from two H3/H4 and two H2A/H2B histone-fold heterodimers using a single structural motif, the four-helix bundle. Each four-helix bundle is formed by the α2 and α3 helices from the adjacent histone-folds. Two H3/H4 dimers interact in a head to head arrangement through an H3/H3 four helix bundle to form an $(H3/H4)_2$ tetramer. An H2A/H2B dimer binds to each half of the $(H3/H4)_2$ tetramer using a four-helix bundle formed by the H4 and H2B histone folds. Several structured N- and C-terminal extensions to the histone-fold regions also contribute to the histone octamer architecture. The αN helix between the N-terminal tail and histone-fold of H3 rests on top of the H4 histone-fold and organizes DNA at the entry/exit site of the NCP. H2A and H2B also contain C-terminal extensions that contribute to the nucleosome core surface. The H2B αC helix extends from the center of the nucleosome disk to the DNA edge opposite the nucleosome dyad, packing against the underlying H2A/H2B histone-fold helices. The H2A C-terminal extension includes a docking domain that interacts with the H2A/H2B histone-fold dimer after which it traverses the nucleosome surface toward the dyad resting on a platform generated by the underlying H3/H4 heterodimer from the opposite side of the octamer.

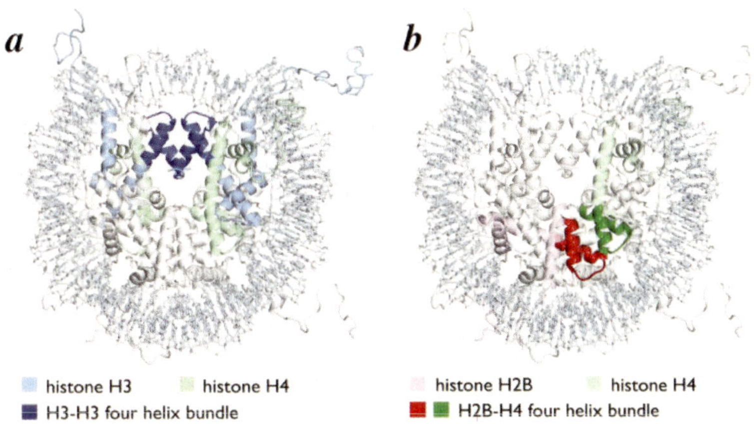

Figure: Histone octamer constructed with four helix bundles. (a) Nucleosome core particle structure highlighting H3–H3 four helix bundle (blue). Remainder of H3 and H4 are shown in light blue and light green, respectively. (b) Nucleosome core particle structure highlighting one H4–H2B four helix bundle (green for H4 and red for H2B). Remainder of H4 and H2B are shown in light green and pink, respectively.

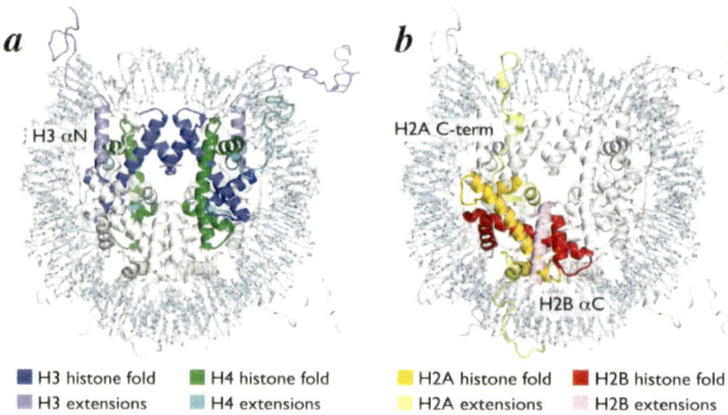

Figure: Histone-fold heterodimers in the nucleosome core particle structure.

In figure (a) Nucleosome core particle structure with central H3/H4 histone-fold tetramer shown in blue (H3) and green (H4). H3 and H4 extensions are shown in light blue and light green, respectively. (b) Nucleosome core particle structure with one H2A/H2B histone-fold dimer shown in yellow (H2A) and red (H2B). H2A and H2B extensions are shown in light yellow and pink, respectively.

The H2A–H2B–H4–H3–H3–H4–H2B–H2A octamer designates a nonuniform path of the nucleosomal DNA. The H2A/H2B dimers bind DNA in two planes perpendicular to the DNA superhelical axis, while the central H3/H4 tetramer forms a diagonal ramp through the nucleosomal dyad, connecting these two planes. Notably, the DNA gyres in neighboring planes align their major and minor grooves, respectively, as they track along the octamer surface. The histone-fold regions of the octamer bind the central ~121 bp of nucleosomal DNA. The remaining ~13 bp at each of the DNA ends is organized by the histone-fold extensions, especially the H3 αN helices. The histone octamer contacts the DNA superhelix at regular intervals projecting an arginine into the minor groove of the neighboring DNA segment. Histone-DNA interfaces are mediated by extensive direct and water-mediate hydrogen bonds, ionic interactions, nonpolar contacts, and the alignment of helix dipoles relative to phosphate backbone ions.

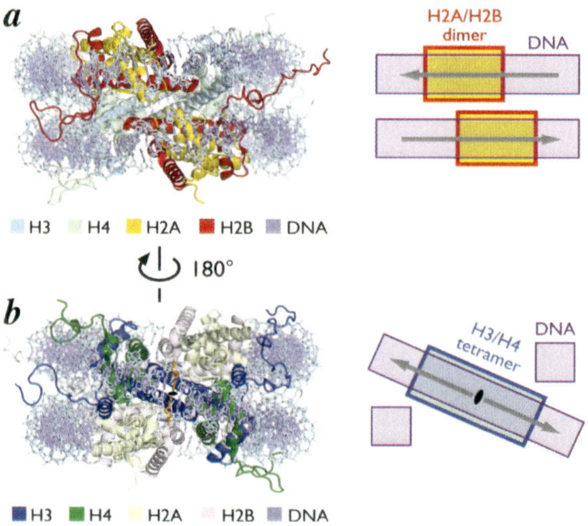

Figure: Histone-fold heterodimers form a ramp for nucleosomal DNA.

(a) H2A/H2B histone-fold heterodimers interact with DNA in two different parallel planes. Structure of NCP viewed from opposite dyad, highlighting H2A and H2B in yellow and red, respectively (left) and scheme of DNA planes (right). (b) H3/H4 tetramer forms a diagonal ramp for DNA connecting two parallel planes. Structure of NCP view from dyad (black oval and orange base pair) with H3 and H4 in blue and green, respectively, (left) and scheme of diagonal DNA ramp (right). Arrows point away from central dyad base pair.

Nucleosome Topology

The ~200 kDa disk-shaped nucleosome core particle has a diameter of approximately 100 Å and height ranging from ~25 Å at the dyad to ~60 Å at the H2B αC helices. It has a multifaceted, solvent accessible surface of about 74 000 $Å^2$. The disk face furthest from the dyad is lined by three parallel ridges, formed centrally by the H2B αC helix and on the sides by the H2B α1 helix and H3 α1 helix together with the H4 N-terminal tail, respectively. These ridges create two intervening grooves, one containing the H2A/H2B acidic patch. The disk face near the dyad contains a central depression overlaying the H3–H3 interface. The complexity of the NCP surface is furthered by the histone N-terminal tails that protrude from the nucleosome surface either outside (H4 and H2A) or between (H3 and H2B) the DNA gyres. These tails, ranging in length from 15 to 36 amino acids, can extend great distances from the NCP, adopt flexible structures, and bind intranucleosomal DNA and/or DNA and histone surfaces in neighboring nucleosomes. The DNA phosphodiester backbone at the perimeter of the NCP presents a highly negative electrostatic surface. An additional negatively charged surface is found in a groove on the H2A/H2B dimer surface that is often referred to as the nucleosome or H2A/H2B acidic patch. Eight acidic residues contribute to the acidic patch, six from H2A (E56, E61, E64, D90, E91, and E92) and two from H2B (E102 and E110). As discussed in detail below, this acidic patch may be a hot-spot for nucleosome binding by chromatin factors. In contrast to the prominent negatively charged surfaces of the NCP disk, the histone tails contain many arginine and lysine residues and carry a strong net positive charge. Overall, the topological and electrostatic complexity of the nucleosome core affords the opportunity for a diverse set of surfaces for nucleosome binding.

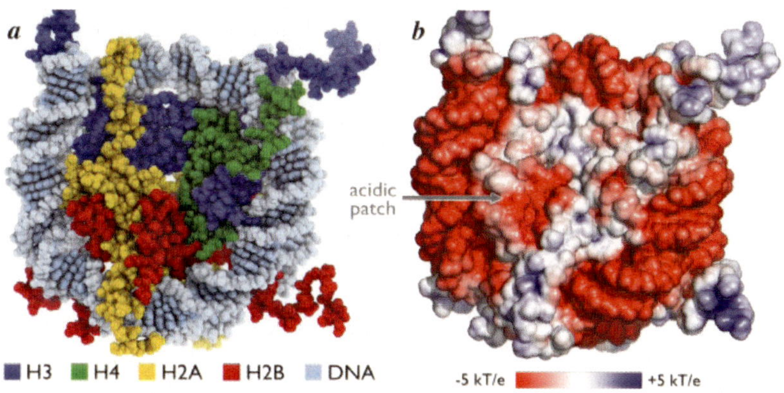

Figure: Surface topology and charge of the nucleosome core particle. (a) Surface of nucleosome core particle viewed down the DNA superhelical axis in space-filling representation. (b) Surface electrostatic potential of nucleosome core particle contoured from −5 to +5 kT/e calculated with ABPS.164 Location of acidic patch is indicated.

Variability in NCP Structure

Since the solution of the core particle containing *Xenopus* histones was reported, crystal structures have been solved using histones from yeast, fly, and man. While minor sequence differences result in small changes to the composition of exposed surfaces and complementary coevolution within the hydrophobic core, the architecture of the complexes remains nearly constant. A myriad of structural studies of histone variants have also revealed some variant-specific roles in the stability and exposed surfaces of the NCP. For example, H2A.Z extends the H2A/H2B acidic patch and causes a subtle destabilization of the H2A/H2B interface with H3/H4. Similar destabilization was observed recently with testes-specific variant H3T. A 2011 structure of the NCP containing the centromeric H3 variant CENP-A revealed that CENP-A nucleosomes only organize the central 121 bp of nucleosomal DNA potentially due to a shortened H3 αN helix. Other structural and biochemical data validate increased opening of the entry exit DNA in CENP-A nucleosomes. This trait is not specific to CENP-A as biochemical and biophysical interrogation of H2A.Bbd nucleosomes show similar opening of the DNA ends in nucleosomes containing H2A.Bbd. Much like histone variants, histone PTMs can induce structural changes in the solvent accessible nucleosome surface and interactions with nucleosomal DNA. Variant- and PTM-specific structural changes can contribute to the ability of nucleosomes to recruit chromatin factors as discussed below. Finally, several structures of NCPs containing different nucleosome positioning sequences have revealed that DNA sequence has effects on nucleosomal DNA structure allowing the octamer to wrap 145–147 bp of DNA.

Structure of Nucleosomal DNA

Nucleosome Positioning Sequences in Nucleosome Structures

Nucleosomes examined by structural studies have for the most part contained four types of DNA sequences: mixed sequence genomic DNA, the 5S RNA coding sequence, the human α-satellite repeat, and the Widom 601 nucleosome positioning sequence. Early structural studies of nucleosomes utilized nucleosomes isolated from naturally occurring sources such as beef kidney and consequently contained mixed sequence genomic DNA. Furthermore, the length of the nucleosomal DNA was variable at 147 ± 2 bp, the result of using micrococcal nuclease to digest long chromatin into nucleosome core particles. Such nucleosome core particles were used in the 7-Å low-resolution crystal structures in 1984. To improve the internal order and diffraction limits of nucleosome core particle crystals, Richmond and colleagues prepared defined length 146 bp nucleosomal DNA fragments containing the 5S RNA coding sequence characterized by Simpson and others. This technical feat employing (now) classical recombinant DNA technology permitted the preparation of milligram quantities of defined sequence and defined length nucleosomal DNA in 1988, resulting in nucleosome core particle crystals which diffracted to ~4.5 Å. While this marked an improvement over the 7-Å diffraction observed for mixed sequence nucleosome core particles, the diffraction was not sufficient for atomic structure determination.

The crystal structure of the nucleosome core particle determined at 2.8 Å by the Richmond group in 1997 contained instead the human alpha satellite centromeric repeat. Use of this nucleosome positioning sequence for reconstituting recombinant nucleosome core particles was first described by Bunick and colleagues in 1995. The 2.8 Å Luger et al. structure revealed important features of how the histone octamer organizes nucleosomal DNA. One such feature was the somewhat surprising finding that the dyad of the nucleosome core particle lines with a base pair and not between base pairs. This means that a nucleosome core particle with 73 bp on either side of the dyad contains 147 and not 146 bp. Previous studies had generally assumed an even number of base pairs in the nucleosome, and this was in fact the basis for using 146 bp of nucleosomal DNA in nucleosome crystallization trials. The occurrence of a base pair at the dyad axis was actually predicted earlier by single base pair resolution mapping of nucleosome positions using site-directed hydroxyl radical footprinting. This study examined recombinant nucleosome reconstituted with the 5S RNA sequence, providing evidence that the placement of the nucleosome dyad on the central base pair is independent of nucleosome positioning sequence. This conclusion has been borne out by all subsequent studies.

Widom 601 Nucleosome Positioning Sequence

Crystal structures of at least 20 nucleosome core particles incorporating variant histones or DNA sequence changes have been determined since the original 1997 structure. The large majority of these utilized the human α-satellite repeat DNA sequence. In contrast, the most popular DNA positioning sequence used in chromatin biochemical studies is the Widom 601 sequence. Lowary and Widom had performed an in vitro selection experiment to isolate synthetic random DNA sequences with high affinity for the histone octamer. The Widom 601 sequence was among the tightest binding sequences found, and its strong nucleosome positioning and high yields in nucleosome reconstitution experiments have made it a favorite among chromatin researchers. Crystal structures of nucleosome core particles containing the Widom 601 sequence were determined on their own and in complex with the RCC1 chromatin factor in 2010. Since the nucleosome core particles pack differently in the 601 nucleosome versus the RCC1/601 nucleosome crystals, the similarity of the structures indicate that the structures are not artifacts of their crystal packing. This is not an insignificant consideration given that the same crystal packing is present in all crystals of nucleosome core particles on their own to date.

The reasons why the Widom 601 sequence is such a strong nucleosome positioning sequence have intrigued many since the sequence was first characterized in 1998, and we are now beginning to understand the mechanistic basis. Nucleosomal DNA must endure an aggregate bend of about 600° in approximately 150 bp, and consequently, DNA sequences that can be bent to contour the histone octamer will be favored. By sequencing chicken nucleosomal DNA, Satchwell et al. showed in 1986 that AA/TT, TA, and AT base steps were favored where the double helix minor groove faces the histone octamer and conversely that GG, GC, and CG base steps were more likely to be found 5 bp away or where the minor groove faces away from the histone. The particular significance of the TA base step in nucleosome positioning was suggested in several experiments, including the analysis of in vitro selected sequences (such as the 601 sequence) which highlighted a 10 bp sequence periodicity of the TA base step. In fact, the 601 sequence contains the TA base step at 5 of the 8 central positions where the DNA minor groove faces the histone octamer (SHL ± 0.5, ± 1.5, ± 2.5, ± 3.5). Crystallographic and biochemical experiments by the Davey laboratory now provide a structural explanation for the significance of this observation. Noting that 4 of the 5 TA base steps are located on the "left" half of the 601 sequence and the remaining one on the "right" half, Davey and colleagues examined the salt stability of nucleosome core particles reconstituted with symmetrized versions of the 601 sequence. Nucleosomes containing the original, asymmetrical 601 sequence dissociated at a salt concentration of 1.26 M, significantly more than the 0.94 M concentration needed to dissociate nucleosome reconstituted with the human α-satellite sequence. However, nucleosome reconstituted with symmetrical DNA based on the left half of the 601 sequence ("601L") were noticeably more stable to salt, while nucleosomes containing the right 601 sequence ("601R") were less stable than the original 601 sequence. These results, as well as the fact that other high affinity in vitro selected nucleosome sequences, such as the 603 and 605 sequence also contain

TA base steps in similar positions, provide evidence for an important role of the TA base step in nucleosome positioning of these high affinity in vitro selected nucleosomes.

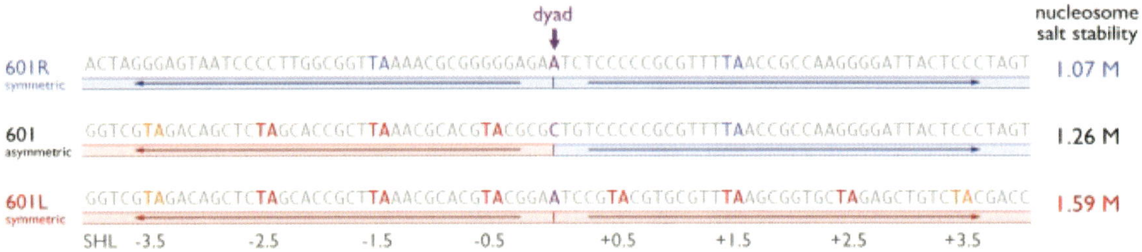

Figure: Scheme of asymmetric and symmetric 601 sequences

Sequences of 601R symmetric, (canonical) 601 asymmetric, and 601L symmetric sequences with H3/H4 TA steps highlighted in red for left half and blue for right half.61 Nucleosome salt stability values (molar monovalent salt) are listed at right and indicate stability as follows: 601L > 601 > 601R. This trend correlates with the number of H3/H4 TA steps: 601L (6), 601 (4), 601R (2). The dyad position is indicated (purple).

Crystal structures of nucleosome core particles containing the 601 and 601L sequences confirm that the TA base steps at the aforementioned central positions directly face the histone octamer. These TA base steps also exhibit significant distortions from ideality particularly in their propeller twist values. The crystal structures help explain the significance of this observation. Histones H3/H4 form the tetramer, which binds DNA around the nucleosomal dyad. The L1/L2 loops, α1/α1 N-terminal ends and L2/L1 loops of histone H3/H4 grip DNA phosphates groups where the minor groove faces the histone octamer and thus constrains the nucleosomal DNA path to these locations (termed "pressure points" by Davey and colleagues). These pressure points occur at SHL ± 0.5, ± 1.5, and ± 2.5 (i.e., 5 bp from the nucleosomal dyad and then again at 10 bp intervals), precisely where TA base steps are located. Fixing these phosphate groups at these pressure points creates stress particularly in the base pairs between the phosphates. Base steps that are more flexible will more easily accommodate this stress through distortions such as propeller twisting within a base pair, rolling between base pairs or sliding one base pair with respect to the adjacent base pair. Since the TA base step is the most flexible of all base steps, it is most able to accommodate the stress created at the pressure points. Thus, we can understand why high affinity nucleosome positioning sequences such as the 601 sequence contain TA base steps where the minor groove faces the histone tetramer.

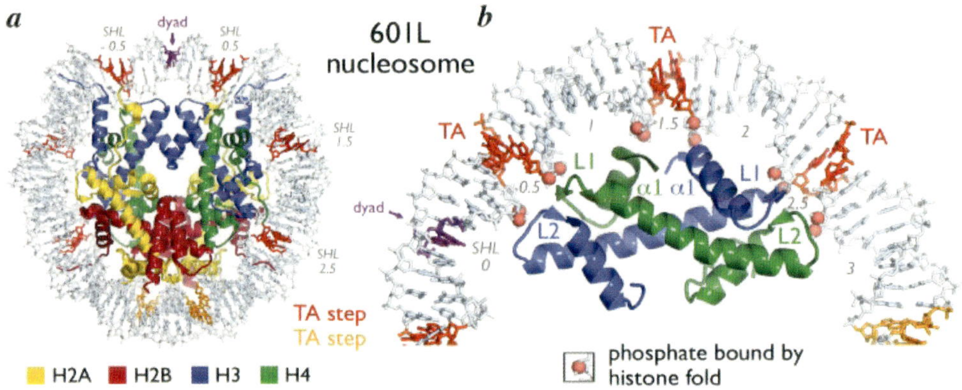

Figure: Location of TA steps in 601L nucleosome core particle structure DN

(a) 601L NCP structure viewed down the DNA superhelical axis with TA steps interacting with H3/H4 and H2A/H2B colored red and orange, respectively. The dyad is indicated (purple). Histones H3, H4, H2A, and H2B are shown in cartoon representation and colored blue, green, yellow, and red, respectively. Nucleosomal A is shown as sticks (light blue). (b) Enlarged view showing one H3/H4 heterodimer bound to DNA containing three TA steps (other histones are not shown for clarity purposes). Backbone phosphates bound to the H3/H4 histone folds are shown in space-filling representation as indicated. Secondary structure elements of dimer are shown.

Why is the TA base step so flexible? The fewer hydrogen bonds between the bases do allow for greater flexibility compared to base steps containing GC base pairs. However, this cannot be sufficient since TT, AA, and AT base steps are not as flexible despite having the same number of Watson–Crick hydrogen bonds. A simple structural explanation is that the stacking of bases is minimal in the TA base step, allowing greater flexibility for roll, propeller twisting and other distortions than the TT, AA, or AT base steps. The methyl group on the thymine base also plays an important role because the relatively bulky methyl group must be accommodated when a T-A base pair is distorted. In the TA base step, the minimal base stacking and the position of the methyl group allow for large roll or propeller twisting without the thymine methyl clashing with other atoms. It is for a similar reason that the eukaryotic transcriptional initiation TATA box is distorted so dramatically upon binding of the TBP TATA binding protein. In contrast, the TT, AA, and AT base steps each offer steric challenges to distortions including roll and propeller twisting. It is worth emphasizing that the geometry of the TA base steps is distinct from the AT base step despite the common alternating A/T sequence. The other base step with fewer constraints to distortions is CA = TG, and this base step has been found to be among the most flexible in protein–DNA structures.

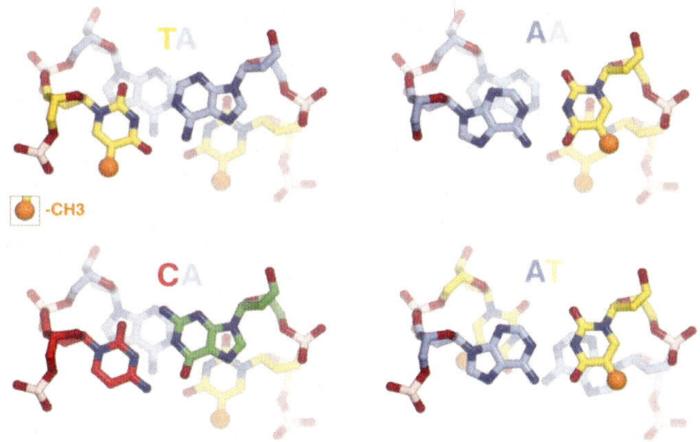

Figure: Minimal base stacking in TA and CA compared to other base pair steps.

TA, CA, AA, and AT base pair steps colored as follows: T = yellow, A = blue, G = green, C = red. The thymine methyl groups are shown highlighted in space-filling representation (dark yellow), all other non-hydrogen atoms shown in sticks representation. The minimal base stacking and the absence of atoms close to the thymine methyl group permit greater flexibility of the TA and CA base pair steps.

The concept that the flexible TA base steps located between critical pressure points in nucleosomal DNA provides an explanation for surprising results of experiments studying the ability of RNA polymerases to progress through a nucleosome. Studitsky and colleagues found that transcriptional elongation by yeast and human RNA polymerase II were blocked by a nucleosome reconstituted with the 601 positioning sequence in one but not the opposite orientation. This blockage was mediated by the H3/H4 histone tetramer since the same effect was observed in the absence of the H2A/H2B dimer. Similar results were obtained for the Widom 603 and 605 nucleosome positioning sequences. These were puzzling findings because the symmetrical nature of the nucleosome structure made it difficult to imagine why a sequence block would function in one orientation but not in the opposite orientation. However, inspection of the DNA sequences with a focus on TA base steps shows a strong correlation between TA base steps at SHL +0.5, +1.5, and +2.5 and the ability to block RNA polymerase II progression. For each of the Widom 601, 603, and 605 nucleosome positioning sequences, RNA polymerase II transcriptional elongation was blocked when the TA base steps are positioned at the pressure points facing the H3/H4 tetramer downstream of the dyad. At first glance, the fact that the block to transcriptional elongation occurs when tight binding to the nucleosome is downstream of the dyad seems counterintuitive. If we imagine RNA polymerase II as an engine peeling off DNA from a one-dimensional histone track, we might expect that TA base steps positioned upstream of the dyad to be more efficient at blocking. The problem, of course, is that this Flatland analysis ignores the three-dimensionality of molecules instead of focusing on how RNA polymerase interacts with the architecture of the three-dimensional nucleosome. The finding that tight binding of nucleosomal DNA to the histone tetramer downstream of the dyad blocks RNA polymerase II indicates the ability of RNA polymerase II to unwrap DNA from the tetramer on the downstream side is a critical aspect of the mechanism of passing through a nucleosome. This insight was exploited by Studitsky and colleagues in a subsequent study where they propose a structural mechanism for this very process.

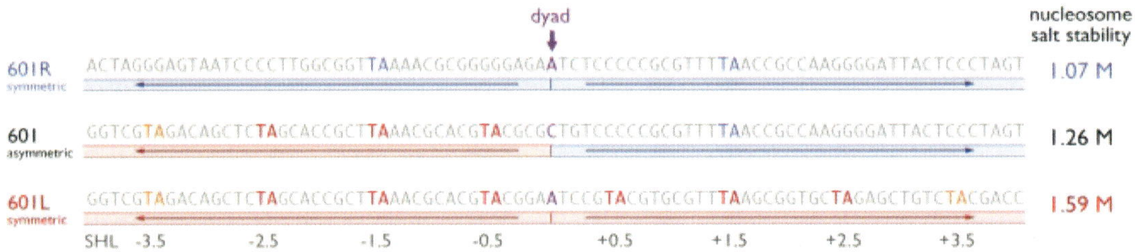

Figure: RNA polymerase II blocking by nucleosome positioning sequences

Sequences of NCP601, NCP603, and NCP605 sequences and their reversed counterparts together with ability to block RNA polymerase II. Multiple TA steps bound to the H3/H4 tetramer downstream (red) of the dyad (purple) blocks RNA polymerase II passage as compared with upstream (blue) of the dyad. TA steps bound to the H2A/H2B dimers are shown in orange.

DNA Stretching in the Nucleosome

The original 2.8 Å crystal structure of the nucleosome core particle contained 146 bp of symmetrized human alpha satellite DNA. Since the nucleosomal dyad lies on a base pair, there was necessarily 73 bp on one side of the dyad base pair and 72 bp on the other side. Thus, the structure was asymmetrical despite the fact that the nucleosomal DNA was symmetrical. To date, every single nucleosome core particle crystallized by itself has packed in the crystal essentially the same way: in space group $P2_12_12_1$ with end-to-end DNA contacts between individual nucleosomes (with the possible exception of the CENP-A centromeric nucleosome where the DNA ends are not visible in the crystal structure). In order for the DNA ends to pack against each other in the crystal, the DNA on the 72 bp side had to stretch by one bp localized around SHL ±2. Subsequent nucleosome core particle crystal structures containing human alpha satellite DNA of variant length or sequence show stretching of one bp around SHL ± 2 and SHL ± 5 but not at other locations. For example, the NCP146b nucleosome, which incorporates a symmetrical version of a different human α-satellite half-repeat, is stretched around SHL -5, while nucleosomes prepared from the original human α-satellite 146 bp DNA and recombinant human histones display stretching at SHL -2 and ±5. In contrast, a 147 bp pseudosymmetric human α-satellite sequence (pseudosymmetric because the symmetry of the two 73 bp halves was broken at the dyad) displayed no DNA stretching, and it is generally accepted that the human alpha satellite sequence forms a 147 bp nucleosome core particle in solution.

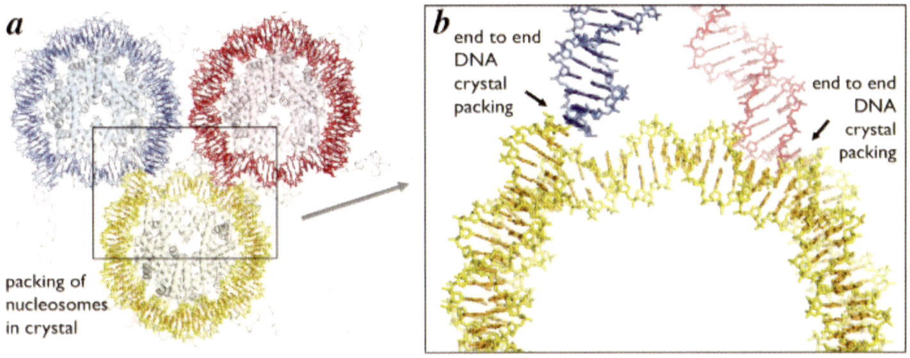

Figure: DNA end-to-end packing in nucleosome core particle crystals

Three nucleosome core particles from one plane of the high resolution NCP crystal structure (PDB ID 1KX5) colored yellow, red and blue. (a) Full and (b) enlarged views of the alignment of the DNA ends from adjacent NCP in the structure. The DNA end-to-end packing exists in all crystals of the nucleosome core particle on its own.

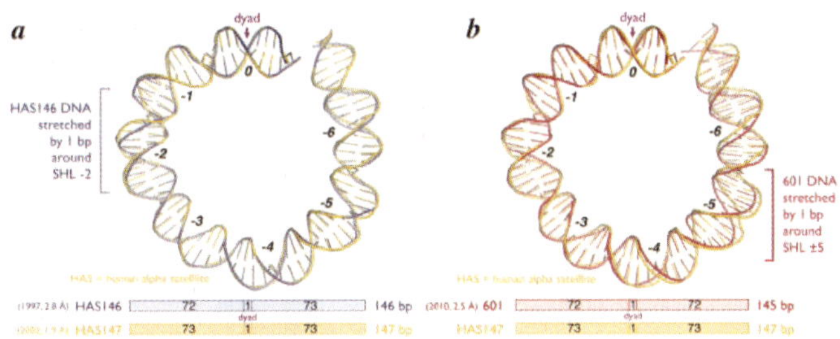

Figure: DNA stretching in nucleosome core particle structures

Cartoon representation of structure of approximately half of the nucleosomal DNA for (a) 146 bp human alpha-satellite (HAS146) (PDB ID 1AOI, blue) and (b) 145 bp 601 (PDB ID 3LZO, red) nucleosome positioning sequences relative to the HAS147 sequence (PDB ID 1KX5, yellow) (top). Stretching of 1 bp is observed at superhelical location (SHL) −2 with the HAS146 sequence and 1 bp each at SHL ± 5 with the 145 bp 601 sequence. SHLs and the dyad = SHL 0 are indicated. The length of DNA wrapped on each side of the NCP for each of the sequences is also shown (bottom).

When we crystallized the chromatin factor RCC1 in complex with the 601 nucleosome, we anticipated that the Widom 601 sequence would also likewise form a 147 bp nucleosome core particle. We therefore employed a 147 bp 601 DNA sequence in our crystallization studies. To our surprise, structure determination showed that the 601 nucleosome in the RCC1/nucleosome structure forms a 145 bp nucleosome core particle due to stretching of DNA by one bp at SHL ± 5. Two lines of evidence indicate that the DNA stretching in the 601 nucleosome did not result from crystal contacts. First, unlike crystals of nucleosome core particles on their own, the RCC1/nucleosome complex does not make DNA end to DNA end crystal contacts. A symmetry related RCC1 does make important crystal contacts with one DNA end, but the DNA end on other end of the nucleosome makes no crystal contacts. Second, the 601 nucleosome core particle on its own (i.e., in the absence of RCC1) crystallized in a different space group ($P2_12_12_1$ vs $P2_1$ for RCC1/nucleosome) and via different crystal packing interactions than in the complex with RCC1. Despite the different crystal packing arrangement, the 601 nucleosome particle on its own also exhibits stretching at SHL ± 5. This stretching can explain why the 601 nucleosome core particle had evaded crystallization for such a long time: the 147 bp 601 nucleosome core particles with its extra bp extending beyond the nucleosome core particle would prevent the canonical DNA end to DNA end crystal packing found in all nucleosome only crystals. We were fortunate that our use of the 147 bp 601 nucleosome not only did not prevent the RCC1/nucleosome from crystallizing but was in fact important for the RCC1-nucleosomal DNA end crystal contact to occur.

Recognition of the Nucleosome Core by Chromatin Factors

The recruitment of macromolecular chromatin factors to genomic loci is controlled at many levels. Factors can be actively sequestered in or excluded from the nucleus. The accessibility of large territories of the genome can be regulated by altering the degree of chromatin compaction. At a more local level, the binding of chromatin factors can be tuned by the positioning of nucleosomes. Some chromatin factors including many transcription factors bind to specific DNA sequences only in nucleosome free regions. On the other hand, many chromatin factors require nucleosomes for binding to chromatin. Obvious examples are histone-modifying and chromatin remodeling enzymes that by definition modify the chemical composition or architecture/location of nucleosomes, respectively. However, many other chromatin factors also bind to nucleosomal regions including high-mobility group proteins, heterochromatin scaffolding proteins, and even viral proteins. As described above, the nucleosome provides a diverse platform for binding of macromolecular chromatin factors. This platform is further diversified by replacement of canonical histones with histone variants and the chemical modification of both histone and DNA components of the nucleosome.

Chromatin factors can bind the nucleosome using one or more of the three following nucleosome surfaces:

1. The histone N- and C-terminal tails;
2. The disk faces of the histone octamer;
3. The nucleosomal DNA.

[Of particular note, the nucleosomal DNA affords the possibility of novel protein–DNA interactions, given its unique curvature as well as the alignment of the nucleosomal DNA gyres.] Much of the research regarding binding to histone tails is centered around histone PTMs. The molecular recognition of histone tails by numerous catalytic domains that establish and remove histone PTMs and diverse protein domains that bind tails modified at specific residues are thoroughly reviewed elsewhere. Due to technical challenges in the structural characterization of the nucleosome core bound to chromatin factors, much less is understood regarding the molecular recognition of the disk surfaces of the histone octamer and nucleosomal DNA. However, recent advances in the cocrystallization of macromolecular chromatin factors bound to the nucleosome core particle have permitted new atomic scale depictions of recognition of the nucleosomal disk. All crystal structures of macromolecules bound to the nucleosome solved to date share one common interaction motif, an arginine bound to the H2A/H2B acidic patch.

H4 N-Terminal Tail

The first instance of a protein segment binding to the nucleosomal H2A/H2B acidic patch was observed in a crystal contact of the 1997 nucleosome core particle structure. In these crystals, residues 16 to 25 of one H4 tail contact the acidic patch on an adjacent NCP forming a charged interaction surface (Notably, the other tail is not resolved N-terminal to residue 20 even though NMR experiments suggest the H4 tail is structured starting with residue 16). The H4 K16 side chain projects into an acidic cavity generated by H2A acidic patch residues E61, D90, and E92 of the neighboring NCP surface. Other positively charged amino acids in the H4 tail also interact with negative side chains in the H2A/H2B acidic patch, including H4 R19-H2A E64, H4 K20–H2B E110 and H4 R23 which contacts both H2B E110 and H2A E56. Similar interactions are observed for H4 K20 and R23 in the crystal lattice of the 1.9 Å NCP structure though the other basic H4 side chains point toward intranucleosomal DNA. A functional role for the H4 N-terminal tail in chromatin structure has been confirmed using nucleosome arrays in solution. Truncation of the N-terminal H4 tail prior to residue 20 or the charge neutralizing acetylation of H4 K16 leads to incomplete cation-mediated compaction of the 30 nm fiber in vitro. The case for a role in chromatin structure is furthered by disulfide formation between spatially adjacent mutant nucleosomes containing H4 V21C and H2A E64C both within a chromatin array and between arrays. It is important to note that the observed structure of the H4 N-terminal tail seen in the crystal lattice may not reflect a native conformation in chromatin fibers. Recent modeling of the H4 N-terminal tail suggests that H4 residues 15 to 20 exhibit a propensity for α helix formation. This allows H4 K16, R17, R19, and K20 to occupy a single helical face, which can be accommodated within the acidic patch groove. Though the evidence for the H4 tail-acidic patch in higher order chromatin structure is compelling, higher resolution structural characterization is required to accurately define the molecular details of this interaction.

Viral LANA Peptide

In 2006, Barbera et al. demonstrated that the nucleosome docking region of the Kaposi's sarcoma-associated herpes virus (KSHV) latency-associated nuclear antigen (LANA) also interacts with the H2A/H2B acidic patch. The KSHV genome is contained within an episome, which is tethered to mitotic chromosomes using the basic N-terminal region of LANA by anchoring to the nucleosomal acidic patch. The 2.9 Å crystal structure of the LANA nucleosome recognition sequence bound to the nucleosome core particle was solved by soaking the peptide corresponding to LANA residues 1–23 into NCP crystals. LANA peptide forms a hairpin that fits in the acidic patch groove between the αC and α1 helices of H2B and makes multiple charged and hydrophobic interactions. The LANA R9 side chain inserts into the acidic patch to form ionic interactions with H2A residues E61, D90, and D92. This acidic patch "arginine-anchor" (our terminology) is shared with all crystal structures of chromatin factors bound to the NCP reported to date and overlaps the H4 K16 binding site observed in the original NCP structure crystal lattice. LANA R9 is critical in nucleosome binding as mutation of this residue eliminates LANA's chromatin association. An additional LANA arginine (R7) forms an ionic interaction with H2B E110 and LANA S10 hydrogen bonds to H2A E64. LANA hydrophobic residues M6 and L8 bind a hydrophobic surface adjacent to the acidic patch including H2A residues Y50, V54, and Y57. Overall the regions of LANA observed to interact with the NCP correlate with those required for the virus to tether its episomal DNA to chromosomes.

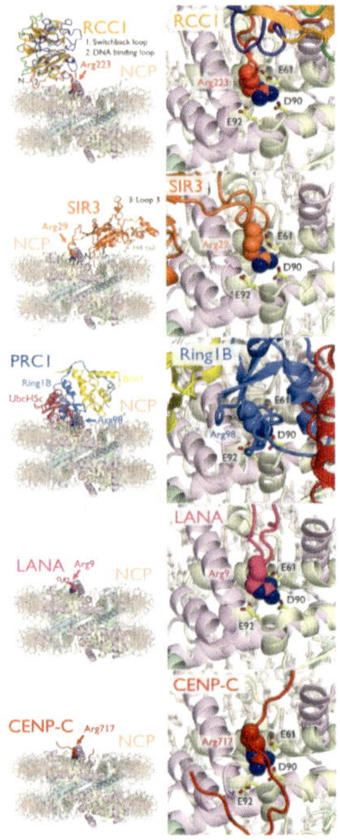

Figure: Nucleosome recognition using the acidic patch arginine-anchor.

From top to bottom, structures of RCC1 (PDB ID 3MVD),52 Sir3 (PDB ID 3TU4),78 PRC1 (PDB ID 4R8P),111 LANA peptide (PDB ID 1ZLA),79 and CENP-C peptide (PDB ID 4INM)107 bound to the nucleosome core particle. Overview of structures as viewed from opposite the dyad (right) and zoomed view of acidic patch (left) with arginine-anchor in space-filling representation and key H2A residues shown as sticks. Locations of RCC1 switchback loop (1), DNA binding loop (2), and N-terminus (N) and Sir3 loop 3 (3) and N-terminus (N) are indicated. Histones H3, H4, H2A, and H2B are shown in cartoon representation and colored cornflower blue, light green, wheat, and pink, respectively. DNA (light pink) is shown as sticks.

Ran Guanine Exchange Factor RCC1

In 2010, we solved the structure of the *Drosophila melanogaster* β-propeller protein RCC1 (Regulator of chromatin condensation) bound to the nucleosome core particle. In contrast to the LANA-NCP structure, the RCC1-NCP structure was solved by cocrystallization of RCC1 on the NCP. RCC1 is a guanine exchange factor for the Ran GTPase (or RanGEF) that establishes a gradient of the GTP bound form of Ran around chromatin. This gradient plays roles in nuclear-cytoplasmic transport, mitotic spindle formation, and formation of the nuclear envelope following mitosis. RCC1 binds to the nucleosome resulting in an increase in its ability to catalyze Ran guanine exchange. Our 2.9 Å structure of the RCC1-NCP complex shows that RCC1 interacts with both the acidic patch and nucleosomal DNA using two β-propeller loops and its N-terminal tail. One loop, termed the switchback loop, binds to the H2A/H2B acidic patch using an intricate network of ionic and hydrogen bonds and van der Waals contacts. The switchback loop contains the RCC1 arginine-anchor residue 223 that binds H2A E61, D90, and E92 nearly identically to LANA R9. Much like LANA, RCC1 uses a second arginine 216 for additional interactions with H2A E61 and E64. Both RCC1 R223 and R216 are important for nucleosome binding in solution. RCC1 S217 forms hydrogen bonds with H2A V45 and E64. An additional hydrogen bond is observed between RCC1 S214 and H2A E64. These ionic and hydrogen bonding interactions are complemented by van der Waals contacts, especially with residues in the H2B αC helix that form a ridge at the edge of the acidic patch.

In addition to the acidic patch, RCC1 also binds to nucleosomal DNA using a distinct β-propeller loop and its N-terminus. The RCC1 DNA binding loop interacts with the phosphodiester backbone across a major groove near SHL ± 6 forming hydrogen bonds or charged interactions with the side chains of K241 and R239. An additional hydrogen bond is formed by the side chain of RCC1 259. The N-terminal tail of RCC1 is also implicated in nucleosome binding. While the N-terminal residues 2–27 are not visible in our RCC1-NCP structure, residues 28 and 29 are positioned to allow the N-terminal tail to enter the major groove of nucleosomal DNA adjacent to the DNA binding loop. Alignment of RCC1 from the RCC1-NCP and RCC1-Ran structures suggest that Ran approaches but does not contact the nucleosome surface. Therefore, either RCC1 or Ran must undergo conformational changes to allow for Ran-NCP interactions to enhance RCC1's RanGEF activity. Further experiments are required to resolve this issue.

Silencing Protein Sir3

In 2011, Armache et al. solved the crystal structure of the BAH (bromo-associated homology) domain of the yeast silent information regulator protein Sir3 bound to the nucleosome core particle. *Saccharomyces cerevisiae* uses SIR (silent information regulator) proteins Sir1, Sir2, Sir3, and Sir4 to establish a transcriptionally repressive chromatin state at telomeres, ribosomal DNA loci and silent mating-type loci.Silencing is thought to be accomplished in part though direct chromatin compaction by Sir3 as demonstrated in vitro. This 3.0 Å structure illustrates one of the most extensive interaction surfaces of the published high-resolution chromatin factor-NCP structures, including 28 Sir3 BAH domain residues and greater than 30 histone residues (and potentially nucleosomal DNA at the BAH domain N-terminus). The structure also suggests that weak self-interactions of the BAH domain observed in the crystal lattice and in solution may contribute to its ability to compact chromatin fibers. Analogous to LANA and RCC1, the Sir3 BAH domain binds to the H2A/H2B acidic patch, but also interacts with the nucleosome in three additional regions: the H4 N-terminal tail, surfaces of H3/H4 in the loss of rDNA silencing (LRS) domain, and the H2B C-terminal helices. Sixteen Sir3 BAH domain residues from loops 2 and 4, strand B5 and the A1 helix interact with H4 tail residues 13–23. This region of the H4 tail is often unstructured in NCP crystals in the absence of crystal lattice contacts. The Sir3 BAH domain-H4 tail interface is predominantly electrostatic in nature owing to the positively charged H4 tail and negatively charged complementary Sir3 surface. Importantly, the H4 K16 side chain forms several hydrogen bonds and ionic interactions with acidic Sir3 side chains, offering a molecular mechanism for the observed loss of Sir3 binding upon H4 K16 acetylation. The nucleosomal recognition surface of the Sir3 BAH domain also includes the α1 helix and L1 loop of H3, the α2 helix and L2 loop of H4 as well as the α3 and αC helices of H2B. These interactions are mediated by the Sir3 BAH loop 3 which becomes structured upon NCP binding and strands B6 and B8. This interaction surface includes the LRS domain residues 76–80 of H3 that are required for silencing in yeast. Much like H4 K16 acetylation, H3 K79 methylation disrupts Sir3 binding. The interactions in this region offer insight into the preference for H3 K79 in the unmodified state due to loss of potential hydrogen bonds with Sir3 BAH side chains. This is further exemplified by two recent structures of the N-α-acetylated Sir3 BAH domain bound to the NCP. The native N-terminally acetylated residue specifically interacts with regions of the Sir3 BAH domain further structuring its loop 3, enhancing contacts with the nucleosome surface in the LRS region and leading to a 30-fold increase in affinity for the NCP. Sir3 BAH also contacts the H2A/H2B acidic patch using the loop 1 region that is unstructured in the absence of the NCP. While electron density is weaker in this region of the Sir3 BAH-NCP structure, it appears that multiple arginines (R28, R29, R30, R32, and R34) line the acidic patch groove to make charged interactions with the NCP surface. Notably, Sir3 R32 occupies an identical binding site to that observed for LANA R9 and RCC1 R223 in a cavity surrounded by H2A E61, D90, and D92. Overall, the multifaceted interaction of the Sir3 BAH with the NCP explains many silencing defects observed upon mutation of both histones and Sir3 in yeast.

Centromeric Protein CENP-C

The crystal structure of the central region of rat centromere protein CENP-C bound to the nucleosome core particle illustrates the ability of a chromatin factor to recognize specific features of a variant histone in the context of the nucleosome. Proper segregation of chromosomes in mitosis requires the mitotic spindle to attach to the kinetochore at the centromere of each chromosome. Centromeric chromatin contains H3 variant CENP-A and a complex of 16 centromeric proteins including CENP-C. The crystal structure of the CENP-C nucleosome binding peptide in complex with a chimeric NCP in which the C-terminal region of H3 was replaced with the LEEGLG solvent exposed sequence of CENP-A was solved by Kato et al. in 2013. In the structure, CENP-C forms an elongated conformation, which contacts the acidic patch and the CENP-A specific regions of the chimeric NCP. CENP-C binds to the acidic patch using two arginines, R717 and R719. The CENP-C R717 arginine-anchor binds in the now characteristic H2A E61, D90, E92 pocket; R719 makes additional ionic interactions with H2A E61 and E64. In the CENP-A specific region of the chimeric nucleosome, CENP-C residue Y725 binds in a hydrophobic pocket created by CENP-A residues I133 and L137. While CENP-A proteins are not highly conserved, they all contain at least one large hydrophobic residue in the CENP-C binding region that is not found in canonical H3. This 3.5 Å structure was validated by extensive NMR experiments using chemical shift perturbations and site-specific incorporation of paramagnetic spin labels.

Polycomb Repressive Complex 1 (PRC1) Ubiquitylation Module

Recently, we solved the crystal structure of the PRC1 ubiquitylation module, containing the Ring1B-Bmi1 ubiquitin E3 ligase RING heterodimer together with the E2 enzyme UbcH5c, bound to its nucleosome core particle substrate at 3.3 Å resolution. PRC1 is a member of the Polycomb group family of complexes and plays a role in transcriptional repression of developmentally regulated genes at least in part through H2A K119 ubiquitylation and intrinsic chromatin compaction. PRC1 contains a RING-type ubiquitin E3 ligase composed of RING domains from Ring1B and Bmi1 that can pair with one of several ubiquitin E2 conjugating enzymes, including UbcH5c. Our structure reveals that all three proteins in the PRC1 ubiquitylation module bind to the nucleosome surface, together contacting all components of the nucleosome core particle. The Ring1B-Bmi1

RING heterodimer forms a saddle over the proximal end of the H2B αC helix, anchored on each side by histone interactions. The RING domain of Ring1B binds the H2A/H2B acidic patch using multiple positively charged side chains including an arginine anchor residue R98. The substantial Ring1B-acidic patch interface contrasts a more modest Bmi1-nucleosome interface. The RING domain of Bmi1 interacts with a smaller acidic surface on the H3/H4 tetramer and forms a cap on the distal end of the H3 α1 helix. The structure is consistent with mutagenesis experiments implicating the H2A/H2B acidic patch, the arginine anchor, and several other basic side chains. In addition to these E3 ligase-histone interactions, the E2 UbcH5c binds to nucleosomal DNA in two positions. Near the DNA end, the UbcH5c antiparallel β-sheet aligns several charged and polar side chains for interaction with the adjacent nucleosomal DNA backbone. UbcH5c binds nucleosomal DNA again at the dyad using basic α3 side chains. Mutagenesis at these novel E2-substrate interfaces diminishes nucleosome binding and activity by the PRC1 ubiquitylation module. Importantly, unlike other histone modifying enzymes structurally characterized to date, the PRC1 ubiquitylation module does not appear to directly recognize its targeted primary sequence. Rather the E3 and E2 components bind to topologically unique nucleosome surfaces distant from the site of catalysis to position the E2 active site directly over the H2A C-terminal tail near the target lysine.

NMR-based Model of HMGN2

As a complementary approach to X-ray crystallography, Bai and colleagues have introduced NMR-based techniques to characterize chromatin factor-nucleosome interactions. The combination of methyl-TROSY and paramagnetic spin label NMR experiments allowed Kato et al. to map the nucleosomal surface bound by high mobility group nucleosomal protein HMGN2. A model for the HMGN2-NCP complex was created based on comprehensive NMR-based restraints. HMGN proteins are chromatin architectural proteins with roles in DNA damage repair, chromatin remodeling, and histone PTMs. They can also directly decompact chromatin and compete for chromatin binding with the linker histone H1. HMGNs share a common N-terminal nucleosome-binding domain with the conserved basic octapeptide sequence, RRSARLSA. Kato et al. observed chemical shift perturbations of labeled side chain methyl groups of H2A L65 and H2B V45 and L103 upon HMGN binding. These residues are in proximity to the H2A/H2B acidic patch. Based on many NMR experimental restraints, a model of the HMGN2-NCP complex was proposed, suggesting arginines in the conserved HMGN nucleosome binding domain (R22, R23, and R26) interact with the H2A/H2B acidic patch. HMGN2 lysines 39, 41, and 42 were also proposed to interact with DNA near the entry/exit from the NCP. The two NCP-binding regions of HMGN2 are separated by a rigid proline-rich linker. Of note, two HMGN2 proteins bind the pseudosymmetry-related faces of the NCP with positive cooperativity (K_d of 1.5 and 0.17 μM for the first and second binding events, respectively, giving a Hill coefficient of 1.4). The authors suggest that HMGN-NCP interactions staple the DNA end to the histone octamer blocking the activity of chromatin remodelers. Moreover, the model explains the release of HMGNs from chromatin during mitosis following phosphorylation of S24 and S28. The negative charge carried on the phosphorylated serines would create repulsive interactions with the neighboring acidic patch.

Acidic Patch Arginine-anchor as a Common Motif for Nucleosome Recognition

All crystal structures of chromatin factors bound to the nucleosome core particle share a common structural motif: an arginine-anchor that binds to a specific cavity generated by H2A E61, D90, and E92 side chains in the H2A/H2B acidic patch. And these are not the only examples of chromatin enzymes and factors relying on the acidic patch for nucleosome binding and/or activity. Ubiquitin E3 ligases RNF168 and BRCA1 exhibit defective ubiquitylation when the acidic patch is mutated and IL-33 binds to the acidic patch, likely in a manner similar to the LANA nucleosome targeting peptide. So why would such a common motif for recognition of a complex as large as the nucleosome exist? The simple answer is that the acidic patch is the most unique region of the nucleosome surface. It is topologically poised for chromatin factor interaction as a deep groove with a complex surface. The width of the groove allows the binding of multiple types of structures including loops (RCC1, Sir3), hairpins (LANA), extended conformations (HMGN2, CENP-C) but can also accommodate helical and β-strand secondary structure elements. In addition, the acidic patch carries the greatest net charge of the solvent exposed region of the histone octamer disk surface. Furthermore, the guanidinium group of the arginine-anchor is optimal for ionic interaction with all three H2A acidic side chains in the shared acidic-patch binding pocket.

Why would other distinct surfaces not be targeted to minimize interference? Luger and colleagues propose that the overlap of binding sites may serve a regulatory role in the determination of chromatin structure.That is, competition for the acidic patch between factors that condense chromatin (H4 tail and Sir3) with other macromolecules (HMGN2, RCC1, etc.) may tune the higher-order state of chromatin. Many questions still remain regarding this emerging paradigm for nucleosome binding. How common is it? Are we seeing it so frequently in biochemical experiments because we know to look for it? Are chromatin factors that bind to the acidic patch just easier to crystallize owing to binding affinities or resultant crystal packing opportunities? How is the binding of multiple chromatin factors to the same nucleosomal surface regulated? Are other acidic patch binders post-translationally modified to tune their binding affinities similar to HMGNs? Currently, the sample size is too small to address most of these questions. However, as described above, recent strides have been made in the structural characterization of chromatin factor-nucleosome complexes. This will provide a foundation for future work to address these unanswered questions and undoubtedly uncover new paradigms for nucleosomal recognition.

Cryo-EM Models of Chromatin Factor-Nucleosome Complexes

In addition to X-ray crystallography and NMR structures, several cryo-EM structures have enhanced our understanding of nucleosome recognition by chromatin enzymes and factors. The chromatin factors studied by cryo-EM to date fall into three functional categories: chromatin remodeling enzymes, histone modification enzyme complexes, and chromatin architectural proteins. They include large and dynamic macromolecules that present difficulties to crystallographers and NMR spectroscopists. Of note, all reported cryo-EM structures of chromatin factor-nucleosome complexes were solved at resolutions greater than ~20 Å. This allows overall architecture to be revealed but precludes molecular description of NCP interactions. Studies using multimodality approaches, pairing cryo-EM with crystallographic characterization of subcomplexes, cross-linking mass spectrometry and/or comprehensive biochemical analysis permit nearly residue-specific understanding of interactions and thus heighten mechanistic insight as compared to cryo-EM reconstructions alone.

ATP-dependent chromatin remodeling complexes can alter the position and composition of nucleosomes by sliding them along DNA, unwrapping nucleosomal DNA, or ejecting/exchanging histone dimers or octamers. There are four principal families of chromatin remodelers: SWI/SNF, ISWI, Mi-2/CHD, and SWR/INO80. Nucleosome binding of representatives of all families except the Mi-2/CHD family have been characterized by cryo-EM. These three families interact with the nucleosome in distinct ways. In 2008, Chaban et al. used negative stain reconstructions of the SWI/SNF family remodeling complex RSC to show that RSC nearly engulfs the NCP, consistent with DNaseI protection experiments and the 2007 Leschziner et al. model docking the NCP into the central cavity of a reconstruction of RSC alone. Interestingly, some density was missing for both nucleosomal DNA and one H2A/H2B dimer, suggesting RSC-mediated remodeling even in the absence of ATP.

Cryo-EM reconstructions of two ISWI family remodeling complexes bound to nucleosomes show less extensive interactions. In a 2009 study, Racki demonstrated that the ACF catalytic subunit, Snf2, binds with 2:1 stoichiometry to the nucleosome. While the ATPase domains and linker DNA are not visible in their reconstructions, biochemical data suggests that the ATPase domain binds the nucleosome at SHL ± 2 and the interaction also involves the H4 N-terminal tail. The authors propose a competition mechanism through which nucleosome spacing is accomplished by competitive sliding by two Snf2 subunits bound to opposite sides of the nucleosome. A different mechanism for nucleosome spacing was suggested for ISWI family member ISW1a by Yamada et al. based on combined cryo-EM and crystallographic data. Cryo-EM reconstructions of ISW1a in the absence of its ATPase domain bound to nucleosomes containing one or two DNA extensions revealed two modes of interactions with linker DNA. Together with a crystal structure of the ISW1a construct bound to free DNA, these cryo-EM reconstructions led to a model in which ISW1a acts as a ruler, using its size and shape to space adjacent nucleosomes.

Two cryo-EM studies also offered insight into the nucleosome binding of SWR/INO80 family remodelers. In 2012, Saravanan generated a model for the 2:1 Arp8:NCP complex using crystal structures of the components and a 21 Å cryo-EM reconstruction. In this model, Arp8 interacts with the H3/H4 surface though the molecular details are unclear. More recently, Tosi et al. used cryo-EM and cross-linking mass spectrometry (XL-MS) to extensively characterize the interaction of the holo-INO80 complex with the NCP. The architecture of the INO80 complex alone shows four domains: the Rvb1/2 dodecamer head, the Ino80 ATPase-Ies2-Arp5-Ies6 neck, the Ino80 N-terminus-Nhp10-Ies1-Ies3-Ies5 body, and the Ino80 HAS-Act1-Arp4-Arp8-Ies4-Taf14 foot. While heterogeneity in the INO80-NCP cryo-EM images prevented proper 3D reconstruction, extensive XL-MS was observed between all four domains of INO80 and the nucleosome disk and tails. Notably, some cross-linking was observed to surfaces of the H2A/H2B dimer that are buried in the NCP structure. These contacts may facilitate opening of the NCP structure required for INO80 mediated H2A/H2B exchange. The authors created a model for the INO80-NCP complex based on the XL-MS data in which the NCP rests on a cradle surrounded by all four domains of the INO80 complex.

NCP bound cryo-EM reconstitutions have also been reported for the *Saccharomyces cerevisiae* Piccolo NuA4 histone acetyltransferase (HAT) complex and the HP1-like heterochromatin protein Swi6 from *Schizosaccharomyces pombe*. Piccolo NuA4 is an H4- and H2A-specific HAT complex that functions alone and as part of the larger NuA4 complex. Piccolo's Esa1 catalytic subunit is unable to bind and acetylate nucleosomes without its accessory subunits Epl1 and Yng2. The 2011 Chittuluru cryo-EM reconstruction shows that Piccolo binds to the NCP with 1:1 stoichiometry using two prongs that contact the NCP opposite to the dyad and over the H4 histone-fold with flexibility observed between Piccolo and the NCP. Subsequent cross-linking experiments place the Esa1 Tudor domain in proximity to nucleosomal DNA and the Epl1 EPcA domain near the N-terminal tail of H2A.

Swi6 is an HP1 ortholog that binds to trimethylated H3 K9 to enable the spreading of heterochromatin. The 2013 25 Å cryo-EM reconstitution of two Swi6 dimers bound to the NCP in open/disinhibited forms reported by Canzio et al. suggests that one Swi6 chromodomain and one chromoshadow domain from each dimer contact the nucleosome near the exit of the H3 tail and at the nucleosomal DNA at SHL ± 5, respectively. The other chromodomain in each dimer is poised for binding H3 K9me3 in a neighboring nucleosome to facilitate spreading of Swi6 across chromatin. While the authors could thoroughly investigate the autoinhibitory function of Swi6, the low resolution of the cryo-EM structure prevented a molecular understanding of the Swi6-NCP interactions.

These cryo-EM studies offer unique views of nucleosome recognition by large and complex chromatin enzymes and factors. At this time, cryo-EM allows for the general architecture of chromatin factor-NCP complexes to be ascertained. The molecular workhorse remains other modalities, including crystallography of subcomplexes, XL-MS and comprehensive biochemistry. Yet, with technological advances in sample preparation together with higher resolution and faster detectors and more powerful image alignment algorithms, cryo-EM holds promise to complement if not rival or surpass crystallographic and NMR methods for atomic-resolution determination of chromatin factor-nucleosome complex structures.

Structural Studies of the Chromatosome and 30 nm Fiber

The structure of the chromatosome and the structure, and even relevance, of the 30 nm fiber are two of the most highly studied, yet controversial, topics in chromatin biology. It is clear linker histone H1 (or H5) binds to the nucleosome core particle and linker DNA and promotes compaction of chromatin arrays into 30 nm fibers. Linker histones contain a central globular domain and unstructured N- and C-terminal extensions. The globular domain gH1 and the C-terminal domain are primarily involved in chromatin binding and compaction. Many years of biochemical, biophysical and computational experiments have led to two distinct classes of gH1-NCP interactions:

1. Symmetric in which gH1 binds at the dyad and interacts with linker DNA extending from both sides of the core particle; and
2. Asymmetric in which gH1 binds near the dyad and interacts with 10–20 bp of linker DNA extending from one side of the nucleosome core. Two recent structural studies offer unique insights into H1 binding in the chromatosome and the 30 nm fiber.

NMR and Cryo-EM Models of the Chromatosome

In 2013, Zhou et al. used extensive NMR measurements to generate a unique residue-specific model for the gH1/NCP complex. The authors first observed chemical shift perturbation of isotopically labeled gH1 residues 37–211 and nucleosome core particles to define the regions of each involved in gH1/NCP complex formation. Then they incorporated paramagnetic spin labels to define distances between regions of gH1 and the NCP to orient the complex. Finally, they performed computational docking to generate models of the gH1/NCP complex. The favored model correlates with asymmetric binding to the nucleosome consistent with strong effects observed with spin labels attached to H3 R37 and H2A T119, as these observations are incompatible with symmetric binding models. Similar results were seen with the H1 tails that are unstructured in the chromatosome. gH1 only interacts with the 10 bp extending from the NCP given that no differences were seen with longer segments of nucleosomal DNA. In the favored computational model, gH1 uses two positively charged surfaces defined by NMR experiments (residues 119–125 and 164–174) to bridge the nucleosome core surface and 10 bp of linker DNA on one side of the NCP. However, the authors could not rule out weaker binding to the other linker DNA. No evidence was seen of gH1 binding histones within the nucleosomal disk. However, gH1 binding imparts structural organization of the H2A C-terminal tail consistent with a direct interaction. This explains the decreased binding of gH1 to H2A.Z containing nucleosomes that have been attributed to divergent C-terminal sequences. In 2014, Song et al. reported an 11 Å reconstruction of the 30 nm fiber reconstituted with histone H1. The density for the gH1 domain showed a 1:1 H1: nucleosome binding with H1 binding asymmetrically near, but off-center from, the dyad and interacting with both entry and exit linker DNAs. These studies provide exciting views of the position of H1 within the chromatosome and bolster growing evidence for the asymmetric binding model. Precise molecular details await higher-resolution structural solutions.

A Cryo-EM Structure of a Two-Start 30 nm Fiber

Traditional dogma holds that the 11 nm unfolded chromatin strand (often referred to as beads on a string) compacts into a 30 nm fiber with side-to-side packing of nucleosomes perpendicular to the fiber axis. The folding of the 30 nm fiber is encouraged by the interiorly positioned linker histones. The 30 nm fiber further condenses into progressively higher-order chromatin states. Most models for the 30 nm fiber fall into two categories:

1. One-start models are solenoidal with sequential nucleosomes connected by bent linker DNA segments arranged along a helical path;
2. Two-start models separate sequential nucleosomes by straight linker DNA in a zigzag patter either longitudinally along the fiber (helical ribbon) or radially across the fiber (crossed-linker).

Studies by two groups in the mid 2000s used reconstituted arrays with defined nucleosome positions to build opposing two- and one-start molecular models. Richmond and colleagues proposed a two-start model based on digestion patterns of short, cross-linked nucleosomal arrays. The two-start model was further supported by a 9 Å tetranucleosome crystal structure showing two stacks of two nucleosomes separated by a zig-zagging pattern of straight linker DNA. This structure was used to generate an idealized model for the two-start crossed-linker-type 30 nm fiber. Of note, the modeled fiber has a smaller diameter owing to the 167 nucleosome repeat length used for the tetranucleosome structure and the dependence of crossed-linker-type fiber diameters on linker length. Meanwhile, Rhodes and colleagues characterized longer H1-containing 30 nm fibers by cryo-EM. They observed similar fiber diameter over widely varying linker lengths characteristic of a one-start solenoid structure. Later modeling also suggested potential two-start solutions to the cryo-EM data in addition to the original one-start model.

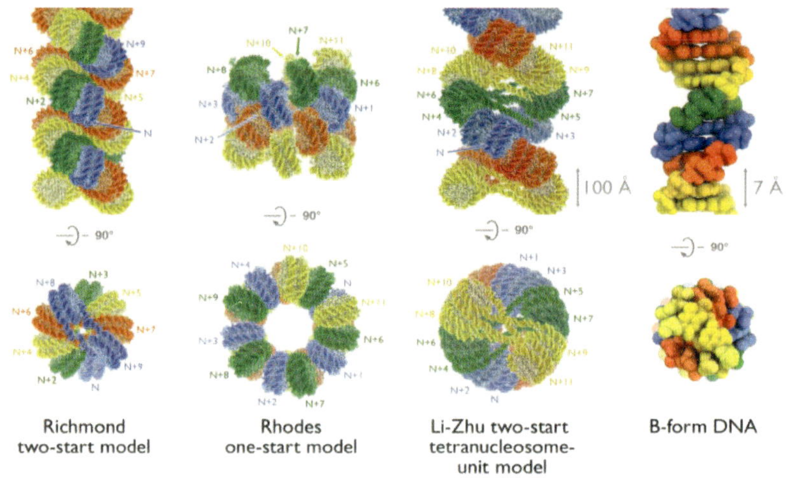

Figure: Models of the 30 nm fiber

Orthogonal views perpendicular to the 30 nm fiber axis (top) and down the axis (bottom) of the Richmond two-start model (left), Rhodes one-start model (center) and Li-Zhu tetranucleosome-unit repeat two-start model (right). The sequence of nucleosomes in each model is indicated. In the Richmond model, each sequential pair of nucleosomes across the fiber is colored similarly. For the Rhodes model, all nucleosomes in the same turn of the solenoid are colored similarly. In the Li-Zhu model, each tetranucleosome repeating unit is colored similarly. Unlabeled nucleosomes in the two-start models are not shown in the bottom views for figure clarity. Linker DNA is not present in the Rhodes model but, given the nature of the solenoidal structure, must be bent. The B-form DNA double helix is shown for comparison (far right). All models shown in space-filling representation and scaled as indicated.

Despite this long-standing hierarchical paradigm for chromatin folding, recent cryo-EM and SAXS measurements with mitotic chromosomes show no evidence of the 30 nm fiber. Rather mitotic chromatin may assume a fractal-like state. Similar experiments with chicken erythrocytes did show evidence of the 30 nm fiber. Altogether, this implies that the 30 nm fiber does exist in certain cell-types and/or cell-cycle stages, but may not be as pervasive as once thought.

In 2014, Song solved 11 Å cryo-EM structures of 30 nm fibers reconstituted with 12 × 177 and 12 × 187 bp of the Widom 601 nucleosome positioning sequence. The structures clearly show a left-handed parallel double helix similar to that proposed by Richmond and colleagues. This structure bears some resemblance to the DNA double helix, although the DNA double helix contains right-handed antiparallel strands. The diameter of the fiber is dependent on small changes in DNA linker length, suggesting a two-start model in which straight linker DNA crossing the central channel determines the fiber diameter. This two-start crossed-linker type model is confirmed by clear density for straight segments of linker DNA crossing the center channel of the fiber. The repeating unit of the fiber is a tetranucleosome with two stacked nucleosomes on opposing sides of the superhelix. The arrangement of nucleosomes places the thinner half of the nucleosome near the dyad close to the interior of the fiber where the fiber diameter is smaller. Save the different linker length, the tetranucleosome repeating unit is very similar to the 9 Å tetranucleosome crystal structure. Unexpectedly, gH1 from neighboring nucleosomes in the 30 nm fiber align with alternating head-to-head arrangements within the tetranucleosomal unit and tail-to-tail arrangements between tetranucleosomal units. The tail-to-tail aligned gH1s interact with one another, imparting an additional twist to the fiber. This results in different internucleosomal contacts between adjacent stacked nucleosomes in the tetranucleosome unit and between tetranucleosome units. Notably, the H4 tail-H2A/H2B acidic patch interaction is plausible between tetranucleosome units, as in the idealized model from Richmond and colleagues. This interaction is not possible within the tetranucleosome unit due to juxtaposition of the H2B αC helix of one nucleosome with the acidic patch H2A α2 helix of the neighboring nucleosome. This phenomenon was also seen in the tetranucleosome crystal structure. While controversy remains regarding the prevalence of the 30 nm fiber in cell-cycle specific structures of chromatin in different cell types in vivo, this model provides a higher resolution view of the 30 nm fiber, which gives new insights into the orientation of H1 within each nucleosome and across the chromatin fiber.

Chromatin

Chromatin is a complex of DNA and proteins that forms chromosomes within the nucleus of eukaryotic cells. Nuclear DNA does not appear in free linear strands; it is highly condensed and wrapped around nuclear proteins in order to fit inside the nucleus.

Chromatin exists in two forms. One form, called euchromatin, is less condensed and can be transcribed. The second form, called heterochromatin, is highly condensed and is typically not transcribed.

Under the microscope in its extended form, chromatin looks like beads on a string. The beads are called nucleosomes. Each nucleosome is composed of DNA wrapped around eight proteins called histones. The nucleosomes are then wrapped into a 30 nm spiral called a solenoid, where additional histone proteins support the chromatin structure. During cell division, the structure of the chromatin and chromosomes are visible under a light microscope, and they change in shape as the DNA is duplicated and separated into two cells.

Euchromatin

Euchromatin is a form of chromatin that is lightly packed—as opposed to heterochromatin, which is densely packed. The presence of euchromatin usually reflects that cells are transcriptionally active, i.e. they are actively transcribing DNA to mRNA. Euchromatin is found in the nucleus of eukaryotes and represents more than 90% of the human genome.

Euchromatin Structure

Before understanding the structure of euchromatin, we should comprehend the different ways in which DNA is packaged in cells.

The DNA in eukaryotic cells is arranged in complexes comprising genes and proteins. These complexes are called chromatin, and they exist in two forms: euchromatin and heterochromatin. Briefly, euchromatin (also known as the *beads-on-a-string structure*) is composed of DNA helices that are condensed at intervals into nucleosomes. Nucleosomes are the basic unit of chromatin and they consist in packaged complexes containing histone proteins around which DNA is wrapped, i.e. nucleosomes are made of DNA coiled around histones. The DNA that connects nucleosomes is known as the linker DNA. Heterochromatin is euchromatin that has been more densely packed into 30-nm fibers. During interphase, heterochromatin is packaged into denser structures—active chromosomes, which are further condensed into denser structures during mitosis and meiosis—metaphase chromosomes. An image of the different chromatin structures can be seen here:

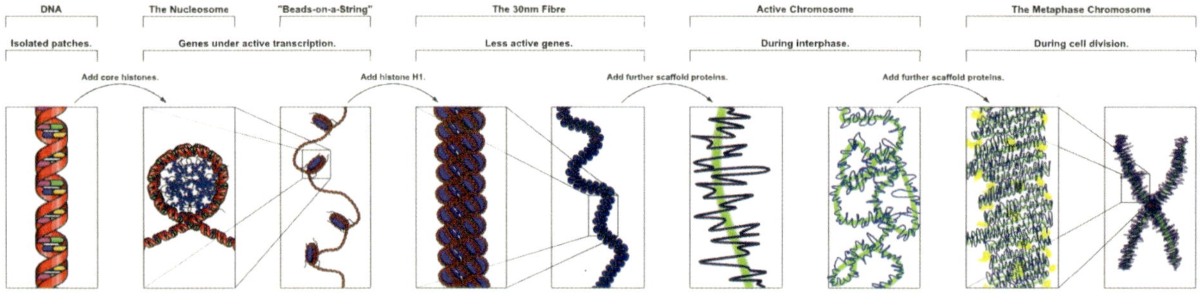

In this figure, the DNA on the left side is condensed into progressively denser structures as we move rightwards, until we reach the densest conformation—the metaphase chromosome that we are used to seeing in micrographs. Note how the double-stranded DNA is wrapped around a center of histone proteins to form nucleosomes, which in turn are part of euchromatin or beads-on-a-string. The third illustration clearly shows why euchromatin is also known as beads-on-a-string, since one can appreciate the linker DNA (string) connecting the nucleosomes (beads).

The main proteins that form chromatin are called histones. Octomers of histones are assembled together to form the nucleosomes: two copies of H2A, two of H2B, two of H3 and two of H4. About 200 base pairs of DNA are wrapped around each nucleosome. Interestingly, histones are thought to act as switches that swap between the different chromatin conformations—euchromatin and heterochromatin—through methylation and acetylation. For instance, a methylated lysine 4 in a part of the histone called *histone tail* seems to induce the euchromatin conformation. This methylated lysine 4 is therefore used as a marker for euchromatin.

Euchromatin Function

Despite being actively researched, the structure of chromatin is still poorly understood although it seems that the cycle in which the cell is at a certain time determines the structure of chromatin. Not surprisingly, the structure of euchromatin provides hints regarding its function and why it is present in transcriptionally active cells. As mentioned above, euchromatin is also called beads-on-a-string because of the resemblance between a necklace of beads connected through a string and the nucleosomes connected through the linker DNA. In this conformation, euchromatin is loose and consequently leaves the linker DNA exposed so that it can be transcribed; this way, RNA and DNA polymerases as well as other proteins can access the DNA. Because of its loose structure, euchromatin is difficult to see under a microscope and appears faintly when stained—in contrast to the easily visible heterochromatin, which is densely packed.

It has been hypothesized that the regulation of the chromatin structure is a way to control gene expression. It is believed that the euchromatic structure is present when genes are *turned on*, that is, when they are being actively transcribed, while the heterochromatic structure is present when genes are *turned off* or inactive. In other words, because euchromatin is present in transcriptionally active cells because of the accessibility to the DNA, folding into heterochromatin may be a way to regulate transcription by preventing the access of RNA polymerases and other regulatory proteins to the DNA. In this line, housekeeping genes, for instance, are always in the euchromatic conformation because they need to be constantly replicated and transcribed to keep the functional activity and survival of the cells.

Euchromatin in Prokaryotes and Some Eukaryotes

Although prokaryotes have a different mechanism to condense DNA, its packaged structure resembles that of euchromatin. It is therefore believed that heterochromatin—the densely packaged chromatin—evolved later, possibly together with the nucleus, to regulate gene expression and to manage large amounts—long strings—of genetic material.

Whereas the DNA in most eukaryotic cells is packaged as described, there are some other eukaryotes that do not conform to this organization. Among these are avian red blood cells and motile sperm cells (spermatozoa), both of which contain chromatin in more densely packaged conformations than most eukaryotes.

Heterochromatin

Heterochromatin is a form of chromatin that is densely packed—as opposed to euchromatin, which is lightly packed—and is found in the nucleus of eukaryotic cells. Whereas euchromatin allows the DNA to be replicated and transcribed, heterochromatin is in such a condensed structure that it does not enable DNA and RNA polymerases to access the DNA, therefore preventing DNA replication and transcription. There are two main types of heterochromatin: constructive heterochromatin and facultative heterochromatin. Heterochromatin represents less than 10% of the human chromatin, with euchromatin accounting for most of it—over 90%.

Heterochromatin Structure

The DNA in eukaryotes is assembled into chromatin, which are complexes made of DNA and proteins. The proteins that form chromatin are called histones, and they are arranged in a way that allows DNA to be wrapped around them. More specifically, the DNA (about 200 base pairs) is coiled around sets of eight histones (octamers) comprising two copies of each of the following: H2A, H2B, H3 and H4. These units made of histones and DNA coiled around them are called nucleosomes. Nucleosomes are in turn connected to one another through DNA strings, also known as linker DNA. In other words, chromatin is the assembly of nucleosomes (DNA and histones) connected by the DNA itself.

The most loosely packaged form of chromatin is called euchromatin, also known as *beads-on-a-string* because of the resemblance between this structure and beads (nucleosomes) held together by a string (DNA). Heterochromatin is a more tightly condensed version of euchromatin and is also known as *30-nm fiber* because the diameter of this helically coiled heterochromatin measures 30 nm. In fact, while G-banding shows very faintly stained euchromatin due to its loose form, heterochromatin is easily seen because it is densely stained due to its denser packaging. Heterochromatin can also be further condensed into active chromosomes and even further into metaphase chromosomes.

The following figure shows the different structural units of DNA packaging in eukaryotic cells:

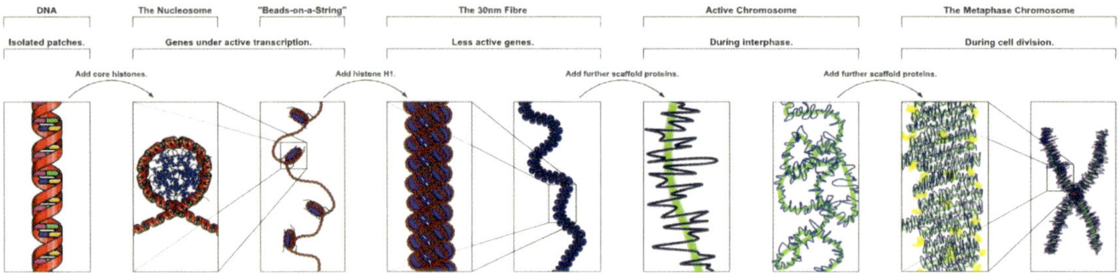

From left to right, double-stranded helical DNA (first illustration) is coiled around histones, forming nucleosomes (second illustration), which constitute the euchromatin or beads-on-a-string structure (third illustration). Euchromatin is further condensed into heterochromatin or 30-nm fibers (fourth and fifth illustrations). The last four illustrations depict more tightly condensed DNA in the form of active and metaphase chromosomes.

Looking at the figure above, we can also appreciate why DNA is in the heterochromatin conformation when it is not being actively replicated or transcribed: the DNA is not exposed and therefore regulatory proteins and polymerases cannot access it. Note the difference between euchromatin and heterochromatin while linker DNA in the euchromatic conformation is exposed and accessible to polymerases and other proteins in order to be replicated and transcribed, the DNA in the heterochromatic conformation is tightly coiled around the nucleosomes and does not allow access to transcriptional elements.

Types of Heterochromatin: Constitutive and Facultative

The structure of heterochromatin can be described in more detail by taking into account its several types. The two main types are constitutive heterochromatin and facultative heterochromatin. These two types can be distinguished based on their features. It has been suggested that other types of heterochromatin also exist and that these other types have mixed features of constitutive and facultative heterochromatin.

Constitutive heterochromatin is the stable form of heterochromatin, i.e. it does not loosen up to form euchromatin, and contains repeated sequences of DNA called satellite DNA. It can be found in centromeres and telomeres, and is usually involved in structural functions.

Facultative heterochromatin, on the other hand, is reversible, i.e. its structure can change depending on the cell cycle, and is characterized by another kind of repeated DNA sequences known as LINE sequences. An example of facultative heterochromatin that changes its structural conformation with the cell cycle is the inactivated X-chromosome (Barr body) of females.

Cell Cycle and Gene Expression

It is not surprising that the way in which the DNA is packaged is related to the cell cycle. When the DNA needs to be copied (replicated) and proteins need to be synthesized (transcription and then translation), the DNA is found in the euchromatin form. When genes do not need to be replicated and transcribed, the DNA is in the heterochromatin form. Furthermore, when the DNA is in the active chromosome form, the cell is in the interphase stage of the cell cycle, and when it is in the metaphase chromosome form, the cell is in dividing, i.e. it is in the mitosis or meiosis stage.

In line with this, it has been proposed that regulating the way in which the DNA is packaged is a way of regulating gene expression. Therefore, housekeeping genes that maintain the functions and survival of the cell are always in the euchromatin form, whereas those that do not need to be expressed are in the heterochromatin form. The means by which this is achieved is by modification of the *histone tail*, a part of the histones that can be acetylated or methylated. Modifying the histone tail results in changes in the packaging of the DNA. For instance, hypoacetylation on the histone tail is associated with the heterochromatic conformation, whereby DNA is not exposed and consequently gene transcription is prevented.

References

- Human-Chromosomes, health: news-medical.net, Retrieved 11 June 2018
- Chromosomes-structure-functions-and-other-details-about-chromosomes-22946: yourarticlelibrary.com, Retrieved 21 June 2018
- What-is-a-telomere: yourgenome.org, Retrieved 31 March 2018
- What-is-the-X-Chromosome, health: news-medical.net, Retrieved 11 May 2018
- Nucleosome-nucleosomes-30: nature.com, Retrieved 08 July 2018
- Heterochromatin: biologydictionary.net, Retrieved 26 May 2018

Chapter 5

Genetic Testing

The determination of human bloodlines and the diagnosis of the vulnerabilities to hereditary diseases is possible through genetic testing. This chapter discusses in extensive detail the varied forms of genetic testing such as carrier testing, predictive testing, DNA paternity testing, genealogical DNA test and preimplantation genetic diagnosis.

Genetic testing is a type of medical test that identifies changes in chromosomes, genes, or proteins. The results of a genetic test can confirm or rule out a suspected genetic condition or help determine a person's chance of developing or passing on a genetic disorder. More than 1,000 genetic tests are currently in use, and more are being developed.

Several methods can be used for genetic testing:

- Molecular genetic tests (or gene tests) study single genes or short lengths of DNA to identify variations or mutations that lead to a genetic disorder.
- Chromosomal genetic tests analyze whole chromosomes or long lengths of DNA to see if there are large genetic changes, such as an extra copy of a chromosome, that cause a genetic condition.
- Biochemical genetic tests study the amount or activity level of proteins; abnormalities in either can indicate changes to the DNA that result in a genetic disorder.

Genetic testing is voluntary. Because testing has benefits as well as limitations and risks, the decision about whether to be tested is a personal and complex one. A geneticist or genetic counselor can help by providing information about the pros and cons of the test and discussing the social and emotional aspects of testing.

Need of Genetic Testing

- A genetic test is generally performed in a particular individual or family for a specific medical purpose.
- There are a number of reasons why a genetic test may be called for, these include:
    - Pre-implantation genetic diagnosis (PGD): Screening an embryo for a genetic disease,
    - Prenatal testing: finding a genetic disease in an unborn baby,
    - Carrier testing: finding out if parents carry a genetic mutation that they could pass onto their future children,
    - Predictive genetic testing: testing an adult for a genetic disease before they have symptoms, usually where the disease runs in the family and they want to find out if they may also be affected,

- Diagnostic genetic testing: making a diagnosis in a patient that is showing symptoms of a known genetic disease,
- Pharmacogenetic testing: determining the best dose or type of medicine to give an individual patient based on their genetics.

Process of Genetic Testing

Genetic tests are performed on a sample of blood, hair, skin, amniotic fluid (the fluid that surrounds a fetus during pregnancy), or other tissue. For example, a procedure called a buccal smear uses a small brush or cotton swab to collect a sample of cells from the inside surface of the cheek. The sample is sent to a laboratory where technicians look for specific changes in chromosomes, DNA, or proteins, depending on the suspected disorder. The laboratory reports the test results in writing to a person's doctor or genetic counselor, or directly to the patient if requested.

Newborn screening tests are done on a small blood sample, which is taken by pricking the baby's heel. Unlike other types of genetic testing, a parent will usually only receive the result if it is positive. If the test result is positive, additional testing is needed to determine whether the baby has a genetic disorder.

Direct-to-consumer Genetic Testing

Direct-to-consumer (DTC) genetic testing is a type of genetic test that is accessible directly to the consumer without having to go through a health care professional. Usually, to obtain a genetic test, health care professionals (such as doctors) acquire their patient's permission and then order the desired test. DTC genetic tests, however, allow consumers to bypass this process and order DNA tests themselves.

There is a variety of DTC tests, ranging from tests for breast cancer alleles to mutations linked to cystic fibrosis. Benefits of DTC testing are the accessibility of tests to consumers, promotion of proactive healthcare, and the privacy of genetic information. Possible additional risks of DTC testing are the lack of governmental regulation, the potential misinterpretation of genetic information, issues related to testing minors, privacy of data, and downstream expenses for the public health care system.

Controversy

DTC genetic testing has been controversial due to outspoken opposition within the medical community. Critics of DTC testing argue against the risks involved, the unregulated advertising and marketing claims, and the overall lack of governmental oversight.

DTC testing involves many of the same risks associated with any genetic test. One of the more obvious and dangerous of these is the possibility of misreading of test results. Without professional guidance, consumers can potentially misinterpret genetic information, causing them to be deluded about their personal health.

Some advertising for DTC genetic testing has been criticized as conveying an exaggerated and inaccurate message about the connection between genetic information and disease risk, utilizing emotions as a selling factor.

Risks and Limitations

The physical risks associated with most genetic tests are very small, particularly for those tests that require only a blood sample or buccal smear (a procedure that samples cells from the inside surface of the cheek). The procedures used for prenatal testing carry a small but non-negligible risk of losing the pregnancy (miscarriage) because they require a sample of amniotic fluid or tissue from around the fetus.

Many of the risks associated with genetic testing involve the emotional, social, or financial consequences of the test results. People may feel angry, depressed, anxious, or guilty about their results. The potential negative impact of genetic testing has led to an increasing recognition of a "right not to know". In some cases, genetic testing creates tension within a family because the results can reveal information about other family members in addition to the person who is tested. The possibility of genetic discrimination in employment or insurance is also a concern. Some individuals avoid genetic testing out of fear it will affect their ability to purchase insurance or find a job. Health insurers do not currently require applicants for coverage to undergo genetic testing, and when insurers encounter genetic information, it is subject to the same confidentiality protections as any other sensitive health information. In the United States, the use of genetic information is governed by the Genetic Information Nondiscrimination Act (GINA).

Genetic testing can provide only limited information about an inherited condition. The test often can't determine if a person will show symptoms of a disorder, how severe the symptoms will be, or whether the disorder will progress over time. Another major limitation is the lack of treatment strategies for many genetic disorders once they are diagnosed.

Another limitation to genetic testing for a hereditary linked cancer, is the variants of unknown clinical significance. Because the human genome has over 22,000 genes, there are 3.5 million variants in the average person's genome. These variants of unknown clinical significance means there is a change in the DNA sequence, however the increase for cancer is unclear because it is unknown if the change affects the gene's function.

A genetics professional can explain in detail the benefits, risks, and limitations of a particular test. It is important that any person who is considering genetic testing understands and weigh these factors before making a decision. Other risks include accidental findings—a discovery of some possible problem found while looking for something else. In 2013 the American College of Medical Genetics and Genomics (ACMG) that certain genes always be included any time a genomic sequencing was done, and that labs should report the results.

Costs

The cost of genetic testing can range from under $100 to more than $2,000. This depends on the complexity of the test. The cost will increase if more than one test is necessary or if multiple family members are getting tested to obtain additional results. Costs can vary by state and some states cover part of the total cost.

From the date that a sample is taken, results may take weeks to months, depending upon the complexity and extent of the tests being performed. Results for prenatal testing are usually available more quickly because time is an important consideration in making decisions about a pregnancy.

Prior to the testing, the doctor or genetic counselor who is requesting a particular test can provide specific information about the cost and time frame associated with that test.

Carrier Testing

In genetics, the term carrier describes an organism that carries two different forms (alleles) of a recessive gene (alleles of a gene linked to a recessive trait) and is thus heterozygous for that the recessive gene. Although carriers may act to convey and maintain recessive genes within a population by passing them on to offspring, the carriers themselves are not affected by the recessive trait associated with the recessive gene.)

Carrier testing is also called carrier screening. It is a type of Genetic Testing. Carrier screening is testing that's done to see whether you or your partner carry a genetic mutation that could cause a serious inherited disorder in your baby. Some of the more common disorders screened for include cystic fibrosis, sickle cell disease, thalassemia, and Tay-Sachs disease, but there are more than 100 others that can be tested for.

Many of these conditions are rare, but one large study found that 24 percent of the patients tested were carriers of at least one mutation. And the average risk of having a child with one of these diseases is higher than that of having a child with Down syndrome or a neural tube defect. What's more, these conditions will not be detected by prenatal tests like CVS and amniocentesis unless you have carrier screening first.

These disorders are recessive, which means that a baby must inherit a defective gene from each parent to have the disease. If you're a carrier of a defective gene for a recessive disorder, that means you have one normal copy of the gene from one of your parents and one defective copy from the other. (Carriers don't usually have any symptoms of the disease.)

If both you and your partner are carriers of a disorder like cystic fibrosis, sickle cell disease, or Tay-Sachs disease, your child will have a 1 in 4 chance of inheriting one defective gene from each of you and being born with the disease.

Process of Genetic Carrier Screening

Ideally, you should have the option of being screened before you try to conceive. Your practitioner should offer it to you at your preconception visit. This way, if you find out that you and your partner are both carriers for a condition, you'll have a wider range of options. You can talk to a genetic counselor who will be able to tell you more about the condition and help you sort out your reproductive choices.

Traditionally, couples have only been offered screening for one or two of the most common mutations if they are determined to be at risk for being a carrier. Risk factors include having a family member with the inherited disorder or who's a known carrier, or being part of an ethnic group at increased risk for the disease.

The problem with this approach is that many people don't belong to distinct ethnic categories.

Many people are mixed race, adopted, or simply can't be sure what ethnicity their ancestors were. So there's no good way of determining who's at risk for being a carrier of any particular mutation.

Instead, you can choose to be screened for a wide range of disease mutations – more than 100 instead of just the one or two you may be "at risk" for. This approach is known as expanded carrier screening. If your caregiver doesn't offer you expanded screening, you can ask for it.

If you opt to be screened, you'll be asked to give a blood or saliva sample. If you're found to be a carrier, your partner will be screened as well. Or both partners may be screened at the same time to get the results faster.

You should be given the option of talking with a genetic counselor before the screening and after you get your results. This person can help you understand your results and your options for planning your family.

You might also consider consulting a medical geneticist, a doctor who is specially trained and board certified in genetics. The American College of Medical Genetics and Genomics offers an online tool for finding genetic services.

Cost

The cost of carrier screening has declined dramatically in recent years, thanks to advances in technology. Since carrier screening is a recommended part of preconception and prenatal care, it's sometimes covered by insurance. On the other hand, some insurance companies consider the testing optional and don't cover it. Out-of-pocket costs vary, but they're typically not more than a few hundred dollars, even without insurance.

Below are common genetic conditions that carrier screening tests for:

Cystic Fibrosis Screening

Cystic fibrosis (CF) is a life-threatening genetic disease. People with CF are prone to breathing difficulties (including lung infections and severe lung damage), digestive problems, and other complications.

If you and your partner are both carriers, the odds are 1 in 4 that your baby will have CF.

No screening test is 100 percent accurate, but if both you and your partner are negative for the CF mutation, your chance of having a baby with the condition is less than half of one percent.

Sickle Cell Screening

Sickle cell disease is a debilitating red blood cell disorder.

High-risk groups include people of African, Caribbean, South or Central American, Mediterranean, Indian, or Arabian descent. According to the National Institutes of Health, 1 in 13 African Americans carries the gene for this disorder.

If you're a carrier, your partner will be offered testing as well. If you and your partner are both carriers (or your partner is a carrier of a related blood disorder), your baby's chance of having sickle cell disease is 1 in 4.

Thalassemia Screening

Thalassemia encompasses a varied group of inherited blood disorders, including some that are relatively mild and others that may cause severe anemia and other serious problems. More than 2 million people in the United States carry the genetic trait for it.

High-risk groups include people of Southeast Asian, Chinese, Indian, African, Middle Eastern, Italian, Greek, and Mediterranean ancestry, as well as anyone with a family history of the disease or a family member who is a known carrier.

People who have the thalassemia trait may have a mild form of anemia. If your initial blood count shows that your red blood cells are small but your iron status is normal, further testing will be done to check for the thalassemia trait.

If you're a carrier, your partner should be offered testing as well. If you're both carriers (or your partner is a carrier for another red blood cell disorder like sickle cell disease), your baby is at high risk for the disease.

Tay-Sachs Screening

Tay-Sachs is a fatal disease of the central nervous system. About 1 in 250 people in the United States carries the genetic trait for it.

Among Central or Eastern European (Ashkenazi) Jews, French Canadians, and Cajuns, the carrier rate is about 1 in 27. Among Irish Americans, it's 1 in 50. You're considered high risk if you belong to any of these groups or have a family history of the disease.

Ashkenazi Jews are also at risk for carrying the genes that cause two other severe nervous system disorders – familial dysautonomia and Canavan disease – and over three-dozen other diseases as well.

If you and your partner both carry the gene for Tay-Sachs, your baby has a 1 in 4 chance of having the disease.

Preimplantation Genetic Diagnosis

Preimplantation genetic diagnosis (PGD) is a procedure used prior to implantation to help identify genetic defects within embryos. This serves to prevent certain genetic diseases or disorders from being passed on to the child. The embryos used in PGD are usually created during the process of in vitro fertilization (IVF).

Process of PGD

Preimplantation genetic diagnosis begins with the normal process of in vitro fertilization that includes egg retrieval and fertilization in a laboratory. Over the next three to five days, the embryos will divide into multiple cells.

Preimplantation genetic diagnosis involves the following steps:

1. First, a couple/few cells are micro surgically removed from the embryos, which are about 5 days developed. After this cell collection, the embryos are safely frozen.
2. The DNA of the cells is then evaluated to determine if the inheritance of a problematic gene is present in each embryo. This process takes at least one full week.
3. Once PGD has identified embryos free of genetic problems, the embryo(s) will be placed in the uterus (usually by an IVF procedure), and the wait for implantation and a positive pregnancy test begins.
4. Any additional embryos that are free of genetic problems are kept frozen for possible later use while embryos with the problematic gene(s) are destroyed. This testing process may take weeks.

Getting from the egg retrieval process to the final results of PGD can take several weeks. If you think about it, this process includes collection, fertilization, 3-5 days of development, 1-2 weeks of testing, and scheduling an appointment to discuss results with your doctor. It is important to keep this in mind if you plan to pursue IVF with PGD so that you know what to expect!

Benefits of PGD

Preimplantation genetic diagnosis can benefit any couple at risk for passing on a genetic disease or condition.

The following is a list of the type of individuals who are possible candidates for PGD:

- Carriers of sex-linked genetic disorders
- Carriers of single gene disorders
- Those with chromosomal disorders
- Women age 35 and over
- Women experiencing recurrent pregnancy loss
- Women with more than one failed fertility treatment

PGD has also been used for the purpose of gender selection. However, discarding embryos based only on gender considerations is an ethical concern for many people.

Technical Aspects

PGD is a form of genetic diagnosis performed prior to implantation. This implies that the patient's oocytes should be fertilized in vitro and the embryos kept in culture until the diagnosis is established. It is also necessary to perform a biopsy on these embryos in order to obtain material on which to perform the diagnosis. The diagnosis itself can be carried out using several techniques, depending on the nature of the studied condition. Generally, PCR-based methods are used for monogenic disorders and FISH for chromosomal abnormalities and for sexing those cases in which no PCR protocol is available for an X-linked disease. These techniques need to be adapted to be performed on blastomeres and need to be thoroughly tested on single-cell models prior to clinical use. Finally, after embryo replacement, surplus good quality unaffected embryos can be cryopreserved, to be thawed and transferred back in a next cycle.

Obtaining Embryos

Currently, all PGD embryos are obtained by assisted reproductive technology, although the use of natural cycles and in vivo fertilization followed by uterine lavage was attempted in the past and is now largely abandoned. In order to obtain a large group of oocytes, the patients undergo controlled ovarian stimulation (COH). COH is carried out either in an agonist protocol, using gonadotrophin-releasing hormone (GnRH) analogues for pituitary desensitisation, combined with human menopausal gonadotrophins (hMG) or recombinant follicle stimulating hormone (FSH), or an antagonist protocol using recombinant FSH combined with a GnRH antagonist according to clinical assessment of the patient's profile (age, body mass index (BMI), endocrine parameters). hCG is administered when at least three follicles of more than 17 mm mean diameter are seen at transvaginal ultrasound scan. Transvaginal ultrasound-guided oocyte retrieval is scheduled 36 hours after hCG administration. Luteal phase supplementation consists of daily intravaginal administration of 600 μg of natural micronized progesterone.

Oocytes are carefully denudated from the cumulus cells, as these cells can be a source of contamination during the PGD if PCR-based technology is used. In the majority of the reported cycles, intracytoplasmic sperm injection (ICSI) is used instead of IVF. The main reasons are to prevent contamination with residual sperm adhered to the zona pellucida and to avoid unexpected fertilization failure. The ICSI procedure is carried out on mature metaphase-II oocytes and fertilization is assessed 16–18 hours after. The embryo development is further evaluated every day prior to biopsy and until transfer to the woman's uterus. During the cleavage stage, embryo evaluation is performed daily on the basis of the number, size, cell-shape and fragmentation rate of the blastomeres. On day 4, embryos were scored in function of their degree of compaction and blastocysts were evaluated according to the quality of the throphectoderm and inner cell mass, and their degree of expansion.

Biopsy Procedures

As PGD can be performed on cells from different developmental stages, the biopsy procedures vary accordingly. Theoretically, the biopsy can be performed at all preimplantation stages, but only three have been suggested: on unfertilised and fertilised oocytes (for polar bodies, PBs), on day three cleavage-stage embryos (for blastomeres) and on blastocysts (for trophectoderm cells).

The biopsy procedure always involves two steps: the opening of the zona pellucida and the removal of the cell(s). There are different approaches to both steps, including mechanical, chemical, and physical (Tyrode's acidic solution) and laser technology for the breaching of the zona pellucida, extrusion or aspiration for the removal of PBs and blastomeres, and herniation of the trophectoderm cells.

Polar Body Biopsy

A polar body biospy is the sampling of a polar body, which is a small haploid cell that is formed concomitantly as an egg cell during oogenesis, but which generally does not have the ability to be fertilized. Compared to a blastocyst biopsy, a polar body biopsy can potentially be of lower costs, less harmful side-effects, and more sensitive in detecting abnormalities. The main advantage of the use of polar bodies in PGD is that they are not necessary for successful fertilisation or normal embryonic development, thus ensuring no deleterious effect for the embryo. One of the disadvantages of PB biopsy is that it only provides information about the maternal contribution to the embryo, which is why cases of maternally inherited autosomal dominant and X-linked disorders that are exclusively maternally transmitted can be diagnosed, and autosomal recessive disorders can only partially be diagnosed. Another drawback is the increased risk of diagnostic error, for instance due to the degradation of the genetic material or events of recombination that lead to heterozygous first polar bodies.

Cleavage-stage Biopsy (Blastomere Biopsy)

Cleavage-stage biopsy is generally performed the morning of day three post-fertilization, when normally developing embryos reach the eight-cell stage. The biopsy is usually performed on embryos with less than 50% of anucleated fragments and at an 8-cell or later stage of development. A hole is made in the zona pellucida and one or two blastomeres containing a nucleus are gently aspirated or extruded through the opening. The main advantage of cleavage-stage biopsy over PB analysis is that the genetic input of both parents can be studied. On the other hand, cleavage-stage embryos are found to have a high rate of chromosomal mosaicism, putting into question whether the results obtained on one or two blastomeres will be representative for the rest of the embryo. It is for this reason that some programs utilize a combination of PB biopsy and blastomere biopsy. Furthermore, cleavage-stage biopsy, as in the case of PB biopsy, yields a very limited amount of tissue for diagnosis, necessitating the development of single-cell PCR and FISH techniques. Although theoretically PB biopsy and blastocyst biopsy are less harmful than cleavage-stage biopsy, this is still the prevalent method. It is used in approximately 94% of the PGD cycles reported to the ESHRE PGD Consortium. The main reasons are that it allows for a safer and more complete diagnosis than PB biopsy and still leaves enough time to finish the diagnosis before the embryos must be replaced in the patient's uterus, unlike blastocyst biopsy. Of all cleavage-stages, it is generally agreed that the optimal moment for biopsy is at the eight-cell stage. It is diagnostically safer than the PB biopsy and, unlike blastocyst biopsy, it allows for the diagnosis of the embryos before day 5. In this stage, the cells are still totipotent and the embryos are not yet compacting. Although it has been shown that up to a quarter of a human embryo can be removed without disrupting its development, it still remains to be studied whether the biopsy of one or two cells correlates with the ability of the embryo to further develop, implant and grow into a full term pregnancy.

Not all methods of opening the zona pellucida have the same success rate because the well being of the embryo and/or blastomere may be impacted by the procedure used for the biopsy. Zona drilling with acid Tyrode's solution (ZD) was looked at in comparison to partial zona dissection (PZD) to determine which technique would lead to more successful pregnancies and have less of an effect on the embryo and/or blastomere. ZD uses a digestive enzyme like pronase, which makes it a chemical drilling method. The chemicals used in ZD may have a damaging effect on the embryo. PZD uses a glass microneedle to cut the zona pellucida, which makes it a mechanical dissection method that typically needs skilled hands to perform the procedure. In a study that included 71 couples, ZD was performed in 26 cycles from 19 couples and PZD was performed in 59 cycles from 52 couples. In the single cell analysis, there was a success rate of 87.5% in the PZD group and 85.4% in the ZD group. The maternal age, number of oocytes retrieved, fertilization rate, and other variables did not differ between the ZD and PZD groups. It was found that PZD led to a significantly higher rate of pregnancy (40.7% vs 15.4%), ongoing pregnancy (35.6% vs 11.5%), and implantation (18.1% vs 5.7%) than ZD. This suggests that using the mechanical method of PZD in blastomere biopsies for preimplantation genetic diagnosis may be more proficient than using the chemical method of ZD. The success of PZD over ZD could be attributed to the chemical agent in ZD having a harmful effect on the embryo and/or blastomere. Currently, zona-drilling using a laser is the predominant method of opening the zona pellucida. Using a laser is an easier technique than using mechanical or chemical means. However, laser drilling could be harmful to the embryo and it is very expensive for in vitro fertilization laboratories to use especially when PGD is not a prevalent process as of modern times. PZD could be a viable alternative to these issues.

Blastocyst Biopsy

In an attempt to overcome the difficulties related to single-cell techniques, it has been suggested to biopsy embryos at the blastocyst stage, providing a larger amount of starting material for diagnosis. It has been shown that if more than two cells are present in the same sample tube, the main technical problems of single-cell PCR or FISH would virtually disappear. On the other hand, as in the case of cleavage-stage biopsy, the chromosomal differences between the inner cell mass and the trophectoderm (TE) can reduce the accuracy of diagnosis, although this mosaicism has been reported to be lower than in cleavage-stage embryos.

TE biopsy has been shown to be successful in animal models such as rabbits, miceand primates. These studies show that the removal of some TE cells is not detrimental to the further in vivo development of the embryo.

Human blastocyst-stage biopsy for PGD is performed by making a hole in the ZP on day three of in vitro culture. This allows the developing TE to protrude after blastulation, facilitating the biopsy. On day five post-fertilization, approximately five cells are excised from the TE using a glass needle or laser energy, leaving the embryo largely intact and without loss of inner cell mass. After diagnosis, the embryos can be replaced during the same cycle, or cryopreserved and transferred in a subsequent cycle.

There are two drawbacks to this approach, due to the stage at which it is performed. First, only approximately half of the preimplantation embryos reach the blastocyst stage. This can restrict the number of blastocysts available for biopsy, limiting in some cases the success of the PGD. Mc Arthur and coworkers report that 21% of the started PGD cycles had no embryo suitable for TE biopsy. This figure is approximately four times higher than the average presented by the ESHRE PGD consortium data, where PB and cleavage-stage biopsy are the predominant reported methods. On the other hand, delaying the biopsy to this late stage of development limits the time to perform the genetic diagnosis, making it difficult to redo a second round of PCR or to rehybridize FISH probes before the embryos should be transferred back to the patient.

Cumulus Cell Sampling

Sampling of cumulus cells can be performed in addition to a sampling of polar bodies or cells from the embryo. Because of the molecular interactions between cumulus cells and the oocyte, gene expression profiling of cumulus cells can be performed to estimate oocyte quality and the efficiency of an ovarian hyperstimulation protocol, and may indirectly predict aneuploidy, embryo development and pregnancy outcomes.

Genetic Analysis Techniques

Fluorescent in situ hybridization (FISH) and Polymerase chain reaction (PCR) are the two commonly used, first-generation technologies in PGD. PCR is generally used to diagnose monogenic disorders and FISH is used for the detection of chromosomal abnormalities (for instance, aneuploidy screening or chromosomal translocations). Over the past few years, various advancements in PGD testing have allowed for an improvement in the comprehensiveness and accuracy of results available depending on the technology used. Recently a method was developed allowing to fix metaphase plates from single blastomeres. This technique in conjunction with FISH, m-FISH can produce more reliable results, since analysis is done on whole metaphase plates.

In addition to FISH and PCR, single cell genome sequencing is being tested as a method of preimplantation genetic diagnosis. This characterizes the complete DNA sequence of the genome of the embryo.

FISH

FISH is the most commonly applied method to determine the chromosomal constitution of an embryo. In contrast to karyotyping, it can be used on interphase chromosomes, so that it can be used on PBs, blastomeres and TE samples. The cells are fixated on glass microscope slides and hybridised with DNA probes. Each of these probes is specific for part of a chromosome, and are labelled with a fluorochrome.

Dual FISH was considered to be an efficient technique for determination of the sex of human preimplantation embryos and the additional ability to detect abnormal chromosome copy numbers, which is not possible via the polymerase chain reaction (PCR).

Currently, a large panel of probes is available for different segments of all chromosomes, but the limited number of different fluorochromes confines the number of signals that can be analysed simultaneously.

The type and number of probes that are used on a sample depends on the indication. For sex determination (used for instance when a PCR protocol for a given X-linked disorder is not available), probes for the X and Y chromosomes are applied along with probes for one or more of the autosomes as an internal FISH control. More probes can be added to check for aneuploidies, particularly those that could give rise to a viable pregnancy (such as a trisomy 21). The use of probes for chromosomes X, Y, 13, 14, 15, 16, 18, 21 and 22 has the potential of detecting 70% of the aneuploidies found in spontaneous abortions.

In order to be able to analyze more chromosomes on the same sample, up to three consecutive rounds of FISH can be carried out. In the case of chromosome rearrangements, specific combinations of probes have to be chosen that flank the region of interest. The FISH technique is considered to have an error rate between 5 and 10%.

The main problem of the use of FISH to study the chromosomal constitution of embryos is the elevated mosaicism rate observed at the human preimplantation stage. A meta-analysis of more than 800 embryos came to the result that approximately 75% of preimplantation embryos are mosaic, of which approximately 60% are diploid–aneuploid mosaic and approximately 15% aneuploid mosaic. Li and co-workers found that 40% of the embryos diagnosed as aneuploid on day 3 turned out to have a euploid inner cell mass at day 6. Staessen and collaborators found that 17.5% of the embryos diagnosed as abnormal during PGS, and subjected to post-PGD reanalysis, were found to also contain normal cells, and 8.4% were found grossly normal. As a consequence, it has been questioned whether the one or two cells studied from an embryo are actually representative of the complete embryo, and whether viable embryos are not being discarded due to the limitations of the technique.

PCR

Kary Mullis conceived PCR in 1985 as an in vitro simplified reproduction of the in vivo process of DNA replication. Taking advantage of the chemical properties of DNA and the availability of thermostable DNA polymerases, PCR allows for the enrichment of a DNA sample for a certain sequence. PCR provides the possibility to obtain a large quantity of copies of a particular stretch of the genome, making further analysis possible. It is a highly sensitive and specific technology, which makes it suitable for all kinds of genetic diagnosis, including PGD. Currently, many different variations exist on the PCR itself, as well as on the different methods for the posterior analysis of the PCR products.

When using PCR in PGD, one is faced with a problem that is inexistent in routine genetic analysis: the minute amounts of available genomic DNA. As PGD is performed on single cells, PCR has to be adapted and pushed to its physical limits, and use the minimum amount of template possible: which is one strand. This implies a long process of fine-tuning of the PCR conditions and a susceptibility to all the problems of conventional PCR, but several degrees intensified. The high number of needed PCR cycles and the limited amount of template makes single-cell PCR very sensitive to contamination. Another problem specific to single-cell PCR is the allele drop out (ADO) phenomenon. It consists of the random non-amplification of one of the alleles present in a heterozygous sample. ADO seriously compromises the reliability of PGD as a heterozygous embryo could be diagnosed as affected or unaffected depending on which allele would fail to amplify. This is particularly concerning in PGD for autosomal dominant disorders, where ADO of the affected allele could lead to the transfer of an affected embryo.

Several PCR-based assays have been developed for various diseases like the triplet repeat genes associated with myotonic dystrophy and fragile X in single human somatic cells, gametes and embryos.

Establishing a Diagnosis

The establishment of a diagnosis in PGD is not always straightforward. The criteria used for choosing the embryos to be replaced after FISH or PCR results are not equal in all centers. In the case of FISH, in some centers only embryos are replaced that are found to be chromosomally normal (that is, showing two signals for the gonosomes and the analyzed autosomes) after the analysis of one or two blastomeres, and when two blastomeres are analyzed, the results should be concordant. Other centers argue that embryos diagnosed as monosomic could be transferred, because the false monosomy (i.e. loss of one FISH signal in a normal dipoloid cell) is the most frequently occurring misdiagnosis. In these cases, there is no risk for an aneuploid pregnancy, and normal diploid embryos are not lost for transfer because of a FISH error. Moreover, it has been shown that embryos diagnosed as monosomic on day 3 (except for chromosomes X and 21), never develop to blastocyst, which correlates with the fact that these monosomies are never observed in ongoing pregnancies.

Diagnosis and misdiagnosis in PGD using PCR have been mathematically modeled in the work of Navidi and Arnheim and of Lewis and collaborators. The most important conclusion is that for the efficient and accurate diagnosis of an embryo, two genotypes are required. This can be based on a linked marker and disease genotypes from a single cell or on marker/disease genotypes of two cells. An interesting aspect explored in these papers is the detailed study of all possible combinations of alleles that may appear in the PCR results for a particular embryo. The authors indicate that some of the genotypes that can be obtained during diagnosis may not be concordant with the expected pattern of linked marker genotypes, but are still providing sufficient confidence about the unaffected genotype of the embryo. Although these models are reassuring, they are based on a theoretical model, and generally the diagnosis is established on a more conservative basis, aiming to avoid the possibility of misdiagnosis. When unexpected alleles appear during the analysis of a cell, depending on the genotype observed, it is considered that either an abnormal cell has been analyzed or that contamination has occurred, and that no diagnosis can be established. A case in which the abnormality of the analyzed cell can be clearly identified is when, using a multiplex PCR for linked markers, only the alleles of one of the parents are found in the sample. In this case, the cell can be considered as carrying a monosomy for the chromosome on which the markers are located, or, possibly, as haploid. The appearance of a single allele that indicates an affected genotype is considered sufficient to diagnose the embryo as affected, and embryos that have been diagnosed with a complete unaffected genotype are preferred for replacement. Although this policy may lead to a lower number of unaffected embryos suitable for transfer, it is considered preferable to the possibility of a misdiagnosis.

Preimplantation Genetic Haplotyping

Preimplantation genetic haplotyping (PGH) is a PGD technique wherein a haplotype of genetic markers that have statistical associations to a target disease are identified rather than the mutation causing the disease.

Once panels of associated genetic markers have been established for a particular disease it can be used for all carriers of that disease. In contrast, since even a monogenic disease can be caused by many different mutations within the affected gene, conventional PGD methods based on finding a specific mutation would require mutation-specific tests. Thus, PGH widens the availability of PGD to cases where mutation-specific tests are unavailable.

PGH also has an advantage over FISH in that FISH is not usually able to make the differentiation between embryos that possess the balanced form of a chromosomal translocation and those carrying the homologous normal chromosomes. This inability can be seriously harmful to the diagnosis made.

PGH can make the distinction that FISH often cannot. PGH does this by using polymorphic markers that are better suited at recognizing translocations. These polymorphic markers are able to distinguish between embryos that carried normal, balanced, and unbalanced translocations. FISH also requires more cell fixation for analysis whereas PGH requires only transfer of cells into polymerase chain reaction tubes. The cell transfer is a simpler method and leaves less room for analysis failure.

Embryo Transfer and Cryopreservation of Surplus Embryos

Embryo transfer is usually performed on day three or day five post-fertilization, the timing depending on the techniques used for PGD and the standard procedures of the IVF center where it is performed.

With the introduction in Europe of the single-embryo transfer policy, which aims at the reduction of the incidence of multiple pregnancies after ART, usually one embryo or early blastocyst is replaced in the uterus. Serum hCG is determined at day 12. If a pregnancy is established, an ultrasound examination at 7 weeks is performed to confirm the presence of a fetal heartbeat. Couples are generally advised to undergo PND because of the, albeit low, risk of misdiagnosis.

It is not unusual that after the PGD, there are more embryos suitable for transferring back to the woman than necessary. For the couples undergoing PGD, those embryos are very valuable, as the couple's current cycle may not lead to an ongoing pregnancy. Embryo cryopreservation and later thawing and replacement can give them a second chance to pregnancy without having to redo the cumbersome and expensive ART and PGD procedures.

Side Effects to Embryo

PGD/PGS is an invasive procedure that requires a serious consideration, according to Michael Tucker, Ph.D., Scientific Director and Chief Embryologist at Georgia Reproductive Specialists in Atlanta. One of the risks of PGD includes damage to the embryo during the biopsy procedure (which in turn destroys the embryo as a whole), according to Serena H. Chen, M.D., a New Jersey reproductive endocrinologist with IRMS Reproductive Medicine at Saint Barnabas. Another risk is cryopreservation where the embryo is stored in a frozen state and thawed later for the procedure. About 20% of the thawed embryos do not survive. There has been a study indicating a biopsied embryo has a less rate of surviving cryopreservation. Another study suggests that PGS with cleavage-stage biopsy results in a significantly lower live birth rate for women of advanced maternal age. Also, another study recommends the caution and a long-term follow-up as PGD/PGS increases the perinatal death rate in multiple pregnancies.

In a mouse model study, PGD has been attributed to various long term risks including a weight gain and memory decline; a proteomic analysis of adult mouse brains showed significant differences between the biopsied and the control groups, of which many are closely associated with neurodegenerative disorders like Alzheimers and Down syndrome.

Challenges

However, PGD also brings with it additional challenges, and for this reason some couples chose not to use PGD. The main challenges of PGD are as follows:

- PGD requires ART treatment, which couples who are fertile do not otherwise need.
- As with all ART treatment, PGD does not provide a guarantee of pregnancy.
- Some types of PGD tests, particularly those that test for specific inherited conditions, require a special test to be designed before ART treatment can commence. This typically takes between three and six months.
- There is a small risk of error with PGD, due to the many technical challenges of testing single cells. Although the risk of error is usually less than 2%, couples who become pregnant using PGD are offered prenatal diagnosis (chorionic villus sampling or amniocentesis) to confirm the PGD diagnosis.
- The effectiveness of PGD depends on the availability of a number of embryos to test. For couples that produce only a small number of embryos, the likelihood of a successful pregnancy is reduced.
- PGD is a relatively new technique. However, current evidence suggests that the risks to the baby are no greater than for other forms of ART. For more information about health risks of ART, please refer to the VARTA brochure, Possible health effects of IVF.
- PGD is expensive. The costs of PGD vary, but the cost of a PGD cycle is typically between two and three times the cost of a standard IVF cycle. At the moment, there is no public funding for PGD.

Five common misconceptions (and little-known facts) about PGS and PGD testing:

Misconception 1: PGS is designed for women of advanced maternal age.

It's easy to understand why so many people believe PGS is geared solely toward women of advanced maternal age. While it's true that an increase in age results in an increased risk for fertility issues, the belief that younger women aren't at risk just isn't true.

Fact: Women of all ages are at risk of having chromosomally abnormal embryos — even women under 30 years of age. Melissa Maisenbacher, a genetic counselor at Natera, explains in an interview that with Day 5 testing, for women under the age of 30, there's a 30 percent risk for each embryo to be abnormal. For women in their late 30s, that number jumps to a whopping 50 percent.

Misconception 2: PGS is only for identifying chromosome abnormalities.

There's no denying that PGS can provide infertile couples with valuable genetic insight. By looking for chromosomal abnormalities, couples can identify and prepare for conditions like Down's syndrome.

Couples who have a family history or are at risk of chromosome abnormality are ideal candidates for PGS, but so are couples that have had multiple failed pregnancies or IVF transfers and simply want to find out why.

Fact: Preimplantation genetic screening can also be used to help infertile couples learn the reasoning behind their infertility. Most people don't know why their embryos either aren't implanting or are resulting in early losses until after genetic testing. Especially for couples dealing with unexplained infertility, PGS can supply some much-needed answers.

There are some who believe PGS is used for gender selection, or so-called "family balancing," but most fertility clinics do not share the gender with patients until they are successfully pregnant following a transfer — and for good reason.

Misconception 3: PGD is "like playing God."

Preimplantation genetic diagnosis is considered by some to be a controversial procedure. PGD is used in conjunction with in vitro fertilization (IVF) to screen for single-cell gene defects that could lead to genetic disorders.

By screening for these genetic conditions, couples affected by an inherited disorder can reduce the risk that their children will also be affected — hence why so many liken the procedure to "playing God."

Fact: This one isn't so black and white, as it plays on other factors, such as faith and morals. However, it's important to remember that so much of what has come about from today's medical technology can be subjected to the same interpretation, from preventative medications to C-sections to ventilators.

Misconception 4: PGD is recommended for individual carriers of single-gene disorders.

For individuals with a family history of single-gene disorders like cystic fibrosis or sickle cell anemia, PGD might seem like the obvious solution. But you may want to think twice before spending an arm and a leg on PGD to assess embryo risk.

Fact: According to Maisenbacher, most couples are not at risk for single-gene disorders. Most individuals are actually carriers of four to six different genetic diseases, but their partner is not usually a carrier for the same genetic disease. And, in the case of most single-gene disorders that PGD is done for, both parents need to be carriers in order for their to be a risk to the child.

Misconception 5: A low-grade embryo results in an unsuccessful pregnancy.

IVF embryos are "graded" to help pick the best for transfer. While it might seem obvious to shoot for mostly grade A embryos, low grade embryos also have the potential to result in a successful pregnancy.

Fact: While there seems to be a relationship between embryo grade and chromosome abnormality, the two don't necessarily correlate.

Advantages

- Improved embryo selection: Only embryos presenting no abnormalities in terms of the number of chromosomes will lead to the birth of a healthy child. Therefore, when working with good embryos, when PGS techniques are applied, we are able to select chromosomally normal embryos and rule out those which would never be capable of leading to the birth of a healthy child even when their appearance would suggest that they are good quality embryos.

- Avoids the transfer of embryos that will not implant: certain chromosomal abnormalities are incompatible with life and prevent the embryo from developing during its early stages and even from implanting in the mother's uterus. PGS means that embryos of this kind can be ruled out, thus optimizing the number of transfers.
- Avoids the transfer of embryos that will lead to pregnancy loss or the birth of children with a variety of syndromes: within the range of possible chromosomal abnormalities, some are less harmful to the embryo and allow it to implant. However, they do stop the pregnancy from developing correctly and can lead to pregnancy loss or the birth of a child with a number of possible syndromes such as Down's syndrome, Patau's syndrome or Edwards' syndrome. PGS means that embryos that will be the cause of situations of this kind can be ruled out.
- Reduces the time required in order to get pregnant; by using PGS, we are able to avoid transferring embryos that will not lead to the birth of a healthy child since they will have been ruled out using the technique. Since we know which embryos will give rise to a full term pregnancy, 'time is not wasted' transferring embryos that will undergo embryo arrest during development and will not lead to the birth of a healthy child.
- Reduces the financial burden: Adding a new analysis to the process could be indicative of an increase in the cost. However, an in-depth knowledge of the characteristics of each embryo means that embryos that would appear to be healthy but which, in fact, are not healthy from a genetics point of view are not frozen and stored. Additionally, the cost of transferring embryos that will not lead to a pregnancy is avoided.
- Positive impact on psychological wellbeing: Using PGS means that the uncertainty that patients go through is reduced. On the one hand, they have the guarantee of the health of their embryo and that the very latest technology has been used in order to ensure this. On the other, the risk of pregnancy loss is reduced and this reduces emotional stress, particularly in the case of patients who have already gone through this.

Disadvantages

- Invasive procedure: PGS means that the embryo needs to be biopsied in order to carry out the genetic test. However, significant progress to reduce the possible damaging effect of the biopsy has been made over the last few years. Carrying out the embryo biopsy on day 5 of development rather than on day 3 has been key to now being able to say that the embryo biopsy does not have a negative impact on embryo viability.
- A cycle with no transfer: In some cases, patients are at a high risk of having abnormal embryos. This is the case, for example, of mothers of an advanced age. When this is the case, it is possible that, following PGS analysis, all the embryos are chromosomally abnormal and not suitable for transfer. As well as the upset caused by calling the treatment off, there is also a significant emotional impact.
- Embryo mosaicism: It is commonly accepted that human embryos have a certain degree of mosaicism. However, diagnosis was difficult. Nowadays, thanks to the development of genetic analysis techniques, we are able to see if there are both normal and abnormal cells in the embryo (mosaic). What needs to be determined is if this affects the embryo in any

way. Numerous lines of research work have been carried out at Institute of Bernabeu in order to evaluate this.

- PGS as a means of screening: PGS analyses the outer section of the embryo with the aim of leaving the part that will give rise to the baby (the internal cell mass) intact because scientific research has shown that there is a significant connection between the two. Therefore, we accept that the biopsy sample taken is representative of the entire embryo.
- A difficult decision: Many couples find taking the decision to analyze their embryos difficult for ethical and emotional reasons. Psychological and professional support is available to patients at our clinics. This serves as a guide but the final decision is always up to the family.

Predictive Testing

Predictive genetic testing is the use of a genetic test in an asymptomatic person to predict future risk of disease. These tests represent a new and growing class of medical tests, differing in fundamental ways from conventional medical diagnostic tests. The hope underlying such testing is that early identification of individuals at risk of a specific condition will lead to reduced morbidity and mortality through targeted screening, surveillance, and prevention. Yet the clinical utility of predictive genetic testing for different diseases varies considerably.

Utility of predictive genetic testing for different diseases varies considerably. We explore here the factors that contribute to this variation and which will dictate the utility of any of these new tests now or in the future.

Methods and Definition of Terms

The definition of utility used here encompasses all aspects of a test (individual and societal) that render it more or less useful in the clinical arena.

Difference from Conventional Medical Testing

Current and Future use

A conventional medical diagnostic test, such as a blood count or an imaging study, defines something about the patient's current condition. Although such information may have implications for the future, its overwhelming utility lies in the information it provides about the patient's current state.

A predictive genetic test, in contrast, informs us only about a future condition that may (or may not) develop. The identified risk is sometimes high—for example, in a positive test for Huntington's disease)—but always contains a substantial component of uncertainty, not only about whether a specific condition will develop, but also about when it may appear and how severe it will be. Predictive genetic tests often carry a further element of uncertainty: the interventions available for individuals at risk are often untested, and recommendations may be based on presumed benefit rather than observations of outcomes.

These uncertainties contrast with the presentation of predictive genetic testing in the popular media, which often fosters an illusion that genetic risk is highly predictable and determinative. An article, for example described a "genetic report card" that would predict a baby's health history at birth. In fact, uncertainties inherent in most genetic tests represent a major limitation to their clinical utility.

Individual versus Family

Whereas conventional diagnostic testing rarely has medical importance for anyone other than the person tested (except in the case of communicable diseases) predictive genetic testing typically has direct implications for family members. Concern for relatives may be an important motivating factor for a patient wanting to undergo such testing; some family members, however, may resist participating in the testing because they prefer not to have information about their genetic risk. The utility of a predictive genetic test will therefore depend on whose point of view is considered.

Utility of Predictive Genetic Testing for Different Diseases

An examination of predictive genetic testing in various diseases helps to identify factors that determine utility. The figure shows the degree of utility for various diseases (ranked according to how clinically useful testing currently is). These diseases are discussed below, from those for which testing is most useful through to those for which testing is least useful or even harmful.

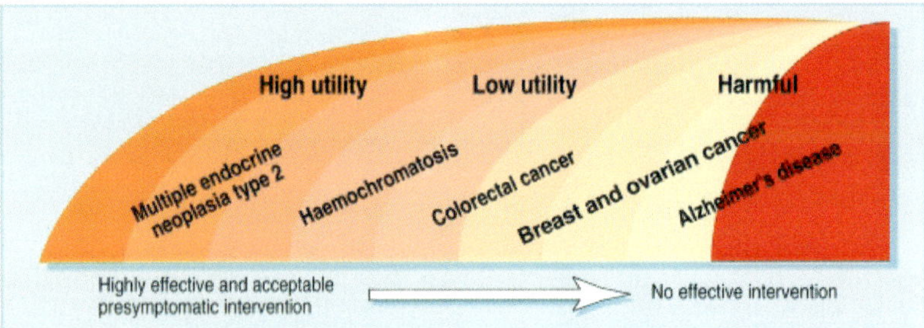

Figure: Utility in predictive genetic testing

Multiple Endocrine Neoplasia Type 2

The rare disorder multiple endocrine neoplasia type 2 results from mutations in the RET proto-oncogene. People with the disorder are almost certain to develop medullary thyroid carcinoma unless they undergo prophylactic thyroidectomy. Studies comparing children with multiple endocrine neoplasia type 2 who underwent thyroidectomy with those who did not, offer compelling evidence that such surgery reduces the likelihood of dying from cancer. Predictive genetic testing makes it possible to identify those who will benefit from surgery.

This example illustrates that when predictive genetic testing strongly predicts a deleterious clinical outcome and an efficacious early intervention exists, it is of high utility. Indeed, such testing for multiple endocrine neoplasia type 2 is the accepted standard of care for individuals at risk.

Haemochromatosis

Haemochromatosis is an uncommon (but not rare) condition of tissue iron deposition, leading to diabetes, cirrhosis, heart disease, arthritis, and gonadal dysfunction. Phlebotomy is a simple and effective preventive treatment, and predictive genetic testing is therefore useful to raise suspicion of this often elusive diagnosis. Testing is less useful for haemochromatosis than for multiple endocrine neoplasia type 2, however, because of low predictive value.

Although excess iron accumulation results from a genetic predisposition, other factors contribute to the development of clinically important iron overload, including sex, diet, and exposure to liver toxins such as alcohol. Thus, the penetrance of the haemochromatosis genotype (the proportion of individuals with genetic susceptibility who will develop the associated clinical condition) is low. The resultant uncertainty limits the utility of predictive genetic testing because preventive action based only on the results of such testing would subject many individuals who would never develop clinical sequelae to unnecessary phlebotomy.

Colorectal Cancer

About 5-10% of colorectal cancer results from inheritance of a few highly penetrant gene mutations that confer a high lifetime risk of the disease. Predictive genetic testing can be useful when family history suggests increased risk—for example, three or more affected relatives, with one in whom the disease was diagnosed before age 50 and is compatible with a diagnosis of hereditary non-polyposis colon cancer. Affected individuals have about a 70% lifetime risk of colorectal cancer. Periodic colonoscopic surveillance of these individuals reduces the development of colorectal cancer by 62% when compared with unscreened controls, showing the utility of predictive genetic testing in this circumstance.

However, hereditary non-polyposis colon cancer involves other cancer risks as well. Affected women have a high risk of endometrial cancer, as well as increased risks of ovarian cancer, other gastrointestinal cancers, and cancers of the ureteral tract. No established surveillance strategies are available for these other cancers. Thus predictive genetic testing provides an established outcome benefit for only one of the risks identified, and therefore although useful, it provides less clear cut benefit than in a condition such as multiple endocrine neoplasia type 2.

Breast and Ovarian Cancer

About 5-10% of breast and ovarian cancers result from the inheritance of mutations in the BRCA1 or BRCA2 gene. Predictive genetic testing for breast and ovarian cancer, as for hereditary non-polyposis colon cancer, can be useful to identify those at increased risk. In both breast and ovarian cancer, however, utility is limited because of considerable uncertainty about the predictive value of the test.

A woman carrying a mutation in the BRCA1 or BRCA2 gene may develop breast cancer, ovarian cancer, both cancers, or neither. Penetrance estimates range from 36-85% for breast cancer and 10-44% for ovarian cancer. Moreover, the age at which cancer occurs is widely variable. These uncertainties probably reflect a combination of factors, including the environment, modifying genes, the nature of a woman's specific mutation, and purely stochastic processes.

The utility of predictive genetic testing for breast and ovarian cancer is further limited by the nature of available surveillance and prevention strategies. Starting mammography at age 25 to 35 is recommended for carriers of the BRCA1 or BRCA2 gene, but the efficacy of this early surveillance is unknown. Because mammography is already widely encouraged for women aged over 40 (in the United States) or 50 (in the United Kingdom), information on genetic susceptibility is less relevant at later ages. Finally, adequate surveillance for ovarian cancer is not available.

Chemoprevention with tamoxifen shows promise for reducing risk of breast cancer, but conflicting data exist. Moreover, chemoprevention increases risk of endometrial cancer and venous thromboembolic disease. Oral contraceptives may reduce risk of ovarian cancer but may also increase risk of breast cancer. Prophylactic oophorectomy and mastectomy are reasonable options for some women and seem to be effective in reducing cancer risk. Such measures carry substantial burdens, however, and mastectomy in particular is not widely accepted by women at risk.

In short, knowledge of an inherited predisposition to breast or ovarian cancer does not lead to simple, straightforward measures to reduce risk, thus limiting the utility of predictive genetic testing.

Alzheimer's Disease

Alzheimer's disease illustrates the potential for predictive genetic testing to cause harm. Measurement of the Apo lipoprotein E genotype can predict risk of developing Alzheimer's disease in people of European descent. Two copies of the apolipoprotein E4 gene (present in 2% of the general population) are associated with a 10-fold increased risk of Alzheimer's disease; one copy is associated with a twofold-increased risk, and the inheritance of an Apo lipoprotein e2 allele is protective. Thus a positive test is an imprecise measure of risk and could result in anxiety, stigmatization, or discrimination. The principle of avoiding harm suggests that currently such testing would generally be unethical because no effective prevention is available.

Factors Affecting Utility

The ideal context, therefore, is a highly predictive test for a disease that is serious and incurable but preventable by means that are imperfect or expensive. The table shows factors affecting utility of predictive genetic testing.

Severity of Disease and Availability of Effective Treatment

The utility of predictive genetic testing declines when a disease is curable. Testing for tuberculosis, for example, makes little sense, even though genetics contributes to susceptibility to the disease. Similarly, as scientific advances make breast or colon cancer curable by increasingly innocuous means, the utility of predictive genetic testing will decline.

Screening and Prevention

Effective and inexpensive screening methods also make predictive genetic testing less useful because these measures can be readily applied to the entire population. Testing for hypertension makes little sense—despite evidence of strong genetic contributors to this condition—because universal screening and treatment are the rule. As the expense of screening rises, predictive genetic testing becomes more appealing. Thus, if magnetic resonance imaging (which is expensive) were shown to be superior to mammography (less expensive) in screening for breast cancer, testing could target those who would benefit most.

Available preventive measures must be either imperfect or expensive for predictive genetic testing to be of high utility. Testing makes sense in women at high risk of breast or ovarian cancer if they are considering oophorectomy or mastectomy: a positive test would confirm risk and support the use of invasive, imperfect interventions. When prevention is simple, however, the value of testing decreases. Vaccination is so cheap, safe, and effective that universal administration is rational. Thus testing has no utility in measles, mumps, or rubella despite evidence of genetic differences in susceptibility to infectious disease. The same would be true if an effective, safe, and inexpensive vaccination existed for breast cancer.

Perceptions of Utility

Family history and experience are important factors in determining how an individual perceives the utility of predictive genetic testing. Figures below shows how a woman's perception of the utility of testing for risk of breast cancer, for example, can vary depending on whether other close relatives have died of the disease or on her own family structure.

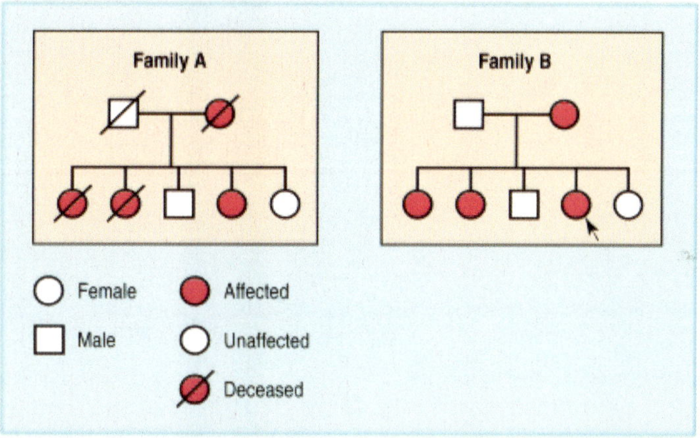

Figure: Families' experiences affect their perceptions of utility of predictive genetic testing

The affected woman in family A, whose sisters and mother died of breast cancer, may perceive chemoprevention or prophylactic surgery favorably, and welcome the guidance that predictive genetic testing can provide in making such decisions. Her counterpart (arrowed) in family B may perceive breast cancer to be a less traumatic disease and feel comfortable with routine surveillance, thus lessening the utility of testing for her.

Cost

The potential demand for genetic tests has many firms seeking to discover and patent information about genetic markers that are correlated (however tenuously) with disease. Policy-makers worldwide are considering the potential impact — both good and bad — of new predictive genetic tests on the cost, quality and equity of health care systems. Policy discussion has tended to focus on the per-unit costs of genetic tests, and particularly on how patent law influences this. However, because the expanded use of health care products and services sold on the basis of test results may be as important in determining the health system impact of predictive genetic testing as the tests themselves, policy-makers should pay close attention to all aspects of complete testing services and their various effects on health system costs.

The ultimate cost impact of a genetic test will depend on how the availability of the test changes health care behaviors. The immediate cost impact is the cost of screening itself. Genetic testing can change screening costs by changing the unit cost per case screened, the number of cases screened, or both. Genetic tests per se are currently very expensive; moreover, appropriate counseling represents a potentially major, and always necessary, component of genetic testing. However, even if the per-unit cost of genetic screening were lower than that of conventional screening, the total cost of the screening program could increase if a genetic test were applied more widely than conventional screening. This might be anticipated if the genetic test were more convenient, or if it were heavily promoted by those who held patents on the tests themselves or on the products sold on the basis of test results.

The long-term financial consequences of predictive genetic testing include changes in the uptake of diagnostic and preventive services. Surveillance activities, undertaken to detect the development of disease early in its progression for those identified as being at high risk, may vary widely. In the case of BRCA1 and BRCA2 testing, or other tests for hereditary forms of cancer, the resulting surveillance could range from inexpensive self-examination to high-cost diagnostic imaging or surgical biopsy. When added to a program cost on a year-after-year basis, increased diagnostic testing or molecular imaging, or both, may be particularly important determinants of the health system impact of a genetic testing service. Preventive activities that are undertaken to reduce the likelihood or severity of disease onset can have similar effects on health system costs. They may range from simple behavior change in order to avoid known environmental risk factors, to costly prophylactic surgical removal of healthy tissue, to lifelong pharmacologic management.

Effective surveillance and preventive activities that are a consequence of positive genetic test results will ideally reduce or avert the treatment costs for the associated disease. In many cases, however, the effectiveness of interventions targeted at individuals with a specific genetic make-up will be difficult to prove in the short term. Moreover, although outcomes should improve when cost-effective surveillance and preventive activities are taken up by appropriate patients, population health outcomes can worsen if false- negative genetic test results reduce the uptake of otherwise cost-effective surveillance or prevention, if genetic testing increases the uptake of unnecessary and ineffective surveillance and prevention activities, or if testing diverts health care resources away from more cost-effective means of promoting health. Attention must therefore be paid to the relative risks and benefits of interventions and behavior changes for patients who are tested versus those who are not, and for patients with positive test results versus those with negative test results.

Two important determinants of the ultimate balance of these cost considerations are the timing of the respective costs and the degree to which the test service is appropriately targeted at candidate populations. The cost of testing is always incurred in the present, and surveillance and prevention costs can begin almost immediately and continue for years, yet saved treatment costs often do not become evident until far into the future. The longer the lag between current spending and future savings, the greater the savings must be to justify current spending from a purely financial perspective. This is because of both the year-after-year nature of surveillance or prevention and the fact that a given level of future savings or costs is worth more today than it is in the future — the further into the future savings arise, the less they are worth investing in today. This "discounting" of future returns to investment reflects interest costs and other economic considerations. If, for example, the interest rate is 5%, the treatment savings required to offset $1 per year of ongoing prevention or surveillance, or both, would need to be about $5 if treatment costs are expected to occur in 5 years, $12 if treatment costs occur within 10 years, $34 for 20 years and $128 for 40 years. When prevention is inexpensive and effective this is not a problem, but the costs of long-term risk management can become substantial for daily pharmaceutical consumption or the routine use of high-tech imaging and surveillance equipment.

The extent to which a test is targeted at high-risk populations is a major determinant of the overall cost of screening. In addition, the precision of targeting affects the nature of the information produced by the test and how that information affects the behavior of tested individuals and their health care providers. Although several factors influence the positive predictive value of a test in practice, one of the most important is the prevalence of the genetic risk factor in the target population. Targeting a screening program toward populations at risk reduces in particular the proportion of "false positives" among positive test results. This improvement in the positive predictive value of the test reduces the unnecessary uptake of prevention and surveillance services.

The impact on health systems will vary considerably across different genetic tests. Policy-makers should be most concerned about the adverse health and economic effects of tests to identify elevated risk for common, multifactorial conditions such as heart disease and cancer. Although only a few of these tests are now in clinical use, several (such as Apo lipoprotein E [ApoE] testing for sporadic Alzheimer's disease) are being pursued strongly by those with a commercial interest in the tests themselves or in the products sold based on test results. On the other hand, genetic tests for strongly hereditary conditions, such as Huntington's disease, are unlikely to generate a large impact because such conditions are rare, and the test can be appropriately targeted to family members. Such tests can even save health care costs by obviating the need for surveillance for those family members confirmed not to have the genetic anomaly.

Genetic testing services are plotted in the figure below along the dimensions of the scope of testing offered (the horizontal axis) and the cost impact per test (the vertical axis). The total health system cost impact can range from significant cost savings to significant cost increases. Quadrant I represents a best case scenario for health care funders, but unfortunately is perhaps the least likely to occur in reality because highly predictive tests generally apply only to rare conditions. Quadrant IV represents a worst case scenario associated with risk factor tests — tests which, despite a relatively low predictive value, have a broad "market" potential for those promoting the tests or the goods or services associated with test results. The shaded slope illustrates where economic assessment would place most current genetic tests: those that have a favorable cost impact per person tested tend to be focused screening programs. As genetic testing services extend beyond rare familial disorders to broader populations, the cost impact will tend to grow and become less favorable due to increased screening costs and reduced predictive power. Even good tests, when applied too broadly or without adequate information and support, can generate large, unwarranted cost impacts.

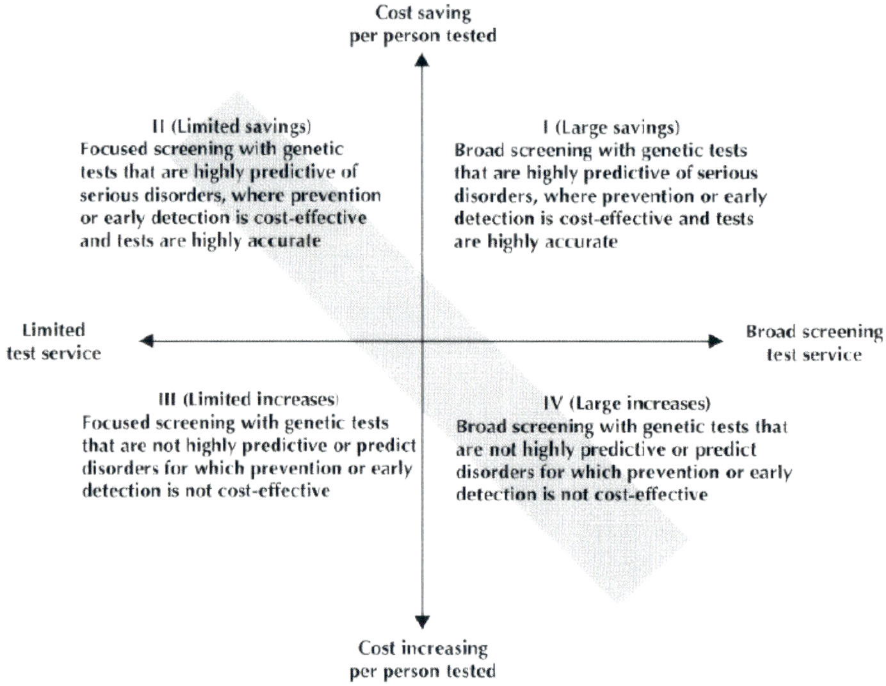

Figure: The cost impact of predictive genetic tests on health systems. Most current genetic tests fall within the shaded slopes.

Because the cost of providing a test may often be relatively small compared with surveillance, prevention and treatment costs, policy-makers have a clear interest in evaluating all aspects of a genetic testing service. Public funders of health care services may have a financial interest in controlling the use and targeting of new predictive genetic tests regardless of who pays for the tests themselves. In many cases, the best way to guide use may be to design well-targeted, publicly funded screening programs rather than allowing access to be determined by private market forces. Policy-makers should focus attention on tests that could have the broadest health system impact. They should also ensure appropriate population targeting, which is at least as important as the technical accuracy of a test in influencing impact on a health system. Wise policy choices can ensure that savings are realized where possible and that use and related costs are justified by associated improvements in health.

DNA Paternity Testing

Paternity testing can determine whether or not a particular man is the biological father of a particular child. This procedure involves collecting and examining the DNA of a small sample of bodily fluid or tissue from a child and the potential father.

DNA is the unique genetic “fingerprint” that makes up a person’s genes and chromosomes. When a baby is conceived, each parent passes on half of his/her DNA to the baby, whose genetic code (DNA) is a shared mix of only its mother’s and father’s DNA. By collecting and examining a small sample of DNA from the baby and the potential father, a paternity test can confirm or disprove that the man is indeed the biological father of the baby.

DNA is present in most of our body's cells. A small sample for testing can be obtained from several bodily sources. The cells that are most commonly tested are obtained from the blood or inside the cheek of the mouth (called buccal cells).

Paternity testing requires a clean sample (a sample free of contaminants and foreign DNA) with a good amount of the participants' DNA. While IDENTIGENE can extract DNA from almost any specimen type, our standard DNA Paternity Test uses easy and non-invasive cheek swabs. IDENTIGENE customers rub the swab against their cheek for about thirty seconds. The swab's special mesh tip (it's not cotton like a Q-Tip) grabs loose cells for DNA testing and typically provides more than enough DNA for a single use, such as a paternity test.

Extracting the DNA

Genetic scientists and technicians use many methods to extract DNA from cells. All methods use three basic steps:

1. Lysing (breaking open) the cells,
2. Separating the DNA from the rest of the cell,
3. Collecting the pure DNA in a single sample, ready for testing.

Using a special chemical process, we break open the collected cells. This separates the DNA from the nucleus and leaves the scientist with a liquid that contains DNA along with other cell parts not needed for testing, such as proteins and lipids. We then separate the DNA from the other cell parts using sophisticated robotics.

The extraction robot uses more chemicals to transfer the DNA mixture to tiny silica-based nano-beads (one billionth of a meter in size); the DNA sticks to the beads while the other cell parts are washed away.

The final step, the elution step, removes the DNA from the beads. The robot collects the pure DNA, which is now ready for the next step in the paternity testing process.

PCR – Amplifying (Copying) the DNA

The DNA scientist puts the extracted DNA into a special solution containing primers. Like toner in a biological copy machine, primers find and make copies of the DNA sample-just those specific regions that we need for a paternity test.

The copy process begins by separating the double-stranded DNA-simply by turning up the heat. As the solution cools, the primers bind to single-stranded DNA, making two copies of the original. We repeat this process (heating and cooling the DNA and the primers) 28 times, making millions of copies of tiny DNA fragments that can now be detected and viewed by a special machine called a genetic analyzer.

Scientists refer to this 'biological copy machine' process as Polymerase Chain Reaction, or PCR.

Measuring DNA for Paternity Testing

The complete PCR process makes copies of 16-18 Genetic Systems (sometimes called markers or loci) to make one DNA Profile: 15-17 markers useful for paternity and one (1) gender marker (used for test-participant verification). Each individual person has different sizes or lengths of DNA fragments at each Genetic System. Special software measures the different sizes of the DNA sections, represented by two numbers (alleles) at each Genetic System on your paternity test report. We then use this information to answer your paternity question.

Paternity Test Reporting

A child's DNA Profile is always a combination of half the father's markers and half the mother's markers. If the tested (possible) father does not share matching markers with the child, then the tested man is excluded as the biological father (he is not the father). If the DNA Profiles do match, the father is not excluded (he is the father) and the probability of paternity is reported (typically greater than 99.99%).

Paternity or Maternity Testing for Child or Adult

The testing is performed by collecting buccal cells found on the inside of a person's cheek using a buccal swab or cheek swab. These swabs have wooden or plastic stick handles with a cotton on synthetic tip. The collector rubs the inside of a person's cheek to collect as many buccal cells as possible. The buccal cells are then sent to a laboratory for testing. For paternity testing, samples from the alleged father and child would be needed. For maternity testing, samples from the alleged mother and child would be needed.

Prenatal Paternity Testing for Unborn Child

Invasive Prenatal Paternity Testing

It is possible to determine who the biological father of the fetus is while the woman is still pregnant through procedures called chorionic villus sampling or amniocentesis. Choronic villus sampling (CVS) retrieves choronic villus (placental tissue) in either a transcervical or transabdominal manner. Amniocentesis retrieves amniotic fluid by inserting a needle through the pregnant mother's abdominal wall. These procedures are highly accurate because they are taking a sample directly from the fetus; however, there is a small risk for the woman to miscarry and lose the pregnancy as a result. Both CVS and Amnio require the pregnant woman to visit a genetic specialist known as a maternal fetal medicine specialist who will perform the procedure.

Non-invasive Prenatal Paternity Testing

Current advances in genetic testing have led to the ability to determine who the biological father is while the woman is still pregnant through a non-invasive method. There is a small amount of fetal DNA (cffDNA) present in the mother's blood during pregnancy. This allows for accurate fetal DNA paternity testing during pregnancy from a blood draw with no risk of miscarriage. Studies have shown that cffDNA can first be observed as early as 7 weeks gestation, and the amount of cffDNA increases as the pregnancy progresses. There are only four companies with headquarters in the United States offering a noninvasive prenatal paternity test: DDC, Viaguard Accumetrics.

DNA Profiling

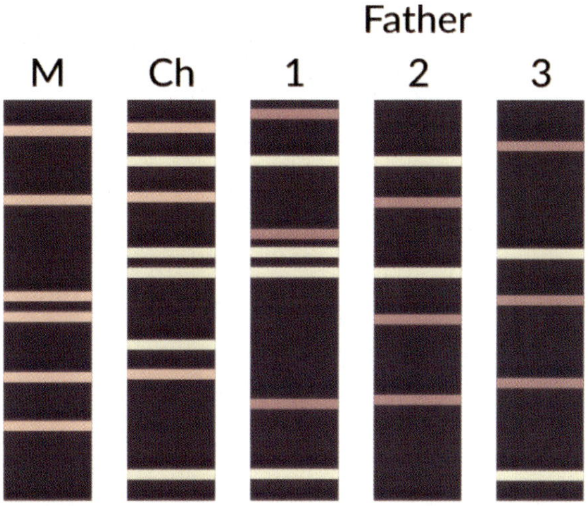

Figure: Example of DNA profiling in order to determine the father of a child (Ch)

Child's DNA sample should contain a mixture of different size DNA bands of both parents. In this case person 1 is likely the father.

The DNA of an individual is the same in somatic (nonreproductive) cell. Sexual reproduction brings the DNA of both parents together randomly to create a unique combination of genetic material in a new cell, so the genetic material of an individual is derived from the genetic material of both their parents in equal amounts. This genetic material is known as the nuclear genome of the individual, because it is found in the nucleus.

Comparing the DNA sequence of an individual to that of another individual can show whether one of them was derived from the other. However, DNA paternity tests are not currently 100% accurate. Specific sequences are usually looked at to see whether they were copied verbatim from one of the individual's genome to the other. If that was the case, then the genetic material of one individual could have been derived from that of the other (i.e. one is the parent of the other). Besides the nuclear DNA in the nucleus, the mitochondria in the cells also have their own genetic material termed the mitochondrial DNA. Mitochondrial DNA comes only from the mother, without any shuffling.

Proving a relationship based on comparison of the mitochondrial genome is much easier than that based on the nuclear genome. However, testing the mitochondrial genome can prove only if two individuals are related by common descent through maternal lines only from a common ancestor and is, thus, of limited value (for instance, it could not be used to test for paternity).

In testing the paternity of a male child, comparison of the Y chromosome can be used, since it is passed directly from father to son.

In the US, the AABB has regulations for DNA paternity and family relationship testing; however, AABB accreditation is not necessary. DNA test results are legally admissible if the collection and the processing follows a chain of custody. Similarly in Canada, the SCC has regulations on DNA paternity and relationship testing; however, this accreditation is recommended, but not necessary.

The Paternity Testing Commission of the International Society for Forensic Genetics has taken up the task of establishing the bio statistical recommendations in accordance with the ISO/IEC 17025 standards. Bio-statistical evaluations of paternity should be based on a likelihood ratio principle - yielding the Paternity Index, PI. The recommendations provide guidance on concepts of genetic hypotheses and calculation concerns needed to produce valid PIs, as well as on specific issues related to population genetics.

Legal Evidence

The DNA parentage test that follows strict chain of custody can generate legally admissible results that are used for child support, inheritance, social welfare benefits, immigration, or adoption purposes. To satisfy the chain-of-custody legal requirements, all tested parties have to be properly identified and their specimens collected by a third-party professional who is not related to any of the tested parties and has no interest in the outcome of the test.

The quantum of evidence needed is clear and convincing evidence; that is, more evidence than an ordinary case in civil litigation, but much less than beyond a reasonable doubt required to convict a defendant in a criminal case.

In recent years, immigration authorities in various countries, such as U.S., U.K., Canada, Australia, France, and others have been requesting immigration petitioners and beneficiaries in a family-based immigration case to voluntarily take the DNA parentage test when primary documents such as birth certificate to prove biological relationship are missing or inadequate.

In the U.S., immigration applicants bear the responsibility of arranging and paying for DNA testing. The U.S. immigration authorities require that the DNA test, if pursued, be performed by one of the laboratories accredited by the AABB (formerly American Association of Blood Banks). Similarly, in Canada, the lab needs to be accredited by the SCC.

The U.S. Department of State and USCIS provide information concerning the DNA parentage test request for immigration purposes.

Although paternity tests are more common than maternity tests, there may be circumstances in which the biological mother of the child is unclear. Examples include cases of an adopted child attempting to reunify with his or her biological mother, potential hospital mix-ups, and in vitro fertilization where the laboratory may have implanted an unrelated embryo inside the mother.

Other factors, such as new laws regarding reproductive technologies using donated eggs and sperm and surrogate mothers, can mean that the female giving birth is not necessarily the legal mother of the child. For example, in Canada, the federal Human Assisted Reproduction Act provides for the use of hired surrogate mothers. The legal mother of the child may, in fact, be the egg donor. Similar laws are in place in the United Kingdom and Australia.

Legal Issues

In the United States, paternity testing is fully legal, and fathers may test their children without the consent or knowledge of the mother. Paternity testing take-home kits are readily available for purchase, though their results are not admissible in court, and are for personal knowledge only. Only a court-ordered paternity test may be used as evidence in court proceedings. If parental testing is being submitted for legal purposes in the U.S., including immigration, testing must be ordered through a lab that has AABB accreditation for Relationship DNA testing.

The legal implications of a paternity result test vary by state and according to whether the putative parents are unmarried or married. If a paternity test does not meet forensic standards for the state in question, a court-ordered forensic test may be required for the results of the test to have legal meaning. For unmarried parents, if a parent is currently receiving child support or custody, but DNA proves that the man is not the father later on, the support automatically stops; however, in many states, this testing must be performed during a narrow window of time if a voluntary acknowledgement of parentage form has already been signed by the putative father; otherwise, the results of the test may be disregarded by law, and in many cases, a man may be required to pay child support, even though the child is biologically unrelated. In a few states, if the mother is receiving the support, then that alleged father has the right to file a lawsuit to get back any money that he lost from paying support. As of 2011, and in most states, unwed parents confronted with a voluntary acknowledgement of parentage form are informed of the possibility and right to request a DNA paternity test. If testing is refused by the mother, the father may not be required to sign the birth certificate or the voluntary acknowledgement of parentage form for the child. For wedded putative parents, the husband of the mother is presumed to be the father of the child. However, in most states, this presumption can be overturned by the application of a forensic paternity test, but in many states, the time for overturning this presumption may be limited to the first few years of the child's life, depending on the law of the state in question.

Personal paternity-testing kits are available. The Standards Council of Canada regulates paternity testing in Canada whereby laboratories are ISO 17025-approved. In Canada, only a handful of labs have this approval, and it is recommended that testing is performed in these labs. Courts also have the power to order paternity tests during divorce cases.

DNA paternity testing for personal knowledge is legal, and home test kits are available by mail from representatives of AABB- and ISO 17025-certified laboratories. DNA Paternity Testing for official purposes, such as sustento (child support) and inheritance disputes, must follow the Rule on DNA Evidence A.M. No. 06-11-5-SC, which was promulgated by the Philippine Supreme Court on October 15, 2007. Tests are sometimes ordered by courts when proof of paternity is required.

In the United Kingdom, there were no restrictions on paternity tests until the Human Tissue Act 2004 came into force in September 2006. Section 45 states that it is an offence to possess without appropriate consent any human bodily material with the intent of analysing its DNA. Legally declared fathers have access to paternity-testing services under the new regulations, provided the putative parental DNA being tested is their own. Tests are sometimes ordered by courts when proof of paternity is required. In the UK, the Ministry of Justice accredits bodies that can conduct this testing. The Department of Health produced a voluntary code of practice on genetic paternity testing in 2001. This document is currently under review, and responsibility for it has been transferred to the Human Tissue Authority.

DNA paternity testing is solely performed on decision of a judge in case of a judiciary procedure in order either to establish or contest paternity or to obtain or deny child support. Private DNA paternity testing is illegal, including through laboratories in other countries, and is punishable by up to a year in prison and a €15,000 fine. The French Council of State has described the law's purpose as upholding the "French regime of filiation" and preserving "the peace of families."

Under the Gene Diagnostics Act of 2009, secret paternity testing is illegal. Any paternity testing must be conducted by a licensed physician or by an expert with a university degree in science and special education in parentage testing, and the laboratory carrying out genetic testing must be accredited according to ISO/IEC 17025. Full informed consent of both parents is required, and prenatal paternity testing is prohibited, with the exception of sexual abuse and rape cases. Any genetic testing done without the other parent's consent is punishable with a €5,000 fine. Due to an amendment of the civil law section 1598a in 2005, any man who contests paternity no longer automatically severs legal rights and obligations to the child.

A paternity test with any legal standing must be ordered by a family court. Though parents have access to "peace of mind" parental tests through overseas laboratories, family courts are under no obligation to accept them as evidence. It is also illegal to take genetic material for a parental test from a minor over 16 years of age without the minor's consent. Family courts have the power to order paternity tests against the will of the father in divorce and child support cases, as well as in other cases such as determining heirs and settling the question involving the population registry. A man seeking to prove that he is not the father of the child registered as his is entitled to a paternity test, even if the mother and natural guardian object. Paternity tests are not ordered when it is believed it could lead to the murder of the mother, and until 2007, were not ordered when there was a chance that the child could have been conceived outside of marriage, making them a mamzer under Jewish law.

Peace-of-mind paternity tests are a "big business" in Spain, partly due to the French ban on paternity testing, with many genetic testing companies being based in Spain.

Peace-of-mind parentage tests are widely available on the internet. For a parentage test (paternity or maternity) to be admissible for legal purposes, such as for changing a birth certificate, Family Law Court proceedings, visa/citizenship applications or child support claims, the process undertaken needs to comply with the Family Law Regulations 1984. Further, the laboratory processing the samples must be accredited by the National Association of Testing Authorities (NATA).

In China, paternity testing is legally available to fathers who suspect their child is not theirs. Chinese law also requires a paternity test for any child born outside the one-child policy for the child to be eligible for a hukou, or family registration record. Family tie formed by adoption can also only be confirmed by a paternity test. A large number of Chinese citizens seek paternity testing each year, and this has given rise to many unlicensed illegal testing centers being set up.

Reverse Paternity Testing

Reverse paternity determination is the ability to establish the biological father when the father of a person, or a suspect, is not available. The test uses the STR alleles in mother and her child, other children and brothers of the alleged father, and deduction of genetic constitution of the father by the basis of genetic laws to create a rough amalgamation. The advantage of this knowledge is the ability to compare the father's DNA when a direct sample of the father is not available.

Genealogical DNA Test

Genealogy dna testing, also known as genetic genealogy, may be able to help you discover more about your ancestral heritage. Some people hail it as the brave new world of genealogical investigation. Others object to it because they say it removes the educational (not to mention, enjoyable) detective work of traditional genealogy research in musty archives and libraries. Others insist it's a load of hoccus pocus nonsense designed to separate fools from their wallet.

Probably the majority simply doesn't know what to think about genealogy dna testing but are curious to know more.

This section is intended primarily to provide information to the latter group. It doesn't pretend to be a learned dissertation on genealogy dna testing. It is merely a primer.

It is aimed at the amateur family historian who is curious about what this scientific development might offer his/her research and wants to know what is available and what to consider before deciding whether or not to proceed with a test.

Need for Genealogical DNA Testing

If you've already unlocked a fair few generations of your ancestry, dna testing may seem an unnecessary luxury. Or, depending on your approach to research or your level of curiosity about your genealogy, dna analysis may offer just the answer you've been waiting for.

Any of the following reasons might make you consider the cost of genealogy dna testing to be within your budget.

- You've hit a brickwall in your paper research.
- You want to see if the genealogy discoveries you've already made stand up to scientific scrutiny.
- You have an unsolved mystery, perhaps about adoptions, rushed marriages or other 'hushed-up' relationships, in your family tree.
- You are satisfied with your knowledge of your more recent heritage (i e during the last few centuries) but you'd like to know about your more distant geographical origin your ethnic origin.
- You want to find out if you are related to others with the same surname.

Surname Studies: DNA and Genealogy Come Together

If you've been a family historian for any length of time, it's possible you've been contacted by someone - a stranger - who shares your surname and wonders if you might be related. After a few cursory questions and answers about immediate family and/or a generation or two of ancestors, both parties usually come to the conclusion that they are not related.

But there is a chance (and, with some names, a likelihood) that you and this stranger are indeed related. The problem is that the 'common ancestor' is further back in time than your combined traditional genealogical research can take you.

This is where surname studies come in, because Y-DNA is passed intact, along with a surname, down the male line.

Surname studies and Y-DNA are, therefore, natural partners in the genealogy DNA testing portfolio.

References

- Permezel, Michael; Walker, Susan; Kyprianou, Kypros (2015). Beischer & MacKay's Obstetrics, Gynaecology and the Newborn. Elsevier Health Sciences. p. 74. ISBN 9780729584050. Retrieved 24 January 2017
- What-is-genetic-testing: yourgenome.org, Retrieved 19 May 2018
- Galanello, Renzo; Origa, Raffaella (2010-05-21). "Beta-thalassemia". Orphanet Journal of Rare Diseases. 5: 11. doi:10.1186/1750-1172-5-11. ISSN 1750-1172. PMC 2893117. PMID 20492708
- "Genetic Testing for Hereditary Cancer Syndromes". National Cancer Institute. National Institute of Health. Retrieved 18 November 2016
- Preimplantation-genetic-diagnosis, infertility: americanpregnancy.org, Retrieved 18 June 2018
- Soliday, F. K.; Conley, Y. P.; Henker, R. (2010). "Pseudocholinesterase deficiency: A comprehensive review of genetic, acquired, and drug influences". AANA Journal. 78 (4): 313–320. PMID 20879632
- Zoloth, Laurie; Holland, Suzanne; Lebacqz, Karen (2001). The human embryonic stem cell debate: science, ethics, and public policy. Cambridge, Mass: MIT Press. ISBN 0-262-58208-2
- Advantages-disadvantages-pre-implantation-genetic-diagnosis-pgd: institutobernabeu.com, Retrieved 25 May 2018
- R, Andorno,. "The right not to know: an autonomy based approach". Journal of Medical Ethics. 30. doi:10.1136/jme.2002.001578. ISSN 0306-6800. PMC 1733927
- John A. Haugen Associates Obstetrics and Gynecology. "The Facts on Prenatal Testing". John A. Haugen Associates Obstetrics and Gynecology. Archived from the original on 2015-04-02. Retrieved 2015-03-26
- Dna-paternity-test-science: dnatesting.com, Retrieved 12 March 2018
- King, Elisabeth (2017). "Genetic Testing: Challenges and changes in testing for hereditary cancer syndromes". Clinical Journal of Oncology Nursing. 21 (5): 589–598. doi:10.1188/17.cjon.589-598
- Genealogy-dna-testing: irish-genealogy-toolkit.com, Retrieved 29 March 2018

Permissions

All chapters in this book are published with permission under the Creative Commons Attribution Share Alike License or equivalent. Every chapter published in this book has been scrutinized by our experts. Their significance has been extensively debated. The topics covered herein carry significant information for a comprehensive understanding. They may even be implemented as practical applications or may be referred to as a beginning point for further studies.

We would like to thank the editorial team for lending their expertise to make the book truly unique. They have played a crucial role in the development of this book. Without their invaluable contributions this book wouldn't have been possible. They have made vital efforts to compile up to date information on the varied aspects of this subject to make this book a valuable addition to the collection of many professionals and students.

This book was conceptualized with the vision of imparting up-to-date and integrated information in this field. To ensure the same, a matchless editorial board was set up. Every individual on the board went through rigorous rounds of assessment to prove their worth. After which they invested a large part of their time researching and compiling the most relevant data for our readers.

The editorial board has been involved in producing this book since its inception. They have spent rigorous hours researching and exploring the diverse topics which have resulted in the successful publishing of this book. They have passed on their knowledge of decades through this book. To expedite this challenging task, the publisher supported the team at every step. A small team of assistant editors was also appointed to further simplify the editing procedure and attain best results for the readers.

Apart from the editorial board, the designing team has also invested a significant amount of their time in understanding the subject and creating the most relevant covers. They scrutinized every image to scout for the most suitable representation of the subject and create an appropriate cover for the book.

The publishing team has been an ardent support to the editorial, designing and production team. Their endless efforts to recruit the best for this project, has resulted in the accomplishment of this book. They are a veteran in the field of academics and their pool of knowledge is as vast as their experience in printing. Their expertise and guidance has proved useful at every step. Their uncompromising quality standards have made this book an exceptional effort. Their encouragement from time to time has been an inspiration for everyone.

The publisher and the editorial board hope that this book will prove to be a valuable piece of knowledge for students, practitioners and scholars across the globe.

Index

A

Adenine, 1, 65, 99, 102, 105, 116, 122, 133-139, 143, 145, 148, 152, 156, 182

Ancient Dna, 57, 65-72

Anti-müllerian Duct Hormone, 187-188, 197

Archaeogenetics, 57, 69, 98

Autosome, 157, 175-178, 194

B

Base Pairing, 105-106, 109, 111, 115, 133, 138, 144, 148

Bioinformatics, 28, 49, 69, 118-119, 141

Blastocyst Biopsy, 232-233

C

Carrier Testing, 224, 227

Centromere, 132, 157, 160-161, 170-172, 181, 212

Chromatid, 160, 164-165, 168, 171-172

Chromomere, 160, 168

Cleavage-stage Biopsy, 232-233, 237

Cline, 21

Coding Dna, 3, 5, 7, 88-89

Colorectal Cancer, 131, 153, 243

Complementary Dna, 111, 119-121

Cytosine, 1, 68, 99, 102-105, 117, 122, 130, 133-136, 138-139, 143-145, 147-149, 156

D

Developing Gonads, 187

Dihydrotestosterone, 187, 196-197

Dna Analysis, 7, 66, 71-72, 255

Dna Damage, 124, 129-131, 142-143, 145-146, 148, 150-151, 153-155, 213

Dna Extraction, 66, 70-72

Dna Paternity Testing, 8, 117, 224, 248, 250, 253

Dna Profiling, 117, 155-156, 251

Dna Repair, 23, 91-92, 108, 113-114, 116, 124, 131, 138, 143-146, 148-151

Down Syndrome, 9, 13, 177-178, 227, 237

E

Epigenetics, 1, 6, 14-15, 21, 42-44

Estrogen, 187, 195, 197-198

Euchromatin, 131, 162-163, 219-223

Excision Repair, 124, 143-144, 148-151, 153-154

G

Gene Loss, 62-63

Genealogical Dna Test, 127, 224, 255

Genetic Architecture, 57-58, 98

Genetic Distance, 33-36, 38, 58

Genetic Recombination, 41, 114-116

Guanine, 1, 65, 99, 102-103, 105, 108, 116, 120, 122, 130, 133-139, 143-147, 149, 210

H

H4 N-terminal Tail, 201, 209, 211, 215

Helicase, 111, 123, 150, 152

Heritability, 1, 50-56

Heterochromatin, 6, 8, 131-132, 163, 185, 208, 215-216, 219-223

Human Evolutionary Genetics, 57

Human Genetic Disorder, 184

Human Genetic Variation, 8, 15, 20-21, 63

Human Genome, 1-10, 12-14, 19-20, 22, 24-26, 31, 36, 45-46, 48, 55-57, 59, 63-65, 83-84, 88-90, 92, 109-110, 115, 118, 139, 141, 179-180, 219, 226

Human Reference Genome, 2, 4-5, 8, 29

K

Karyotype, 19, 73, 122, 158, 162, 184, 196

L

Lampbrush Chromosome, 168

Ligase, 111, 114, 149-152, 154, 212-213

M

Microarray, 22, 25, 29, 32

Mitochondrial Dna, 14, 21, 37, 58, 66, 68, 73, 99, 124, 127-130, 251

Mitochondrial Inheritance, 127-129

N

Noncoding Dna, 3, 5, 110

Noncoding Rna, 3, 5-6

Nuclear Dna, 14, 58, 68, 99, 109, 122, 124, 128-130, 219, 251

Nuclease, 114, 151, 166, 202

Nucleosome, 112, 166-167, 198-220, 223

Nucleosome Core, 166-167, 198-204, 206-213, 216-217

P
Personal Genome, 10, 22
Pigmentation, 17, 40, 44, 85-86, 91
Polymerase, 6, 27, 67, 70-72, 103, 110-111, 113, 115, 117, 119, 121-123, 144, 149-152, 154, 191, 205-206, 234, 237, 249
Population Genetics, 23, 70, 252
Preimplantation Genetic Diagnosis, 224, 229-230, 233-234, 239
Protein-coding Gene, 4-5, 13

S
Selective Sweep, 1, 39-41, 83-85
Selfish-gene Theory, 79, 81, 83
Single Nucleotide Polymorphism, 15, 19, 21, 52, 65
Somatic Structural Variation, 22-23
Split Read, 28-29
Sry Gene, 183, 187, 190-192, 195
Supercoiling, 100, 106-107, 114, 165-166

T
Telomere, 108-109, 157, 161, 173-175, 223
Thymine, 1, 99, 102, 105, 117, 122, 133-136, 138-139, 143-144, 147-150, 154, 156, 205
Topoisomerase, 111
Transposons, 7-8
Turner Syndrome, 9, 13, 180, 182-184

V
Variome, 45-46, 56

X
X Chromosome, 13, 64, 91, 163, 176, 178-182, 184, 186, 193-195

Y
Y Chromosome, 3, 58, 64, 73, 157, 170, 176, 179-187, 190-192, 194-195, 251